Studies in biological control

THE INTERNATIONAL BIOLOGICAL PROGRAMME

The International Biological Programme was established by the International Council of Scientific Unions in 1964 as a counterpart of the International Geophysical Year. The subject of the IBP was defined as 'The Biological Basis of Productivity and Human Welfare', and the reason for its establishment was recognition that the rapidly increasing human population called for a better understanding of the environment as a basis for the rational management of natural resources. This could be achieved only on the basis of scientific knowledge, which in many fields of biology and in many parts of the world was felt to be inadequate. At the same time it was recognised that human activities were creating rapid and comprehensive changes in the environment. Thus, in terms of human welfare, the reason for the IBP lay in its promotion of basic knowledge relevant to the needs of man.

The IBP provided the first occasion on which biologists throughout the world were challenged to work together for a common cause. It involved an integrated and concerted examination of a wide range of problems. The Programme was co-ordinated through a series of seven sections representing the major subject areas of research. Four of these sections were concerned with the study of biological productivity on land, in freshwater, and in the seas, together with the processes of photosynthesis and nitrogen fixation. Three sections were concerned with adaptability of human populations, conservation of ecosystems and the use of biological resources.

After a decade of work, The Programme terminated in June 1974 and this series of volumes brings together, in the form of syntheses, the results of national and international activities.

INTERNATIONAL BIOLOGICAL PROGRAMME 9

Studies in biological control

EDITED BY

V. L. Delucchi

Professor of Entomology
Swiss Federal Institute of Technology, Zurich

CAMBRIDGE UNIVERSITY PRESS
CAMBRIDGE
LONDON · NEW YORK · MELBOURNE

CAMBRIDGE UNIVERSITY PRESS
Cambridge, New York, Melbourne, Madrid, Cape Town,
Singapore, São Paulo, Delhi, Tokyo, Mexico City

Cambridge University Press
The Edinburgh Building, Cambridge CB2 8RU, UK

Published in the United States of America by Cambridge University Press, New York

www.cambridge.org
Information on this title: www.cambridge.org/9780521281645

First published 1976
First paperback edition 2011

A catalogue record for this publication is available from the British Library

Library of Congress Cataloguing in Publication data
Main entry under title:
Studies in biological control.
  'International Biological Programme, 9.'
  English, French, Russian, and Spanish.
  Bibliography: p. 000
  Includes index.
  1. Insect control – Biological control. 2. Tetranychidae – Biological
control. 1. Delucchi, Vittorio L. 11. International Biological Pro-
gramme. [DNLM: 1. Pest control,   Biological.
SB975 S933] SB933.3.S78   632'.7   75-16867

ISBN 978-0-521-20910-6 Hardback
ISBN 978-0-521-28164-5 Paperback

Contents

Contents

Table des matières

Table des matières

Содержание

Содержание

Contenido

Contenido

Collaborators

Editorial board

B. P. Beirne, Pestology Centre, Department of Biological Sciences, Simon Fraser University, Burnaby 2, British Columbia V5A 1S6, Canada

V. L. Delucchi (Chairman), Department of Entomology, Swiss Federal Institute of Technology, Universitätsstrasse 2, 8006 Zürich, Switzerland

C. B. Huffaker, Division of Biological Control, University of California, 1050 San Pablo Avenue, Albany, California 94706, USA

D. F. Waterhouse, Division of Entomology, Commonwealth Scientific and Industrial Research Organization (CSIRO), Canberra City, ACT 2601, Australia

Working group coordinators

M. A. Bateman, Division of Entomology, Commonwealth Scientific and Industrial Research Organization (CSIRO), Zoology Building, University of Sydney, Sydney, NSW, Australia 2006 (fruit flies core project)

P. DeBach, Department of Entomology, University of California, Riverside, California 92502, USA (scale insects core project)

C. B. Huffaker, Division of Biological Control, University of California, 1050 San Pablo Avenue, Albany, California 94706, USA (spider mites core project)

N. W. Hussey, Department of Entomology, Glasshouse Crops Research Institute, Worthing Road, Rustington, Littlehampton, Sussex BN16 3PU, UK (spider mites core project)

M. Mackauer, Pestology Centre, Department of Biological Sciences, Simon Fraser University, Burnaby 2, British Columbia V5A 1S6, Canada (aphid core project)

D. Rosen, Department of Entomology, Faculty of Agriculture, The Hebrew University, PO Box 12, Rehovot, Israel (scale insects core project)

M. J. Way, Imperial College Field Station, Silwood Park, Ascot, Berkshire SL5 7PY, UK (aphid core project)

K. Yasumatsu, Entomological Laboratory, Faculty of Agriculture, Kyushu University, Fukuoka 812, Japan (rice stem-borers core project)

List of collaborators

Contributors to the volume

D. P. Annecke, Plant Protection Research Institute, Private Bag X134, Pretoria, South Africa

L. C. Argyriou, Phytopathological Institute 'Benaki', Kyphissia, Athens, Greece

M. A. Bateman, Division of Entomology, Commonwealth Scientific and Industrial Research Organization (CSIRO), Zoology Building, University of Sydney, Syndey NSW, Australia 2006

R. L. Blackman, British Museum (Natural History), Cromwell Road, London

J. Boczek, Agricultural University of Warsaw, 02-975 Warsaw-Ursynow, Poland

E. F. Boller, Swiss Federal Research Station, 8820 Wädenswil, Switzerland

L. Bonnemaison, Station centrale de Zoologie, Institut National de la Recherche Agronomique (INRA), Route de Saint-Cyr, 78 Versailles, France

G. L. Bush, Department of Zoology, University of Texas, Austin, Texas 78712, USA

L. E. Caltagirone, Division of Biological Control, University of California, 1050 San Pablo Avenue, Albany, California 94706, USA

D. J. Calvert, Division of Biological Control, University of California, 1050 San Pablo Avenue, Albany, California 94706, USA

F. Chaboussou, Station de Zoologie Agricole, Institut National de la Recherche Agronomique (INRA), Domaine de la Grande-Ferrade, 33 Pont-de-la-Maye, France

D. L. Chambers, Insect Attractants, Behavior and basic Biology Research Laboratory, US Department of Agriculture (USDA), Gainesville, Florida 32601, USA

E. Collyer, Division of Entomology, Department of Scientific and Industrial Research, Private Bag, Nelson, New Zealand

Z. T. Dabrowski, Agricultural University of Warsaw, 02-975 Warsaw-Ursynow, Poland

P. DeBach, Department of Biological Control, University of California, Riverside, California 92502, USA

V. Delucchi, Department of Entomology, Swiss Federal Institute of Technology, Universitätsstrasse 2, 8006 Zürich, Switzerland

V. F. Eastop, British Museum (Natural History), Cromwell Road, London

A. P. Economopoulos, 'Democritus' Nuclear Research Centre, Aghia Paraskevi, Attikis, Greece

E. A. Elbadry, Faculty of Agriculture, Ain Shams University, Cairo, Egypt

D. L. Flaherty, Division of Biological Control, University of California, 1050 San Pablo Avenue, Albany, California 94706, USA

B. S. Fletcher, Division of Entomology, Commonwealth Scientific and Industrial Research Organization (CSIRO), Zoology Building, University of Sydney, Sydney, NSW, Australia 2006

G. N. Foster, Department of Zoology, The West of Scotland Agricultural College, Auchincruive, Ayr KA6 5HW, UK

R. H. Gonzalez, Plant Production and Protection Division, FAO, Viale delle Terme di Caracalla, 00100 Roma, Italy

P. Grison, Institut National de la Recherche agronomique (INRA), Station de Recherches de Lutte Biologique, 78 La Minière par Versailles, France

W. Helle, Laboratory of experimental Entomology, University of Amsterdam, Linnaeusstraat 2B, Amsterdam, The Netherlands

H. J. Herbert, Canada Department of Agriculture, Research Station, Kentville, Nova Scotia, Canada

J. Hobart, Department of Agriculture and Forest Zoology, University College of North Wales, Bangor, Gwynedd, LL57 2DG UK

I. Hodek, Institute of Entomology, Czechoslovak Academy of Sciences, 7 Vinicna, Prague 2, Czechoslovakia

S. C. Hoyt, Tree Fruit Research Center, Washington State University, Wanatchee, Washington 98801, USA

M. D. Huettel, Insect attractants, behaviour, and basic biology Research Laboratory, Agricultural Research Service, USDA, Gainesville, Florida 32604, USA

C. B. Huffaker, Division of Biological Control, University of California, 1050 San Pablo Avenue, Albany, California 94706, USA

N. W. Hussey, Department of Entomology, Glasshouse Crops Research Institute, Worthing Road, Rustington, Littlehampton, Sussex BN16 3PU, UK

C. E. Kennett, Division of Biological Control, University of California, 1050 San Pablo Avenue, Albany, California 94706, USA

J. E. Laing, Division of Biological Control University of California, 1050 San Pablo Avenue, Albany, California 94706, USA

F. Leclant, Laboratoire de Recherches de la Chaire d'Ecologie Animale et de Zoologie Agricole, Ecole Nationale Supérieure Agronomique (ENSA), 34060 Montpellier, France

C. S. Li, Entomology Section, Animal Industry & Agriculture Branch, Dept. of Northern Australia, P.O. Box 5150, Darwin, Australia 5794

H. J. B. Lowe, Plant Breeding Institute, Trumpington, Cambridge, UK

M. Mackauer, Pestology Centre, Department of Biological Sciences, Simon Fraser University, Burnaby, British Columbia V5A 1S6, Canada

J. A. McMurtry, Department of Biological Control, University of California, Riverside, California 92502, USA

V. Moericke, Institut für Pflanzenkrankheiten, Universität Bonn, 53 Bonn 1, Federal Republic of Germany

E. Niemczyk, Research Institute of Pomology, ul. Pomologiczna 18, Skierniewice, Poland

List of collaborators

T. Nishida, Department of Entomology, College of Tropical Agriculture, University of Hawaii, 2500 Dole Street, Honolulu, Hawaii 96822, USA

E. R. Oatman, Department of Biological Control, University of California, Riverside, California 92502, USA

W. P. S. Overmeer, Laboratory of experimental Entomology, University of Amsterdam, Linnaeusstraat 2B, Amsterdam, the Netherlands

M. D. Pathak, International Rice Research Institute, Manila, the Philippines

R. J. Prokopy, RT. 1. Bailey's Harbor, Wisconsin, USA

L. Readshaw, Division of Entomology, Commonwealth Scientific and Industrial Research Organization (CSIRO), Canberra City, Australia

G. Remaudière, Service de Lutte biologique contre les Insectes, Institut Pasteur, 28 Rue du Docteur Roux, 75015 Paris, France

J. G. Rodriguez, Department of Entomology, College of Agriculture, University of Kentucky, Lexington, Kentucky 40506, USA

D. Rosen, Department of Entomology, Faculty of Agriculture, The Hebrew University, PO Box 12, Rehovot, Israel

G. H. L. Rothschild, Division of Entomology, Commonwealth Scientific and Industrial Research Organization (CSIRO), Canberra City, Australia

G. E. Russell, Plant Breeding Institute, Trumpington, Cambridge, UK

K. H. Sanford, Canada Department of Agriculture, Research Station, Kentville, Nova Scotia, Canada

G. J. Snowball, Division of Entomology, Commonwealth Scientific and Industrial Research Organization (CSIRO), Mrs Macquarle's Road, Sydney NSW, Australia 2000

M. Tanaka, Kurume Branch, Horticultural Research Station, Ministry of Agriculture and Forestry, Mii-machi, Kurume-City, Fukuoka 830, Japan

T. Torii, Institute of Biological Control, Faculty of Agriculture, Kyushu University, Fukuoka 812, Japan

H. F. van Emden, Department of Agriculture and Horticulture, University of Reading, Earley Gate, Reading RG6 2AT, UK

D. F. Waterhouse, Division of Entomology, Commonwealth Scientific and Industrial Research Organization (CSIRO), Canberra City, Australia

M. J. Way, Imperial College Field Station, Silwood Park, Ascot, Berkshire SL5 7PY, UK

N. Wilding, Entomology Department, Rothamsted Experimental Station, Harpenden, Hertfordshire AL5 2JQ, UK

T. Wongsiri, Division of Entomology and Zoology, Department of Agriculture, Ministry of Agriculture & Co-operatives, Bangkhen, Bangkok 9, Thailand

K. Yano, Entomological Laboratory, Faculty of Agriculture, Kyushu University, Fukuoka 812, Japan

K. Yasumatsu, Entomological Laboratory, Faculty of Agriculture, Kyushu University, Fukuoka 812, Japan

1. Introduction

V. L. DELUCCHI

IBP efforts in biological control have coincided with the increasing concerns about the detrimental side-effects of pesticides, the ever-increasing costs of pest control and the deterioration of environmental quality. In recognition of the need for a more ecological approach in the control of noxious animals and plants than hitherto, research was reoriented to give greater emphasis to cultural practices, the use of plant resistance, and biological control in its broadest sense. This reorientation revealed the existence of extensive gaps in knowledge, not only of the interactions between populations of organisms but also of simple aspects of the behaviour of arthropods which have been considered economically important pests for decades. Moreover, the internationalisation of phytosanitary problems has shown that it is often advantageous to use comparable techniques of investigation throughout the world. The coordination and stimulation of biological control research under the aegis of IBP happened, therefore, at a decisive moment in the development of new control concepts and contributed significantly to identifying research needs and to standardising research methods, and in integrating the efforts of isolated scientists.

This special IBP volume on biological control and related approaches is the result of collaborative studies carried out by several hundred entomologists in more than 30 countries over a period of seven years. It covers five pest groups (Chapter 3 to 7) of major importance to primary crop productivity and human welfare. It does not represent the total IBP contribution to biological control, for in addition to these five groups or core projects selected for international coordination there were several other projects that could not be included in this volume because they were only of national significance.

The need for broad biological approach to fruit fly control

Fruit flies, one of the core projects, constitute an economic problem of worldwide importance. Crop losses resulting from their direct attacks are incalculable. In addition, they pose a serious deterrent to the introduction of important crop species or of new, high-quality varieties. Their impact on agricultural development is comparable with that of tsetse flies in tropical Africa. Fruit fly problems are particularly serious in developing areas where multivoltine polyphagous species occur and their control by insecticides is inefficient because of the existence of extensive reservoirs of wild hosts. The need to apply biological control methods in those areas is urgent.

Satisfactory control of some fruit flies has been achieved in tropical islands by introducing exotic parasites (van den Bosch & Haramoto, 1953; Clausen, Clancy & Chock, 1965) and in warm, temperate, continental areas by inundative releases of indigenous parasites (Monastero, 1967, 1969). In the latter case it was demonstrated that this biological control method is cheaper than control by chemicals and gives better results.

However, complete economic control of fruit flies by the use of parasites cannot be assured and this is a main reason why genetic methods have found great support and enthusiasm. The sterile-insect technique has achieved eradication of fruit flies in certain situations and in others has proved to be valuable in reducing natural populations below levels attained by other control techniques. The most recent effort of this kind concerns the 'medfly', *Ceratitis capitata* (Wied.), on the island of Procida, in the gulf of Naples, where complete control has been achieved for two consecutive seasons (1972 and 1973) through programmed releases of sterile insects, despite the continuous immigration of flies from the mainland (Cirio, personal communication).

A broad biological approach to fruit fly control, including the conservation of resident natural enemies and their mass release, introduction of exotic species, and mass release of sterile insects, is now being applied in Greece against the olive fly, *Dacus oleae* (Gmel.), under the aegis of FAO. The anticipated effect from these techniques in conjunction with the implementation of cultural practices and other non-biological control measures should contribute decisively to reducing substantially the losses caused by the olive fly.

Most of the biological control projects on fruit flies have been cooperative efforts by representatives of national and/or international agencies. Such projects generally entail a complex of operations and require research efforts of many kinds of specialists, as is clearly indicated in Chapter 3 of this volume.

The manipulation of the environment to control aphids

Aphids are also important pests. Some species are of world-wide distribution and many are known to be vectors of viruses causing plant disease. *Macrosiphum euphorbiae* Thomas, *Brevicoryne brassicae* L. and *Myzus persicae* Sulz. are examples of polyphagous species that attack their host plants almost everywhere they grow (van Emden *et al.*, 1969).

M. persicae is of particular importance because it attacks a great number of economic crops and is able to transmit a large number of virus diseases. It was consequently selected to receive special attention in this collaborative IBP study. The intensive use of organophosphate has increased the resistance of *M. persicae* to these chemicals and thus made its control increasingly difficult and expensive. The first observations on its resistance were made at the beginning of the IBP effort and *M. persicae* appeared to be a particularly suitable subject for biological control research. Attempts were made to solve this problem by developing a broad integrated control programme, utilising not only the impact of parasites, predators and pathogens, but also that of other natural mortality factors. However, the action of all these factors is usually insufficient for control of *M. persicae* in most economic crops and their impact occurs too late to lower the aphid populations to tolerable levels, especially at the time of peak virus spreading.

A new strategy was developed which offers great prospects. This strategy is to reduce the aphid's numbers on non-crop vegetation where the natural or semi-natural environment is more favourable than crop environments to the preservation of natural enemies. In addition, reductions are also contemplated by selection of aphid-resistant strains of the crop host plants and by cultural practices which increase the efficiency of resident parasites and predators.

Unfortunately, the quantitative effects of various factors which help to control *M. persicae* are only partly understood. Much research is needed to bridge the numerous information gaps in the population biology of this aphid. An outline for future work on its natural enemies was suggested by van Emden *et al.* (1969).

The importance of biological control of rice stem-borers

The need for international coordination in research on lepidopterous stem-borers of tropical graminaceous crops has often been emphasised. More than 20 years ago Jepson (1954) stated that 'the creation of a small international subcommittee of workers from all continents might be a useful first step' in defining the problems and stimulating an exchange of information on these stem-borers. This became a reality ten years later within the framework of IBP, although the activities of the IBP working group were restricted to rice. In 1970 the International Organization for Biological Control of Noxious Animals and Plants (IOBC) created a similar working group with a much wider scope, including all lepidopterous stem-borers of tropical graminaceous crops, thus implementing Jepson's suggestion. Most of the Lepidoptera associated with graminaceous crops are oligophagous and the dynamics of their populations are intimately connected with a variety of host plants in a given environment. However, experience has proved that with very extensive monocultures, such as rice, the limitation to the single crop facilitates the development of international collaborative efforts.

Rice is the staple food of about one third of the world population. As with other graminaceous crops, rice has long been considered a crop of relatively low financial returns and therefore largely a subsistence crop in peasant economy. At that stage of agricultural development lepidopterous stem-borers were already identified as pests and their control was achieved through cultural practices, for instance by cutting infested shoots, destroying stubble, burning, flooding, ploughing after harvest, rotation, alteration of planting dates, and using resistant varieties. It is generally accepted that natural enemies were at that time an essential element of the rice ecosystem and were able to efficiently control stem-borers in many ecological situations (Yasumatsu, 1967a).

In most of the developing parts of the world rice production has not evolved beyond the subsistence stage and yields remain low, whereas in developed countries yields have been considerably increased but can be maintained only by heavy applications of pesticides. As a result, stem-borers have become resistant to several insecticides and their control increasingly difficult (Waterhouse, 1967), and the former action of natural enemies has been almost entirely suppressed. This situation is not unique for rice. Experience with other crops, as for instance cotton, has shown that the optimal approach for pest control is never to rely solely on a single method (Li, 1972). One possible component alternative to the use of pesticides is biological control. In Asia rice stem-borers are attacked by 75 species of parasitic insects, four species of parasitic nematodes, and about 70 species of predators (Yasumatsu, 1967b). Other natural enemies are known from other continents (Rao, 1965) and others doubtless remain to be discovered. Many data are available today on the natural enemies of rice stem-borers (Yasumatsu & Torii, 1968; Torii, 1971c), but no attempt has yet been made to evaluate the true impact of the natural enemies on rice stem-borer population changes. There is some evidence that egg parasites are an important mortality factor, but this might be only apparent (Rothschild, 1970) as their mass release has often failed to give expected results. The ecological needs of the most important natural enemies are not well understood and nobody knows the effects of cultural practices on them or what might be done to increase their effectiveness (Nishida & Wongsiri, 1972). Chapter 5 deals with this limited state of our knowledge and suggests avenues of research whereby the information gaps may be closed.

Armoured scales: a group of insects suitable for biological control

The armoured scales and their relatives represent an area where the greatest experience in biological control has been gained. On a world basis about 40 % of the pest species which have been controlled biologically by introduction of exotic natural enemies have been coccids. These insects are considered to have certain biological attributes which commonly make them more suitable than most other pests to control by natural enemies (DeBach, 1964, 1971b). They are also easily transferred from one area to another and they often attack high-value crops that have a world-wide distribution. The early success with biological control of *Icerya purchasi* (Mask.) in more than 30 sub-tropical and warm temperate countries led to particular emphasis on biological control of scales and mealybugs. There have been successes on all continents, from the humid tropics (against the coconut scale, *Aspidiotus destructor* Sign.) to the arid zones (against the date palm scale, *Parlatoria blanchardi* T.T.) and are described in many publications. With the inception of this IBP effort scale insects were therefore regarded as an ideal subject to strengthen international cooperation in biological control.

Because much of the background information that can be readily put to use relates to diaspine scales, the IBP project has been restricted to this family. Biological control of diaspine scales, and notably those attacking citrus, has not only produced a large direct benefit to growers, but has also had a major impact on the development of certain disciplines and concepts, as for instance systematics and several aspects of ecology. The interpretation of many specific and sub-specific entities among *Aphelinidae*, *Encyrtidae* and other taxa of parasites, and more rarely among their hosts, has resulted from biological control research and this has put in a new light the taxonomy of parasitic species in general. From the ecological standpoint, studies of the natural enemy component has clarified in many situations the true role played by antagonistic organisms and contributed to the definition of criteria for the application of conventional biological control (DeBach, 1972) which were already indicated by Silvestri (1922).

There were many exchanges of cultures of parasites and predators of armoured scales between countries of all continents during the IBP project (cf. Chapter 6). It will be noted that California, because of its excellent facilities and the experience of its research workers, has established a centre for receipt and culture, and at the same time a source, of natural enemies for biological control of scale insects for the whole world. This kind of arrangement is most rational considering that accurate taxonomic processing of natural enemies is required whenever an introduction is envisaged.

Spider mites, an induced problem

The final core project discussed in this volume is on phytophagous mites, particularly the tetranychids or spider mites. It is well known that mites do not commonly cause extensive damage in situations that are not greatly influenced by man. Before World War II mite outbreaks were in general of only local importance and mainly caused by weather and cultural practices altering the roles of efficient natural enemies or the vigour of the host plants.

During the last 25 years the modifications in natural enemy efficiency and crop plant conditions have become intense and general, so that the control of these mites has become increasingly difficult. The intensive use of broad-spectrum insecticides and fungicides against key pests and changes in cultural practices (especially the use of fertilisers) have considerably modified, almost everywhere, the primitive environment and created conditions favourable for mite increase. The natural regulation of mite abundance has been disturbed by removing or impeding natural enemies, and the intrinsic power of increase of the mites has often been augmented by induced changes in plant metabolism. The resulting situation led to the introduction of acaricides in spray programmes and, as a consequence, to the development of resistance of pest mites to these chemicals, making mite control by chemicals increasingly expensive and finally almost impossible in some instances. In this way mites became serious pests of a wide variety of major crops and similar situations were experienced in many parts of the world almost simultaneously.

This international character of the spider mite problem was of fundamental importance for consideration in IBP. Another important aspect has been the need for a comprehensive approach to pest control of a given crop (Hoyt, 1969a; Readshaw, 1971) that would lead to the restoration of efficient predators and, at the same time, to the control of the other pests by techniques that preserve the natural enemies of the mites. Impressive results have been achieved for the protection of crops in glasshouses (Hussey & Bravenboer, 1971) and others are anticipated for several important field crops.

2. Definition and planning of the project

D. F. WATERHOUSE

The specific stimulus for the inclusion of a project in the general area of biological control within IBP arose from a proposal in the submission by the Australian Academy of Science to the IBP General Meeting in Paris in July 1964. This item, which had been prepared by D. F. Waterhouse, was entitled 'An international approach to the biological control of rice pests'. The Paris meeting endorsed the proposal under the broad heading Biological Control (control of biota by other biota), recommended interaction with the International Organization for Biological Control of Noxious Animals and Plants (IOBC/OILB), the Commonwealth Institute of Biological Control (CIBC) and other relevant bodies, and recommended a strengthening through IBP of cooperative research on biological control on an international basis. It also emphasised that adequate ecological studies were essential as a part of the programme.

The single entomologist at the Paris meeting, P. Grison (France), was appointed to Sectional Committee E (later to become UM, Use and Management of Natural Resources). This Sectional Committee met in Rome in February 1965 and requested P. Grison (as convenor), K. Yasumatsu (Japan) and D. F. Waterhouse to develop plans for a Working Group. Considerable activity occurred in ensuing months, much of it by correspondence. This activity involved, *inter alia*, individuals representing IOBC, CIBC and the newly formed International Advisory Committee on Biological Control. Many proposals were received from these and other leading biological control workers and a meeting of an enlarged Working Party on Biological Control took place in Rome in September 1965. In addition to P. Grison (Chairman), K. Yasumatsu and D. F. Waterhouse, the following were present: B. P. Beirne (Canada), C. B. Huffaker (USA) and, as observers, E. Biliotti (France), V. Delucchi (FAO), M. Laird (WHO), and F. Wilson (Australia/UK). I. A. Rubtzov (USSR), who had been appointed to this Working Party, was unable to attend.

The following broad policy guidelines were established by the Working Party:

(i) The research programmes undertaken should be basic in character, and should be concerned strictly with the control of biota by other biota. Thus, the ecological effects of the integration of chemical and biological methods of control should not be included. Where projects selected for the

7

programme had applied significance, the investigations were to be directed towards the elucidation of basic problems connected with the project rather than towards the achievement of biological control. Thus the establishment of new natural enemies in a country was to be incidental to any programme and not the immediate main objective of it.

(ii) Each project selected was to be of a kind that called for, and would greatly benefit from, simultaneous study by a number of countries. Each project was to be regarded essentially as a cooperative research enterprise, the main lines and methods of which were planned by agreement in advance. To this end the programme was to allow for the possibility of research being conducted in both temperate and tropical areas and in all the main geographical regions of the world.

(iii) Because the research undertaken was to be based essentially on national funds, which were likely to be very limited in many countries, it was seen to be highly advantageous to most countries if some of the projects selected could be fitted readily into their existing research programmes. This would facilitate collaboration in the programme and tend to make additional funds easier to obtain. However, this consideration was not to be allowed to undermine the essentially basic character of each project.

(iv) The programme was to be restricted to as few projects as possible, although to be influenced by the need for reasonably wide international collaboration. It was clear that too extensive a programme would reduce the concentration of effort and the depth to which the research could be pursued. It was desirable that at least one of the projects selected be a basic research study relevant to important pest problems of a major food source in developing countries.

Even with these guidelines the selection of a small number of projects from the long list available was an exceedingly difficult one. Many excellent and worthwhile proposals were either too limited in scope or too applied in nature, or were not viable without more finance than was likely to become available. After considerable deliberation it was decided that research on the following five 'core' projects would comprise in the programme:

An aphid of world importance (the green peach aphid, *Myzus persicae*).
Spider mites of the genus *Tetranychus*.
Rice pests, especially stem-borers.
Codling moth and other apple and pear tortricids.
Fruit flies of major economic importance.

To assist in the endorsement of these proposals by SCIBP it was decided that, for each core project, there would be prepared (*a*) a brief statement of the problem, (*b*) comments on the nature of the biological control approach and (*c*) information on possible IBP coordination and activation of a larger programme than existed at that stage.

A scientist who was an expert in the field concerned was then nominated as convenor for each core project. The Working Party further decided that each convenor was to have the responsibility for contacting other workers in the nominated field throughout the world, of forming a representative project committee and of developing specific plans for action and collaboration within his project. These latter plans were to be submitted to the Working Party for endorsement.

The Working Party also made tentative arrangements for comprehensive reviews to be prepared of the ecology, population dynamics and natural enemies of the groups of pests involved in the five core projects. These reviews were to be aimed at summarising existing knowledge and at outlining profitable fields of future work, especially collaborative work.

The Working Party also requested arrangements to be made for the production of a methodology manual with special reference to these five groups of pests so that population sampling, life tables, natural enemies etc., would be approached in as uniform a manner as possible by collaborating countries.

The foregoing proposals were warmly endorsed by the UM Committee at its subsequent meeting in Paris in November 1965.

The extent to which these plans were implemented will become evident in subsequent chapters of this book, so need no further elaboration here. There are, however, several additional matters that are appropriate to add to the historical record.

One is that J. Weiser (Czechoslovakia) was later added to the Working Party; another that, with the replacement on the UM Committee in 1968 of P. Grison by V. Delucchi, the latter automatically became Chairman of the Working Party; and a third that it became necessary to substitute one of the core projects with a replacement proposal. This occurred because it was not found possible to enlist the assistance of a suitable convenor of the project on codling moth and other apple and pear tortricids. The replacement, selected in 1968, was entitled 'Armoured scale insects (Diaspidinae) of major economic importance'.

Meetings of the UM Working Party on Biological Control were held in Tokyo (August 1966), Moscow (August 1968), Canberra (August 1971) and San Francisco (September 1973). General policy matters were discussed at each of these meetings, guidance and encouragement given where appropriate, SCIBP funds sought to permit meetings and other activities of the core project committees, and progress reports prepared which are incorporated in the central records of IBP.

9

3.   Fruit flies

Coordinator: M. A. BATEMAN

The Working Group on tephritid fruit flies was activated in September 1968. The first meeting was held in Rome, Italy, and attended by experts from 12 countries, and a series of collaborative activities was planned, including three research projects and the preparation of certain literature reviews and methodology manuals.

The aims of the group were re-examined at the second meeting held in Wädenswil, Switzerland, in September 1971, which was attended by 35 experts from 17 countries. Several of the original projects had by that time been completed, others were encouraged to continue, and some were abandoned. Certain projects were initiated at this meeting and some of these have since been completed.

A great deal of research was either initiated and organised by the IBP Working Group or was strongly stimulated as a result of association and collaboration between members. Most of the major advances in our knowledge and understanding of the biology of the Tephritidae, which occurred during the period the group was active, were in the areas covered by this report.

Most of the work was not directly related to the development or improvement of control methods, but it has provided a great deal of background information of the kind that is likely to prove useful for this purpose. The section on life tables and mortality factors, for example, goes some way towards filling considerable gaps in our knowledge of the causes of death during the pupal stage, and of the importance of such mortality factors as parasites, predators and disease. Such information will be essential for the eventual development of integrated control techniques for orchard pests. The sections on marking pheromones and colour attraction present data that could be utilised directly in control systems that seek to minimise the use of insecticides. Pheromone-like substances that deter oviposition in susceptible fruit have obvious potential in such systems, and colour traps without insecticides are already being used for localised control of the European cherry fly, *Rhagoletis cerasi* (L.), in Switzerland (Remund, 1971). Information on movements (p. 33 ff.) is useful for the efficient application of any control method, but in future years fruit fly control is likely to trend more and more towards the sup-

Contributors: M. A. Bateman, E. F. Boller, G. L. Bush, D. L. Chambers, A. P. Economopoulos, B. S. Fletcher, M. D. Huettel, V. Moericke, R. J. Prokopy.

pression of entire breeding populations (i.e. 'area' control) and in this context such information is indispensible. The section on sexual behaviour describes the recently discovered sex-attractant pheromones and provides background information for the development of new control methods based on interference with the mating process. The section on the population genetics of tephritids outlines the application of the technique of gel electrophoresis to fruit flies and describes its usefulness in studies on such topics as the effects of insecticides on gene frequencies, the delineation of host and geographic races, and the quality control of insects being mass-reared for sterile release or other biological control programmes.

Control methods in current use against tephritids throughout the world are of three main types: (i) cover-sprays with persistent insecticides, (ii) bait sprays based on protein hydrolysate preparations or more specific male attractants, and (iii) systemic insecticides. These techniques are effective against most pest species and are usually regarded as providing adequate protection for commercial orchards. But in most countries the major problem lies in the vast numbers of fruit trees growing wild or in backyard situations. The insects breeding in this fruit are rarely subjected to control procedures. They multiply virtually unchecked and are responsible for the destruction of immense quantities of fruit, as well as for the need for expensive quarantine and trade barriers.

The only solution to this problem would seem to be some form of broad area control system directed against entire populations rather than against small local aggregations. One such system is biological control. There have been numerous attempts to control tephritid populations by the introduction of hymenopterous parasites, but the results, while useful and encouraging in some instances, have rarely been spectacular. One real difficulty with the biological control of these insects is the extremely low economic threshold. For table fruit, for example, there is demand for a perfect product, and fruit to be exported to an area where fruit flies do not occur must be completely pest-free. Parasites alone cannot meet such stringent demands.

An alternative area control system is the sterile insect release method which has been shown to be effective, at least with tropical species, in 'island' situations (Steiner *et al.*, 1965a, 1970). Suppression or eradication of entire breeding populations has also been achieved with attractants such as protein hydrolysates or specific male lures (Steiner *et al.*, 1965a; Bateman, Friend & Hampshire, 1966). Research is continuing in several parts of the world on methods of reducing the present high costs of these techniques, and of making them more efficient, but this work is retarded by a lack of comprehensive and reliable ecological information.

From the beginning the working group recognised that serious gaps exist in our knowledge of the population ecology of the tephritids. However, it was impractical to organise a world-wide study in this field because of the long-term nature of the work and its inherent technical difficulties. Participants were encouraged to initiate their own ecological studies wherever possible, particularly field studies on populations, because of their importance in the development and improvement of control techniques.

There are, in fact, certain important and fundamental differences between different tephritid species in population structure. The family is divided into two major groups of species: the univoltine forms which usually have a winter diapause and inhabit the cool temperate regions of the earth, and the multivoltine forms which have no obvious diapause and inhabit the more tropical regions. Both forms are distributed in extensive regional populations fragmented into mosaics of smaller local populations, but there are marked differences between the two in the permanence of these local populations and in their interrelationships.

The tropical species tend to form local populations that are transient. Typically, a population will begin in a local area where host fruit is beginning to ripen. It will multiply rapidly and flourish for a limited time (usually until the fruit starts to diminish) often producing extremely high numbers of offspring. But after a limited time (usually when the supply of host fruit is declining) the population disintegrates. The individuals disperse and scatter, many to join new populations in other parts of the region where the environment is again favourable. Such species tend to be strong flyers with high capacities for dispersal and frequently a distinct post-teneral or juvenile dispersal phase. Observed flight ranges extend from 25 km to 150 km (Christenson & Foote, 1960; Shaw *et al.*, 1967) and there is a great deal of movement between populations throughout the region. Capacities for increase are high because of high fecundity and rapid development. Typically, generations overlap and are relatively short.

The temperate species, on the other hand, tend to form more permanent and stable local populations which inhabit the same areas of vegetation through successive years, with the univoltine life cycle synchronised with the regular appearance of larval food each summer. They are relatively weak flyers with low capacities for dispersal. Observed flight ranges are often less than 1 km, and no definite post-teneral dispersal phase has been reported. Capacities for increase are extremely low, relative to the tropical species, as fecundity is low and development slow. Numbers achieved are never extremely high, and there is little interchange between local populations.

In the terminology of MacArthur & Wilson (1967), the tropical pest species are predominately '*r* selected', whereas the temperate species are predominately '*k* selected'. The former are selected for colonising ability. Polyphagous individuals that disperse widely, reproduce rapidly, and exploit the environment to leave the maximum number of offspring, are strongly favoured. In temperate regions, on the other hand, fewer varieties of fruit are available and they ripen over a shorter period. Wide dispersal is hazardous and selection

13

favours highly specialised monophagous or oligophagous individuals which are adapted for co-existence with their fellows in relatively confined aggregations. This leads to the development of mechanisms for the efficient utilisation of resources and the avoidance of competition. For example, temperate species typically lay only one egg at each oviposition site, and some species mark the fruit with a pheromone to deter oviposition by other females (see p. 23 ff.).

An understanding of these different adaptive strategies is of some importance in the formulation of appropriate control measures for the various pest tephritids, especially where area control techniques are concerned. On theoretical grounds, classical biological control using introduced hymenopterous parasites is likely to be more effective against the k strategists – the temperate species – with their more stable and predictable annual populations, than against the highly mobile and less predictable r strategists. The sterile insect release method may also be expected to be more useful against k strategists, because the individual demes can be attacked separately and there will be little likelihood of rapid re-colonisation of disinfested areas. Methods involving the use of widespread baiting systems, on the other hand, may be more effective against the more tropical r selected species, because their greater activity and mobility will increase the probability that they will move into the effective range of the attractants. Risks of re-infestation will, o course, always be greater for r selected species, regardless of which control system is used.

The material achievements of the working group include the following:

(i) Collaborative projects completed on pupal mortality and natural enemies, marking pheromones, and colour attraction.

(ii) Projects on micro-organisms, attractants and population genetics partly completed.

(iii) A review article 'The ecology of fruit flies' published (Bateman, 1972) and one on zoogeography in preparation.

(iv) A manual of methodology 'The use of gel electrophoresis with tephritids' completed (Bush & Huettel, 1972).

(v) One 'Idea booklet' on 'Fruit Fly Movement' completed and circulated to members.

(vi) World lists of fruit fly workers (containing names, addresses and research interests) circulated to members and updated each year.

(vii) *Fruit Fly News Bulletin* circulated to members.

(viii) Exchange of reprints and research information.

At least as important as these, however, were the contacts and friendships established between the 50 or so active members of the working group. Fruit fly workers throughout the world were, for the first time, welded into a single cohesive entity with agreed aims and common goals, relatively modest though they were. Of particular importance was the bringing together of the two major groups of workers concerned with the temperate and tropical tephritids. Prior to IBP the gulf which separated these groups was far broader than was warranted by the taxonomic distances between the two groups of species. The bridging of that gulf has now been accomplished and the mutual stimulation and collaboration has helped to bring renewed impetus and improved perspective to fruit fly research throughout the world.

Life-table studies and pupal mortality

The need for intensified ecological studies on economically important fruit flies was recognised by the IBP Working Group. One area in particular in which the need for more information was emphasised, was in the mortality factors that act upon the stages of fruit flies associated with the soil – especially the pupal stage. Since most presently available information deals with parasitism it was agreed to emphasise two aspects on which less data were available, namely predation and the influence of abiotic factors. A research programme encompassing a wide range of tephritid species and geographical areas was initiated in November 1969 and a manual of methods was distributed to the participating specialists. The participants, fruit fly species studied, and the geographic areas involved were as follows:

Arambourg, Y.	*Dacus oleae* (Gmel.)	France (south)
Bateman, M. A.	*D. tryoni* (Frogg.)	Australia (south east)
Boller, E. F.	*Rhagoletis cerasi* (L.)	Switzerland (north west)
Cameron, P. J.	*R. pomonella* (Walsh)	Quebec, Canada
Cavalloro, R.	*D. oleae* (Gmel.)	Italy (north)
Cirio, U.	*Ceratitis capitata* (Wied.)	Italy (central)
Nakagawa, S.	*D. dorsalis* (Hend.)	Hawaii, USA (Hilo)
Rivard, I.	*R. pomonella* (Walsh)	Quebec, Canada
Sigwalt, B.	*C. capitata* (Wied.)	Tunisia
Vallo, V.	*R. cerasi* (L.)	CSSR (Slovakia south)

Materials and methods

Groups of ten freshly formed pupae of the fruit fly species under investigation (either from laboratory cultures or field-collected material) were placed at depths of 0, 2.5 and 5 cm in the soil under selected host trees, and either protected by cylindrical barriers made of 10 mm or 1 mm mesh screening or left unprotected (Fig. 1). Every experimental unit thus consisted of 3 × 10 pupae, and there were 36 units per tree laid out according to a design that included as variables, direction (tree quadrants) and distance from tree-trunk (4 quadrants × 3 distances × 3 levels of protection = 36 units). The units, which were placed carefully in the ground to avoid disturbing the surrounding vegetation, were exposed to the influence of abiotic and biotic mortality factors for increasing periods depending on pupal development time. The time interval

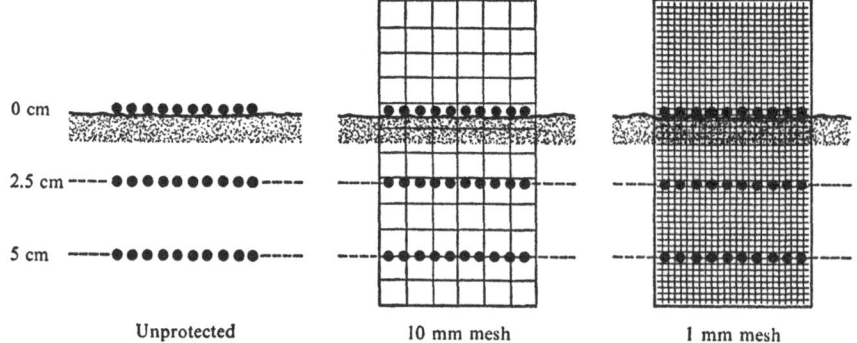

Fig. 1. Diagrammatic presentation of mechanical barriers used to protect fruit fly pupae from predators on the soil surface and deeper soil strata.

between the burial of the pupae to the emergence of the adults was divided into six sampling periods: every day (in the case of *D. dorsalis* in Hawaii), every week (*D. tryoni* in Australia) or every month (species in the temperate zones).

After exposure, pupae from the three different soil depths were separated and the soil examined microscopically for fragments of pupae that had been attacked by predators. For every unit and soil depth, the numbers of pupae that were healthy or had produced adults, and of those that were diseased, desiccated, parasitised, damaged by predators, or missing (presumed to have been taken by predators) were recorded. As far as possible the experiments were repeated at different times of the year, to see if there were seasonal changes, and in different places to assess the effects of different soil types, plant cover, host varieties, etc.

Results

The pattern of predation

Examples of the data obtained (Fig. 2) show the general pattern of predation occurring under the various experimental conditions. The average amount of predator action is expressed as percentages of pupae destroyed with increasing exposure time, at the soil surface and at the pooled soil depths of 2.5 and 5 cm. The differences in the amount of predation are due to the characteristics of the bioclimatic or edaphic regions, and to the time of the year when the experiments were done. The latter especially, had a profound influence on the activity of predators.

Results from a detailed study of predation on the pupal stage of *R. cerasi* in Switzerland in 1963 to 1965 are presented in Fig. 3 and can be considered as representative of many univoltine fruit fly species of the temperate zone (Boller, 1966).

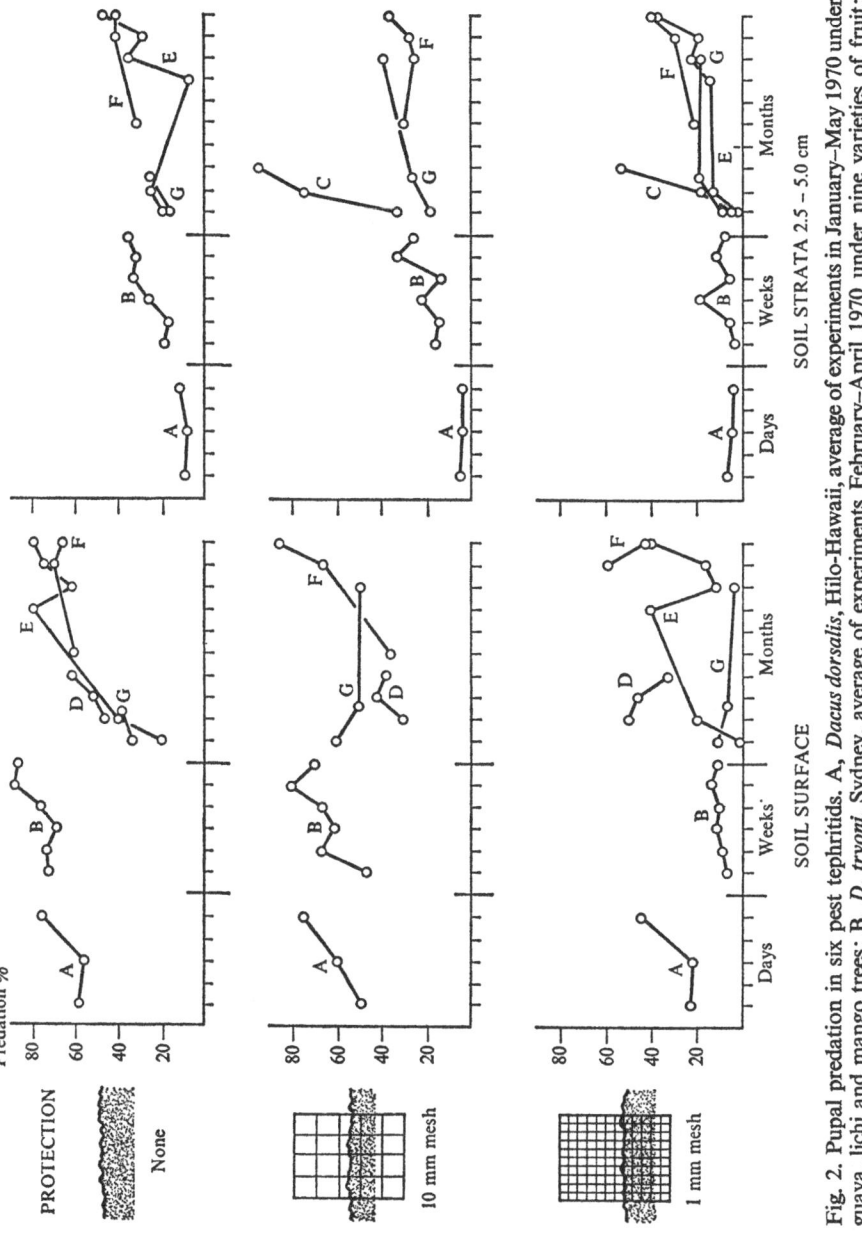

Fig. 2. Pupal predation in six pest tephritids. A, *Dacus dorsalis*, Hilo-Hawaii, average of experiments in January–May 1970 under guava, lichi and mango trees; B, *D. tryoni*, Sydney, average of experiments February–April 1970 under nine varieties of fruit; C, *D. oleae*, Antibes, October 1970–February 1971; D, *Ceratitis capitata*, Rome, October 1970–March 1971; E, *Rhagoletis pomonella*, Quebec, September 1970–July 1971; F, *R. cerasi*, Bratislava, August 1970–April 1971; and G, *R. cerasi*, Switzerland, August 1970–April 1971.

17

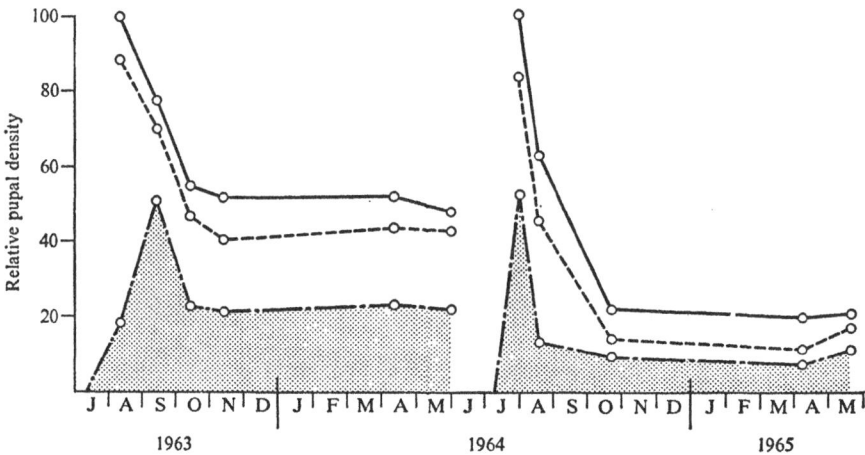

Fig. 3. Changes in the relative density of pupae of *Rhagoletis cerasi* in two consecutive years in north-west Switzerland: —, total pupae; --- intact pupae; —·— parasitised pupae.

Table 1. *Components of pupal mortality (%) assessed at the completion of the pupal stage*

				R. cerasi[d]			
				1963/64		1964/65	
Mortality factor	C. capitata[a]	D. tryoni[b]	R. pomonella[c]				
	A	A	A	A	B	A	B
Predation	38.0	38.7	10.1	51.6	51.6	79.5	79.5
Disease	12.1	1.5	6.2	4.8	10.0	2.1	10.0
Desiccation	14.4	4.7	—	—	—	—	—
Parasitism	30.9	0	29.0	21.8	50.1	11.7	33.3
Other	—	8.8	—	4.5	20.8	1.1	15.0
Total pupal mortality	95.4	53.7	45.3	82.8	—	94.4	—

[a] Cirio, personal communication; [b] Bateman & Elliott (1971); [c] Cameron (1971); [d] Boller (1966).

A, relative percentages of pupal mortality; B, percentages of pupae destroyed by the individual mortality factor if it had acted alone on the initial pupal population.

The components of pupal mortality

The calculation of the percentages contributed by the individual components to the total pupal mortality (Table 1) presented some problems, especially where pupae remained a long period of time in the ground and where various mortality factors acted in sequence. An earlier life-table study in Switzerland (Boller, 1966) had shown that the only parameter that could be measured with accuracy at each sampling was the decrease, due to predation, in pupal density

18

Table 2. *Life-tables for two fruit fly species*

| | R. pomonella (Quebec)* (Cameron, 1971) 1970/71 | | | | R. cerasi (Switzerland)* (Boller, 1966) | | | |
| | Early varieties | | Late varieties | | 1963/64 | | 1964/65 | |
Stage	lx	100qx	lx	100qx	lx	100qx	lx	100qx
Egg	100	5.3 (0–9.5)	100	0.8 (0–2.6)	100	30.0	100	44.9
Larva$_{I-II}$	94.7	81.5 (79.7–83.8)	99.2	74.1 (55.4–86.3)	70	5.0	55.1	5.0
Larva$_{III}$	17.5	73.2 (45.8–89.0)	24.9	90.0 (88.2–100)	66.5	35.0	52.3	58.3
Pupa	4.7	53.7 (41.2–75.3)	2.5	82.8 (45.8–100)	43.2	82.8	21.8	94.3
Adult	1.8		0.4		7.4		1.3	

lx, number of individuals entering stage, expressed in relative figures; 100*qx*, percentage mortality occuring during given life stage.
* Both studies were carried out on individual trees in unsprayed abandoned orchards.

between two population estimates. To determine how many diseased or parasitised pupae were eliminated by predators (a factor that would – if neglected – lead to an overestimation of the importance of the predators) the following sequence of mortality factors that influenced the numbers of pupae was established: predators, diseases (including desiccation), parasites and prolonged diapause. Accordingly, the results in Table 1 are presented in two ways: (i) as percentages calculated on the basis of actual numbers of pupae observed (*A*), and (ii) as percentages of pupae destroyed by the individual mortality factor if it had acted alone on the initial pupal population (*B*) (Morris, 1957).

Importance of pupal mortality

Polyvoltine species with overlapping generations present major problems for the establishment of classical life-tables, and it is not surprising, therefore, that the three life-table studies on fruit flies that are currently in progress are on univoltine species. The data available for *R. pomonella* (Cameron, personal communication) and *R. cerasi* (Boller, 1966; Boller & Remund, personal communication) are presented in Table 2 and allow some preliminary conclusions to be reached on the importance of pupal mortality in the total mortality occurring during the entire life cycle.

For simplicity, the *R. pomonella* data are expressed as averages of eight life tables derived from eight individual trees. The components of the mortalities of the individual life stages are omitted, but the ranges of the mortalities observed are given in parentheses beside the calculated means.

Influence of abiotic factors

No clear picture was gained from a study of the influence of moisture and soil-type on pupal mortality. Bateman (personal communication) found that, during a prolonged dry period, more pupae of *D. tryoni* died beneath the soil, apparently from desiccation, than on the soil surface where they received slightly more moisture as dew and light rain. Cirio (personal communication) found pupal mortality was 50 % higher in clay than in volcanic soil, but only following rain. Cavalloro (personal communication) and Sigwalt (personal communication) observed that the depth at which pupation occurred increased with decreasing moisture content of the soil.

Discussion

The data summarised in Fig. 2 show some interesting patterns of pupal predation, but since the climate differed from place to place and the observations were carried out during different seasons, comparisons between the curves must be made with caution.

The pattern of predation in polyvoltine species (curves A, B, C, D)

The curves labelled A (Fig. 2) show the patterns of predation on *D. dorsalis* at Hilo, Hawaii (twentieth parallel of latitude). Mortality due to predation reached a very high level within a few days of exposure of the pupae on the surface of the soil. Apparently it was caused mainly by predators with body widths above 1 mm, and it frequently reached 100 % after an exposure time of only 24 hours. The effects of these unidentified surface-feeding predators were greatly diminished by 1 mm mesh barriers. Predation was practically zero in the soil at depths of 2.5 and 5 cm.

A similar pattern was observed for *D. tryoni* (curves B) in Australia (thirty-fifth parallel). The amount of predation was strongly influenced by both mesh size and depth of burial, the mortality increasing with increasing accessibility of the pupae to predators. The interaction between these two factors was so high that they cannot be considered independently. There was a general trend towards increasing mortality caused by predation with longer exposure time, but the time during the season when the pupae were buried, and the variety of fruit tree under which they were buried appeared to have little influence. Ants were probably the most effective surface predators, although most species were too large to enter the fine mesh containers. The difference between the coarse mesh and unprotected categories suggests that some predators were deterred even by the 10 mm mesh. Most underground predators were excluded by the fine mesh, but again there was an increase in predation with increasing accessibility. The effect of mesh size was not as great underground as it was on the surface (Bateman, personal communication; Bateman & Elliott, 1971).

The results of the study on *D. oleae* in France (Arambourg, personal communication) are given in curves G. Under these Mediterranean conditions the predators were more active during winter than during the hot dry summer at a depth of 5 cm where all observations were made. Apparently the predators prefer deeper zones in summer than in winter. There are indications that ants (especially the abundant *Crematogaster* spp.) are the most important predators. This was confirmed by observations of ants transporting pupae of *D. oleae* in Northern Italy (Cavalloro, personal communication). Mites (*Epierius* spp.) were observed in the pupal samples and a myriapod, *Monotarsobius crassipes* L., is also suspected of being a predator. Mortality due to factors other than predators seemed to be low during the summer but increased with increasing duration of the pupal stage in the winter.

The pattern of predation in univoltine and oligovoltine species (curves E, F, G, Fig. 2)

R. pomonella and *R. cerasi* have similar patterns of predation. Both of these species were studied between the forty-fifth and forty-eighth parallels, they both normally produce one generation per year, and both spend some 10 months in diapause in the pupal stage. There is, however, a difference of about two months between the two species in the time of pupation. The long period that the pupae are available to the predators compensates for the reduced activity of the latter in the cooler climate of the temperate zones.

Investigations on *R. cerasi* showed that up to 70 % of the pupae disappeared between July and early October, that the pupal density remained more or less constant throughout the winter, and that it sometimes decreased again slightly in spring (Fig. 3) (Boller, 1966). Curves G and F of Fig. 2 indicate that more pupae were attacked on the soil surface than beneath the soil, and that 1 mm mesh provided a barrier against some of the major predators. Surprisingly, a considerable degree of predation occurred at 5 cm depth even in the fine mesh containers. Laboratory tests have indicated that the predominant ant species (especially *Myrmica laevinodis* Nyl.) can discover the pupae only by chance, and then are not able to transport or open them. On the other hand carabid beetles and myriapods devoured a large number of pupae in the laboratory. Carabids are frequently abundant under cherry trees and some myriapods were found during the examination of buried unprotected pupae and those in 10 mm mesh containers.

Serological tests failed to identify the predators in nature (Burki & Boller, 1969) and the identity of the predators acting upon pupae protected by the 1 mm mesh is unknown. We suspect that certain mite species are involved, but it remains to be determined whether these actually entered the intact puparia, or merely took advantage of injury caused by some other agent.

Observations on *R. pomonella* (curves E of Fig. 2) indicate that predation is pronounced on the surface and that it also reaches considerable levels underground at depths where the pupae are normally found. The evidence indicates that pupae on the soil surface were mostly killed by larger predators, whereas the influence of small predators increased deeper in the soil. Carabids, crickets and ants are considered to have been the main predators (Cameron, 1971; Rivard, personal communication; Monteith, personal communication).

Despite large differences in the results we can recognise climatic effects on predation rates for *Dacus* spp. Differences between exposed and buried pupae are most evident in warmer climates. On the other hand, small organisms appear to be more important than larger predators in the temperate zones.

The components of pupal mortality

The data available (Table 1) indicate that predation is, in most cases, the predominant cause of pupal mortality. This is particularly so in warmer areas where polyvoltine species are exposed for relatively short periods. However, additional mortality agents become increasingly important with increasing exposure, in species that undergo a pupal diapause. Parasites and physical factors such as desiccation are certainly important, whereas bacterial and fungal diseases seem to play only a minor role.

Life-table studies (Table 2) demonstrate that important regulatory processes occur during the pupal stage. Although in *R. pomonella* the high mortalities observed in egg and larval stages show no great variance, pupal mortality is suspected of having a regulatory function (Cameron, 1971). In *R. cerasi* also, predation seems to be relatively constant, although considerable variation occurs in mortality caused by parasites. In this species, however, if the overall generation mortality is compared with natality, it seems clear that the reproductive processes are more important for population regulation (Boller, 1966).

Conclusion

Despite the diversity of conditions, which makes a direct comparison of results difficult, it is evident that mortality caused by predators (which has never been studied in such a systematic manner before) is important in regulating fruit fly densities. This is highly relevant to future control programmes, especially in those such as sterile-insect release programmes where fruit fly pupae may be taken to the field. Although few predators have been identified, and despite the possibility that artificial increase of predator

Table 3. *Species of Tephritidae which tend to disperse their eggs uniformly among available oviposition sites*

Species	References
Aciurina ferruginea (Doane)	Tauber & Tauber (1967)
Carpomyia vesuviana Costa	Batra (1954)
Dacus dorsalis Hend.	Newell & Haramoto (1968)
D. oleae (Gmel.)	Martin (1948)
D. tryoni (Frogg.)	Monro (1967)
D. tsuneonis Miyake	Miyake (1919)
Epochra canadensis Loew	Severin (1917)
Eurosta solidaginis (Fitch)	Uhler (1951)
Gonioglossum wiedemanni Meigen	Silvestri (1920)
Rhagoletis alternata Fall.	Persson (1963)
R. basiola (Osten Sacken)	Balduf (1959)
R. cerasi (L.)	Häfliger (1953)
R. cingulata (Loew)	Caesar & Spencer (1915)
R. completa Cresson	Boyce (1934)
R. fausta (Osten Sacken)	Caesar & Spencer (1915)
R. pomonella (Walsh)	LeRoux & Mukerji (1963)
R. ribicola Doane	Jones (1937)
R. suavis (Loew)	Gambrell (1931)
Zonosemata electa (Say)	Foott (1968)
Z. vittigera (Coq.)	Goeden & Ricker (1971)

populations might not prove feasible, we now have a fuller understanding of the causes of pupal mortality and its importance in the population dynamics of tephritids.

Significance of fly marking of oviposition site

A number of parasitic Hymenoptera tend towards uniformity of egg dispersion among available insect hosts, and in several of these species, a marking substance deposited by the female following oviposition has been found to deter egg-laying by subsequent females (Salt, 1937; Greany & Oatman, 1972; Vinson, 1972). Likewise, in at least 20 species of tephritids (Table 3), there is some evidence of a tendency to disperse eggs uniformly among available oviposition sites, and recent findings have revealed that in *D. oleae* and *R. pomonella* this egg dispersion is also mediated by a marking substance laid down following oviposition (Cirio, 1971; Prokopy, 1972a). It has also been observed in a number of tephritid species (Table 4) that copulation is initiated at or very near the oviposition site. Recent evidence indicates that a marking substance deposited at the oviposition site influences mating behaviour in at least one of these species, *R. pomonella* (Prokopy & Bush, 1972).

In view of their potential importance to integrated control, these findings prompted an investigation to determine how widespread among pest tephritids was this behaviour of marking the oviposition site.

Table 4. *Species of Tephritidae for which there is evidence that mating is initiated at or near the oviposition site*

Species	References
Aciurina ferruginea (Doane)	Tauber & Tauber (1967)
Ceratitis capitata (Wied.)	Feron (1962)
Dacus tsuneonis Miyake	Miyake (1919)
Eurosta solidaginis (Fitch)	Uhler (1951)
Eutreta spp.	Stoltzfus & Foote (1965)
Rhagoletis cerasi (L.)	Bush and Boller (personal communication)
R. completa Cresson	Boyce (1934)
R. fausta (Osten Sacken)	Prokopy (personal communication)
R. pomonella (Walsh)	Prokopy, Bennett & Bush (1971)
R. suavis (Loew)	Brooks (1921)
Urophora cardui L.	Zwölfer, Englert & Pattullo (1970)
U. siruna-seva (Hering)	Zwölfer (1969)
Zonosemata electa (Say)	Peterson (1923)
Z. vittigera (Coq.)	Goeden & Ricker (1971)

Influence of oviposition-site marking on oviposition behaviour

Six tephritid species have been investigated. In *D. oleae*, Cirio (1971) observed that, during oviposition, while the ovipositor is still inserted into the olive, the female olive fly sucks up the juice exuding from the puncture. Immediately following oviposition, the female withdraws her ovipositor into her abdomen and then moves about the fruit, periodically touching her proboscis to the fruit surface and depositing fluid. Nearly all females subsequently visiting olives marked with fluid deposited by a previous female refrained from ovipositing in them but readily oviposited in adjacent unmarked olives. Although Fiestas Ros de Ursinos *et al.* (1972) found that the odour of olives is attractive to olive flies, the constituent that attracts the flies is probably different from that involved in deterring oviposition. Cirio demonstrated that the deterrent is present in the water-soluble portion of the juice and not in the oil-soluble portion, suggesting that rain might act to wash some or most of it away.

Prokopy (1972*a*) observed that, immediately following oviposition, a female *R. pomonella* circles around the host fruit for about 30 seconds, dragging her fully extended ovipositor on the fruit surface behind her. Experiments in the field and laboratory showed that during this act of ovipositor-dragging, the female deposits on the fruit surface a pheromone which acts to deter subsequent attempts at oviposition. Contrary to the situation in *D. oleae*, the juice exuding from a punctured host fruit, when spread over the fruit surface, did not act to deter subsequent probing. The degree to which other *R. pomonella* females arriving on pheromone-marked fruit were deterred from probing was found to depend upon the size of the fruit and the strength of the internal drive of the females to oviposit. Where the fruit was small (e.g. cherries of 15 mm diameter) and the drive to oviposit moderate (as among the field population), there was almost total deterrence for the four days of the experiment. However, where the fruit was large (apples of 55 mm diameter) or the drive to oviposit strong, as among laboratory flies deprived of oviposition sites, marking by only one female was not a strong deterrent. Prokopy (1972*a*) suggested that the area of fruit surface marked following a single oviposition was related to the amount of food or space required by a single larva. In further laboratory studies, he showed that fewer *R. pomonella* females attempted to oviposit in an artificial fruit previously occupied by a large number of males than in a clean fruit. This suggests that, whatever its nature, the marking pheromone is produced not only by females, but, at least to some extent, by males as well.

Cirio (1973) observed that, following oviposition, *R. completa* females drag their ovipositors over the surface of walnuts for an average of 22 seconds, and demonstrated that this deposits a marking pheromone. He found in laboratory tests that females attempted oviposition much less often in walnuts recently marked by one to three previous females, than in adjacent unmarked walnuts or in walnuts whose surface had been swabbed with juice from the husk. The marking pheromone lost its effectiveness after five days and it could be washed off with water. Cirio (1973) also showed that chemo-physical alteration of the walnut husk by the developing larvae was an even more potent deterrent to repeated oviposition than was the marking pheromone.

No evidence has been found for the existence of oviposition deterrents in three other tephritid species which were investigated. Fletcher (personal communication) offered small (2.5 cm diameter or less) cherry guava fruits to females of *D. tryoni* in an orchard and found that they attempted oviposition just as often into fruits that had received previous single or multiple ovipositions as into control fruit that had not been probed. Nonetheless, a tendency towards uniformity in dispersion of eggs among available fruits has been reported to occur in *D. tryoni* and it has been suggested that aggressive interactions between flies on the host fruit may be one mechanism by which this is achieved (Pritchard, 1969). Nakagawa (personal communication) offered guavas and papayas to females of *Ceratitis capitata* and *D. dorsalis* both in natural conditions and in large outdoor cages, and found that females of both species attempted oviposition just as often in fruits which had recently received a large number of ovipositions as in fruit that had received none. Interestingly, females of *D. tryoni*, *C. capitata*, and *D. dorsalis*, like females of *R. pomonella* and *R. completa*, engage in at least some dragging of the ovipositor across the fruit surface following oviposition. One can only speculate as to the possible biological significance of such dragging in these three species: (i) some pheromone may indeed be deposited, but it is not an oviposition deterrent; (ii) early in the evolutionary history of these species pheromone was deposited during the dragging process, but such is no longer the case; or (iii) the dragging may serve to clean the ovipositor following oviposition, as was suggested by Feron (1962).

Influence of oviposition-site marking on mating behaviour

The possible influence of oviposition-site marking on mating behaviour has received even less attention than its influence on oviposition behaviour.

In the only published study on this subject, Prokopy & Bush (1972) found that a pheromone deposited by mature *R. pomonella* females on the surface of real and artificial fruit caused arriving males to remain two to five times longer there than on control fruit not previously exposed to any flies. The male-arresting effect of the pheromone was found to largely dissipate within a few hours after deposition. It was also found that mature males and immature females deposited a male-arresting substance on fruit but in lesser amount than that deposited by mature females. Recent evidence (Prokopy, personal communication) indicates that females arriving on such female or male-marked fruit spend less time there than on unmarked fruit. Prokopy & Bush (1972) proposed that the male-arresting pheromone communicates to the male the current or recent presence of other apple maggot flies, especially mature females, in the area, thus increasing the probability of a male's encountering a female if he remains in the vicinity. It is also suggested that the male-arresting pheromone might stimulate sexual aggressiveness in males. The chemical relationship of the male-arresting pheromone to the pheromone deterring repeated oviposition in *R. pomonella* is unknown.

There is some evidence that marking of the oviposition site may have an influence on assembly for mating in three other tephritid species. Thus, males of *R. cerasi* have been found to spend about twice as much time on artificial fruit recently exposed to mature females and males as on unexposed fruit (Boller, personal communication). *R. completa* females have been found to spend a longer time on fruit recently occupied by males (and a shorter time on fruit recently occupied by females) than on unmarked fruit (Cirio, personal communication). Finally, observations by Zwölfer (personal communication) indicate that males as well as females of *Urophora cardui* (L.) are arrested on their host plant, the thistle *Cirsium arvense* (L.), by a marking pheromone deposited there by the males.

Importance of oviposition-site marking to control

As mentioned by both Cirio (1971) and Prokopy (1972*a*), the discovery of oviposition deterrent substances seems to offer great potential in terms of new approaches to control. Of course, any realisation of this potential depends first upon the isolation, identification, and synthesis of the active components.

As yet, none of the tephritid oviposition-site marking substances have been identified.

It is envisaged that, once synthesised, the active component(s) of the oviposition-deterring substance could be sprayed on susceptible trees. The amount of substance and the frequency of application needed to achieve effective deterrence would depend, at least to some extent, upon the strength of the oviposition drive of the females in the vicinity. Selected unsprayed sites within or near the orchard could be equipped with appropriate traps to attract and capture the females deterred from ovipositing on the sprayed trees. Similarly, the synthetic arresting substances for males or females might be sprayed on selected trees equipped with appropriate traps to capture the aggregating flies. Such an integration of oviposition-site marking substance and attractive traps might lead to an acceptable level of population suppression.

Conclusion

The discovery that females of *D. oleae*, *R. pomonella*, and *R. completa* mark their oviposition sites with fruit juice or pheromones which deter repeated oviposition, and that females and/or males of *R. pomonella*, *R. cerasi*, *R. completa* and *U. cardui*, mark oviposition sites with pheromones which influence assembly for mating, is of considerable importance to a better understanding of the biology of these species, as well as to possible new approaches to their control. The fact that no evidence was found for the existence of oviposition deterrents in the three other species of tephritids investigated (*D. tryoni*, *D. dorsalis* and *Ceratitis capitata*) should not be taken to mean that marking of the oviposition site is confined to only a few species in this family. Indeed, because of their tendency to distribute eggs somewhat uniformly among available oviposition sites, or their tendency to initiate mating at or near oviposition sites, many of the species listed in Tables 3 and 4 may be found to exhibit some sort of oviposition-site marking or recognition behaviour.

Oviposition deterrents may be most prevalent among those species that, like *D. oleae* and *R. pomonella*, deposit a single egg per cavity per oviposition and that either infest relatively small fruits or are gall-formers. But the demonstration of an oviposition-deterring pheromone in *R. completa*, a species that lays many eggs per cavity per oviposition illustrates that this phenomenon can also occur in other kinds of species.

Response to colour stimuli

In recent years it has been found that adults of certain tephritid species (e.g. *R. pomonella* and *R. cerasi*) are often strongly attracted to surfaces coloured yellow, green, or orange, in marked preference to similar surfaces coloured

27

white or various shades of grey (Prokopy, 1972*b*). It is supposed that these responses to colour have biological significance, the attraction to green, for example, being probably related to the fact that the flies find food and oviposition sites on or near green leaves. Besides colour attraction there is a reaction to specific shapes such as spheres, spheres with spikes, etc. (Prokopy, 1968; Zwölfer, 1968) but this is outside the scope of this report.

During the second meeting of the IBP Working Group it was decided to initiate a study of the prevalence of colour responses among fruit flies in various parts of the world, and to attempt to determine the colour which elicits the strongest reaction. The best place for suspending colour traps in trees in different seasons, the sex ratio of the flies responding, and the state of development of their ovaries, were also to be determined. In addition, a comparison was proposed between the attractiveness of traps depending upon colour with that of traps of various other kinds. The aim was to explore the possibility of developing a new method of reducing fruit fly populations, or of forecasting or monitoring infestations.

Participants in this study:

M. A. Bateman and B. S. Fletcher (1)	Sydney, Australia	*D. tryoni*
E. Bohlen (2)	Agadir, Morocco	*C. capitata*
P. Fimiani (3)	Portici, Italy	*C. capitata*
V. Moericke (4)	Bonn, Germany	*Platyparea poiciloptera* Schr.
S. Nakagawa (5)	Hawaii, USA	*C. capitata* *D. cucurbitae* Coq. *D. dorsalis*
R. Rivard (6)	Quebec, Canada	*R. pomonella*
A. Tominić (7)	Split, Yugoslavia	*C. capitata* *D. oleae*
R. Traboulsi (8)	Fanar, Lebanon	*D. oleae*
V. Vallo (9)	Ivanka pri Dunaji, Czechoslovakia	*R. cerasi*

Information attributable to an individual participant is identified in the text by the number following his name in this list.

Methods

Five colours were tested: (i) Fluorescent yellow; Day Glo Saturn Yellow, hue similar to ripe lemons. (ii) Yellow; enamel, hue similar to sunflowers. (iii) Fluorescent green; hue similar to green leaves but more intense. (iv) Orange; enamel, hue similar to ripe oranges. (v) Foil; aluminium foil (neutral colour with high reflectance).

Traps: cardboard sheets 15×20 cm and slotted in the middle were fitted together in pairs to form X-shaped, four-winged traps. Four traps were placed in a tree, one per side in each cardinal direction (N, E, S, W) about 2 m above the ground and as close to the periphery of the crown as possible. The wings of the trap were coated with sticky entrapping material, usually bird tanglefoot.

Flies were collected from the traps on the third day after setting up and subsequently three times per month (weekly in the case of *D. tryoni*) during the season. Records were kept of the number of each sex recovered from each of the eight surfaces of the traps. These surfaces were numbered from one to eight progressing clockwise from the north facing wing.

Results and discussion
Colour versus aluminium foil

Six fruit fly species were attracted to coloured traps in much greater numbers than to foil. The two exceptions were *D. cucurbitae* (5) whose catches on foil only surpassed those on fluorescent yellow in July when mostly males were captured (Table 5), and *P. poiciloptera* (4) which was consistently more attracted to foil.

Comparisons between hues

The most attractive pigment was fluorescent yellow, which surpassed yellow either due to its different hue, its greater reflectance of the yellow wave lengths, or both. In *D. tryoni* the difference between fluorescent yellow and yellow did not reach statistical significance (1). Fluorescent green was inferior to both yellow pigments, as was orange, indicating that fluorescence is not a critical factor. The preference for yellow over orange may have been influenced by the lower reflectance of the orange pigment.

Trap placement and orientation

In five tests the traps were placed in the cardinal positions on the trees (Table 6). Highest catches occurred in the south position for *R. pomonella* (6), *R. cerasi* (9), *C. capitata* (2) and *D. dorsalis* (5) and in the west position in *D. oleae* (8) and *C. capitata* (5). For *D. cucurbitae* (5) no positional effect was discerned. Positions which resulted in particularly low catches were east for *D. oleae* (8) and *D. dorsalis* (5) and west for *R. pomonella* (6). In tests with *D. oleae* (8) (Table 7), the walls of the traps facing east caught most of the flies in all tree positions and with all colours. The difference was greatest in traps on the west side of the tree, but the tendency was the same on the east side. The inner surfaces of the traps caught more flies than the outer, particularly on the west side of the tree. By contrast, no recognisable trends relative to position were observed in tests with *D. cucurbitae* (5).

Table 5. Responses of fruit flies to colour traps

Species	Country and author*	Dates of experiment	Totals caught	%♀	Percentage responding to each colour FY	%♀	NY	%♀	FGr	%♀	NO	%♀	ALU	%♀
C. capitata	Morocco (2)	29 Sept–14 Oct	1839	53	45.7	56	16.9	41	19.7	56	14.1	52	3.7	—
	Italy (3)	11 Jul–25 Oct	1165	75	31.8	73	30.9	77	18.8	58	12.4	77	6.0	82
	Hawaii (5)	8 Jun–20 Sept	53	30	43.4	33	34.0	33	—	—	15.1	—	7.5	—
	Hawaii (5)	4 Oct–27 Dec	81	56	27.2	32	54.3	64	14.8	—	—	—	3.7	—
	Yugoslavia (7)	7 Sept–26 Sept	324	62	54.7	66	36.1	48	—	—	9.5	58	0	—
	Yugoslavia (7)	7 Sept–26 Sept	359	45	40.9	35	34.5	46	—	—	25.0	58	0	—
D. oleae	Lebanon (8)	13 Jul–16 Oct	7133	25	41.8	21	34.7	30	21.3	23	—	—	2.1	34
R. pomonella	Canada (6)	11 Jul–23 Aug	428	60	47.0	56	18.2	65	17.7	58	13.7	62	3.4	60
R. cerasi	CSSR (9)	25 May–6 Jul	5214	47	36.6	43	33.7	47	11.7	53	17.9	53	0.1	—
D. dorsalis	Hawaii (5)	8 Jun–20 Sept	45	33	44.4	—	42.2	—	—	—	11.1	—	2.2	—
	Hawaii (5)	4 Oct–27 Dec	33	45	42.4	3	36.4	7	21.1	—	—	—	0	—
D. cucurbitae	Hawaii (5)	8 Jun–20 Sept	274	3	31.4	3	21.9	7	—	—	9.1	0	37.6	2
	Hawaii (5)	4 Oct–27 Dec	24	42	45.8	—	29.2	—	20.8	—	—	—	4.2	—
P. poiciloptera	Germany (4)	20 May–19 Jun	253	61	13.1	64	19.9	61	17.1	51	18.3	68	31.5	59
D. tryoni	Australia (1)	15 Mar–29 Mar	3754	48	32.6	46	34.5	45	16.0	49	16.3	47	0.6	62

FY, fluorescent yellow; NY, normal yellow; FGr, fluorescent green; NO, orange; ALU, silver aluminium foil.
* See authors listed on p. 28.

30

Table 6. *Distribution of catches of fruit flies on different sides of the trees*

Species and author	Percentage of flies caught on different quadrants of trees				
	North	East	South	West	Totals
C. capitata (2)	21	26	33	20	1839
(5)	26	23	21	30	53
D. oleae (8)	30	15	19	36	7133
R. pomonella (6)	26	24	43	7	428
R. cerasi (9)	19	28	35	18	5214
D. dorsalis (5)	33	9	45	13	45
D. cucurbitae (5)	26	26	21	27	274

Table 7. *Catches of* D. oleae *in Lebanon* (8) *on the four wings of the traps distributed according to wing areas facing East, South, West and North*

	Flies caught on trap walls facing cardinal directions									
	Position of traps in the trees									
	East		West		North		South		Total	
Trap wall	Σ	%	Σ	%	Σ	%	Σ	%	Σ	%
1+4 (facing E)	327	32	1063	40	776	37	510	37	2676	38
3+6 (facing S)	226	21	686	27	450	21	264	19	1626	23
5+8 (facing W)	282	27	508	19	459	21	260	19	1509	21
2+7 (facing N)	212	20	339	14	439	21	332	25	1322	18
Total	1047		2596		2124		1366		7133	

Σ = total flies caught; % = percentage on the two walls facing each cardinal direction.

Sex ratio

All coloured traps caught both sexes, but in quite variable ratios (Table 5). The following summarises the results:

(i) More females than males – *C. capitata* (2, 3, 5, 7) *R. pomonella* (6) and *P. poiciloptera* (4).

(ii) Equal ratios – *R. cerasi* (9) and *D. tryoni* (1).

(iii) More males than females – *D. cucurbitae* (5), *D. oleae* (7, 8) and *D. dorsalis* (5). Some interactions between tree position and sex ratio were found. In *D. oleae* (8), for instance, more males were caught in the west and north.

Table 8. *Comparison of catches of fruit flies on fluorescent yellow colour traps with the catches of traps of other types*

Species and author		Trap	No. of traps	Catch per trap	Relative attractiveness (FY = 1)	% ♀
C. capitata	(2)	FY	8	109	1.6	56
		Nad	12	178		0
C. capitata	(3)	FY	8	46	17.9	75
		Nad	4	823		0
C. capitata	(5)	FY	8	3	11.3	33
		McPh	4	34		89
C. capitata	(5)	FY	8	2.7	123.3	32
		McPh	4	333		82
C. capitata	(7)	FY	1	156	0.3	67
		McPh	1	52		63
D. oleae	(7)	FY	1	90	0.28	35
		McPh	1	25		80
R. pomonella	(6)	FY	8	24	0.8	56
		Sph	8	19		26
D. dorsalis	(5)	FY	8	2.5	4	25
		McPh	4	10		73
D. dorsalis	(5)	FY	8	1.8	16.6	43
		McPh	4	30		76
D. cucurbitae	(5)	FY	8	11	1.4	3
		McPh	4	15		40
D. cucurbitae	(5)	FY	8	1.4	4.6	36
		McPh	4	6.5		42
P. poiciloptera	(4)	Alu	1	24	0.7	33
		WTT	1	17		41
D. tryoni	(1)	FY	8	153	3.1	46
		ST	3	468		48

FY, fluorescent yellow; Nad, Nadel trap with trimedlure; McPh, McPhail trap with diammoniumphosphate; Sph, spherical sticky trap; Alu, aluminium foil; WTT, white test-tube trap; ST, suction trap.

Comparisons with traps of other designs

The coloured X-shaped sticky traps were compared with the McPhail invaginated trap (5, 7), the Nadel trap (2, 3), a 24″ (60 cm) suction trap (1) and a spherical sticky trap (6) (Table 8). Odour-baited traps and the suction trap were usually superior to the coloured traps. Other visual traps were fairly similar in attractiveness to the X-traps. Results with *C. capitata* were extremely varied. McPhail traps baited with diammoniumphosphate caught from 0.3 to 120 times and Nadel traps baited with trimedlure from 2 to 18 times more flies than the X-traps, even though the Nadel trap attracted only males. For *P. poiciloptera*, artificial asparagus stalks (test tubes 1.6 cm diameter and 15 cm long, of different colours) were compared with coloured traps. The optimal tube, which was white, caught 17 flies, and the optimum trap (aluminium) caught 24 flies. A white cylinder of 10 cm diameter caught only one fly.

The following explanations are proposed for the variation within and between captures in colour and odour traps:

(i) The distance of attraction of the two types of stimulation differs, that of odour being greater. Thus, more flies move into the area of odour stimulus than of colour.

(ii) A higher proportion of flies in a population are responsive to odour than to colour.

(iii) The test method may not accurately measure colour response; for example, some flies may respond to a colour but fail to alight on the trap, as has been observed in the case of aphids.

Ovarian development

With the exceptions of *D. cucurbitae* (5) and *D. tryoni* (1) the majority of the female flies captured were gravid, although there were always some females without eggs. In *D. tryoni*, the result probably reflected the high proportion of juvenile flies in the population when the experiment was carried out (1).

Application of colour traps

Fluorescent yellow X-traps for survey and control are proposed or have been applied in tests with *R. cerasi* and *R. pomonella* (Remund, 1971; Russ *et al.*, 1973) and *C. rosa* Karsh (Monty, personal communication). These are species which respond readily to visual cues, and the experiments described here on the influence of trap position may assist in the more efficient use of traps. Other species are characteristically responsive to odour, but the simultaneous use of colour may or may not enhance the efficiency of the odour-baited traps. Indeed, coating a McPhail trap with yellow pigment decreased captures of *D. oleae* (Orphanidis & Soultanopoulos, 1962) and of *Anastrepha ludens* (Loew) (Chambers, personal communication).

Adult movements

Laboratory studies

Laboratory studies on the movements of insects should be designed to indicate comparative relationships and should not be expected to provide measurements directly applicable to field behaviour. They can give a better picture than can field studies of the relative effects of controlled variables, such as light, temperature, humidity, wind movement, colour, olfactory stimuli etc., or of differences such as are associated with laboratory strains, genetic or

3   DSI

physical markers, irradiation doses etc. Normally the studies are of two kinds: (i) propensity tests, which measure the capacity of flies to change from a resting to a flying state, and (ii) flight tests which measure in-flight parameters such as number and length of flights, and respiration.

Two procedures now in use for studies of flight propensity in fruit flies were described at the second meeting of the Fruit Fly Working Group. Haisch (personal communication) described a balance beam actograph for measuring the flights of insects confined within a cage. Departure of the insect from the wall changed the cage weight sufficiently to cause a mark to be recorded on a strip chart. Thus, the number of flights in a test period could be recorded and compared under a variety of experimental conditions, and flies subjected to treatments such as irradiation could be compared.

A startle-test device developed in Honolulu measures the propensity for flight, or irritability (or, reciprocally, the sedentary tendency) of flies. The test flies are confined within an opaque container, approximately 1 litre in volume, and allowed time to acclimate. The chamber is fitted with a butterfly valve at the top, which, when opened, allows light to enter and the flies to escape into an upper chamber the interior of which is coated with sticky entrapping material. The sudden exposure to light stimulates the more irritable flies into flight and they are caught and counted. As an example of results with such a unit, Schroeder, Chambers & Miyabara (1973) found that laboratory-reared *C. capitata*, irradiated with 10 krads two days before adult eclosion, were approximately 50 % less prone to initiate flight than untreated wild flies.

The essential factor of devices used to measure flight propensity is that the flies be at rest with tarsal surface contact; the tendency of the fly to terminate this contact by flight is then measured.

Laboratory studies of flight ability using flight mill systems have been reported. In these, the fly is suspended from a rotor arm which pivots about a central hub. The arm is rotated by the insect as it flies. The data derived from the revolutions of the rotor indicate rate of flight and the length and number of flight periods. In all of the flight mill systems yet devised, once the fly loses tarsal contact it cannot recover it at will. Thus, the tarsal reflex causes thresholds for flight to remain at lower levels than for an insect which is free to rest. For this reason suspended flight positions are not recommended for examination of flight propensity unless a mechanism for normal resting posture of the fly is incorporated.

Chambers & O'Connell (1969) reported a simple mill system for testing flight of the Mexican fruit fly, *A. ludens*. It incorporated a magnetically suspended rotor but no automatic data recording system. Subsequently it was improved with more stable rotors and with fully automated data processing (Chambers & Sharp, personal communication). This unit and a similar one developed by Boller (personal communication) which incorporates strip-chart data recording, were described at the second meeting of the IBP

Working Group. Their use was recommended for quality control tests of flies produced and sterilised in preparation for sterile release control programmes.

Sharp, Chambers & Haramoto (1975) examined the flight performance of *D. cucurbitae* and *D. dorsalis* in flies of different sexes and ages. Boller & Remund (1971) made similar comparisons using *R. cerasi* and *D. oleae*, and also included an examination of the effects of the available energy levels, and the weights of the individual flies. The flight mill data revealed differences attributable to sex. For example, females of all of the species tested reached their maximum flight ability earlier than males; however, 16-day-old males recorded the maximum flight rates and distances for the *Dacus* spp., and it appeared that flies between one and two weeks old had the greatest flight ability, indicating, for example, that flies one week old might be most suitable for release in field studies. Standardisation of the amount of food given to flies used in such comparative experiments is strongly recommended.

Sharp & Chambers (personal communication) also compared the flight ability of irradiated (20 krads two days prior to adult emergence) and normal *D. dorsalis* males and females. In young flies, whose performance is poorer than older flies, irradiation enhanced flight performance. However, detrimental effects of irradiation were increasingly manifest after four days, indicating that flies treated for a sterile release programme should not be held longer than necessary. It should be noted that at the sterilising dose, 10 krads, only slight detrimental effects could be found at any age, and at 20 krads the effects were neither pronounced nor universal. However, Sharp (1972), studying wingbeat frequencies, clearly showed harmful effects of 20 krads on both sexes, regardless of age. Thus, it appears that physical change due to radiation can be observed, but it is not great enough, except at very high doses, to be clearly measured in flight ability studies. This may be partly caused by the low threshold for flight induced by tethering and it indicates the value of additional tests on flight propensity.

Boller & Remund (1971) reported flight mill tests showing that unmarked *R. cerasi* flew more slowly than flies marked with fluorescent powders or with samarium chloride, but for longer periods of time. Also *D. oleae* reared on an artificial diet showed lower flight capability than native flies. Schroeder, Mitchell & Miyabara (1974) found that marking with a dye increased flight ability but decreased flight propensity in *D. cucurbitae*.

Yates (1969) employed very simple miniature flight mills using as rotors aluminium strips balanced on pins set in cork bases. Up to 20 of these small devices could be observed simultaneously in an array. Yates (1969), and also Fletcher (1974*a*) found that peak flight duration in *D. tryoni* occurred when the flies were one to three days of age, and declined steadily thereafter. This is an interesting contrast to the data for the other *Dacus* spp. described above.

Field studies

Most field data on fruit fly movements have been obtained by releasing numbers of marked individuals at a central point surrounded with traps. Although useful information on dispersive potentials of various species has been obtained, it is difficult to derive quantitative data because of the artificial situation created. Neilson (1971), Boller, Haisch & Prokopy (1971) and Fletcher (1973) have obtained data on movements using natural populations, and other studies are now in progress.

The movements of tephritid fruit flies appear to be of two main types: dispersive (migratory) flights and non-dispersive (trivial) flights (Bateman, 1972). There are considerable differences among the various species in the extent and importance of these two types of movement.

The multivoltine tropical and sub-tropical species

Studies on *Dacus* pest species, particularly *D. dorsalis*, *D. cucurbitae*, and *D. tryoni*, indicate that the adults are very mobile and that a variety of dispersive movements occur, including post-teneral flights, movements away from overwintering sites, host seeking, and response to adversity. The movements of *D. oleae*, *Anastrepha suspensa* (Loew), *A. ludens* and *C. capitata* probably fall into these same general categories.

The post-teneral flights of juveniles are preceded by feeding to build up glycogen reserves and appear to be an important phase of adaptive dispersal. In *D. tryoni*, about 75 % of newly emerged male and female flies left a study area orchard in the first week after emergence (Fletcher, 1973, 1974*a*). Most species have a fairly long preoviposition period which, even under highly favourable conditions for maturation, may last for 7–10 days. Towards the end of this period, the flights presumably become oriented towards host plants, but the physiological mechanism that brings this change about is not known.

Some individuals may cover large distances during this post-teneral phase. Iwahashi (1972) recorded long distance flights of *D. dorsalis* between islands of the Ogasawara group in Japan, and De Murtas, Enkerlin & Cirio (1972) noted similar movements of *C. capitata* between islands off the west coast of Italy. Teneral males of *D. tryoni* (Fletcher, 1974*a*) that were marked and released were subsequently recaptured up to 24 km away. Juvenile adults have been found to leave emergence areas even when these appear highly favourable (Bateman & Sonleitner, 1967; Fletcher, 1973; and others).

A similar dispersive phase, which may have the same underlying physiological basis, occurs in the spring when adults leave their overwintering sites. During the late autumn *D. tryoni* stops breeding, at least in the southern part of its range, the ovaries regress, and the adults congregate in sheltered sites where they remain throughout the winter. When higher spring temperatures allow maturation to proceed, the flies disperse, presumably in search of hosts. Here also, there may be a period of adaptive dispersal prior to host seeking because it was observed that the majority of flies left an overwintering site in early spring, even though it contained more host trees with ripening fruit than did the surrounding areas (Fletcher, personal communication).

Searching for hosts may also result in a significant amount of dispersal. The movement of various species of fruit flies into areas where new host crops are ripening and their disappearance from areas where crops are diminishing has been observed. The immigration and emigration of large numbers of *D. tryoni*, including many gravid females, in an orchard, concurrent with the appearance and disappearance of ripe fruit was recorded by Bateman (1968) and Fletcher (1973).

As the mature adults move from host to host, they may cover considerable distances, particularly when suitable hosts in the area become scarce. Mature males of *D. tryoni* have been recaptured almost 24 km from their release point and circumstantial evidence suggests that both males and females frequently travel up to 64 km in search of oviposition sites when fruit is not available locally (Fletcher, 1974*b*). Similar long distance movements have also been observed in *A. ludens* (Shaw *et al.*, 1967) and in *D. dorsalis, D. cucurbitae*, and *C. capitata* (Steiner, 1969).

Studies on *D. oleae* have indicated that this species also moves reasonable distances, although so far there are no records of long range movements. This may be due, in part at least, to the lack of a powerful lure similar to those that exist for some other *Dacus* spp. Economopoulos *et al.* (personal communication) found that when irradiated *D. oleae* were released two days after emergence, most stayed close to the point of release, although a few reached the farthest traps at a distance of 3 km. Traps on non-host trees did not catch any released or wild flies.

Flies frequently travel through areas which do not contain host fruit. Fletcher (1974*b*) trapped large numbers of *D. tryoni* in areas remote from breeding sites. Because of the lures used, only males were trapped but it seems unlikely that such movements would be limited to one sex only. Similar observations had been made earlier by Steiner & Lee (1955) on *D. dorsalis*.

Besides leaving areas where fruit is diminishing, flies also move when climatic or other factors become unfavourable. In the autumn when leaf fall reduces the amount of shelter, adults of *D. tryoni* leave many of their breeding areas and move into overwintering sites. Experiments on the movement of *D. dorsalis* in the Ogasawara Islands also suggested that poor habitat conditions encouraged long distance dispersal (Iwahashi, 1972).

When mature flies reach areas where there is ample fruit, movements tend to become non-dispersive, and show daily patterns associated with requirements for food, shelter and oviposition sites. Mature adults of *D. tryoni* tended to remain in parts of an orchard where ripe or ripening fruit was present, although there was some movement between the orchard and surrounding vegetation (Sonleitner & Bateman, 1963). Lopez and Chambers (personal communication) observed movement of *A. ludens* between adjacent orchards of oranges and grapefruit as fruit availability varied between them. Sigwalt *et al.* (1968) were unable to find any evidence of a daily rhythm of movements of *C. capitata* adults between a grove of orange trees and the trees of a windbreak nearby.

At least in the colder regions of their distribution, adults also have very restricted movements during the winter months. Non-dispersive movements were observed in an overwintering population of *D. tryoni* in Camden, NSW, Australia. Most individuals remained in the overwintering site from late May until the end of August, although on warm days flies frequently made short local flights from tree to tree. At the same place, individuals of another species of tephritid, *Dirioxa pornia* (Walk.), were observed occupying individual leaves, in close proximity to each other, on which they stayed almost continuously throughout the winter (Fletcher, personal communication). In Pakistan, Syed (1972) reported overwintering aggregations which often contained several hundred individuals of various species, including *Dacus dorsalis*, *D. cucurbitae*, *D. scutellaris* Bezzi, *D. diversus* Coq., and *D. hageni* de Meij. Individuals left these clusters as temperatures increased, to feed on nearby food sources. As spring approached, the clusters disappeared as the flies dispersed.

The temperate univoltine species

Movements of these species tend to be non-dispersive or limited to short-range dispersive flights, probably because the adults normally emerge in areas where hosts are abundant, at a time when oviposition sites are available. There is very little information about long-distance dispersal flights, although there is evidence that they occur in some species when the immediate surroundings are unfavourable.

Neilson (1971) found that movements of the apple maggot, *R. pomonella*, were fairly limited, recaptured flies having travelled a maximum of only 0.9 km. Maxwell (1968) found that sticky barriers facing an orchard caught comparatively large numbers of *R. pomonella* suggesting that there was quite a lot of dispersal between orchards. When flies were released in an area without host trees some individuals were captured in an orchard over 1.5 km away (Maxwell & Parsons, 1968).

Boller *et al.* (1971) studied the movements of *R. cerasi* in a cherry orchard in Switzerland and found that very few marked individuals left the area within a 0.1 km radius of their release point as long as suitable host fruit was available. A few individuals travelled up to 0.5 km when the trees were separated by open fields. However, when unusual meteorological conditions produced a scarcity of host fruit, flies left the main areas of infestation, which had remained fairly static for several years, and migrated to other areas so that the distribution of flies became more uniform (Boller, 1974). Similar movement patterns have been observed in the blueberry maggot *Rhagoletis mendax* Curran in Michigan, and *R. pomonella* in Wisconsin (Boller, personal communication).

Sexual behaviour of pest tephritids

The family Tephritidae contains four main subfamilies, Dacinae, Trypetinae, Tephritinae and Oedaspinae. Many generalisations are possible about sexual behaviour in the Dacinae, Tephritinae and Oedaspinae, which appear to be natural groupings of species with close biological relationships, but the Trypetinae contains diverse genera with diverse sexual behaviour. The Dacinae and Trypetinae include the major economic pest species and consequently their sexual behaviour is better known than that of the other subfamilies. Even so, most of our knowledge is based on studies of artificially reared flies in laboratory conditions. Critical studies on natural populations remain to be done. A comprehensive review of the literature has been prepared by Tychsen (1972). The IBP Working Group encouraged further study of the sexual behaviour of fruit flies because new information in this field may lead to the development of improved methods of control.

The sexual behaviour of the Dacinae

In the Dacinae nearly all of the information on sexual behaviour has come from studies of pest species of the genus *Dacus*.

Males of this genus mate frequently but the females appear to have a refractory period after mating during which they become unreceptive to males. In *D. tryoni* the majority of females are monogamous, although some of them will mate again several weeks after the first mating (Tychsen, 1972). In *D. dorsalis* the females mate more frequently, often at four to five day intervals (Christenson & Foote, 1960). Reports on the frequency of remating in *D. oleae* vary considerably, depending upon the origin of the flies, their age, previous sexual activity, dietary factors, environmental conditions and the mating pressure from males (Tzanakakis, Tsitsipis & Economopoulos, 1968; Zouros & Krimbas, 1970a; Cavalloro & Delrio, 1971; Economopoulos, 1972). Economopoulos (1972) concluded that wild females which survived for $2\frac{1}{2}$ months would mate up to six or more times whereas laboratory-reared females surviving a similar time would mate only four or five times. Tzanakakis *et al.* (1968) suggested that a male factor might be transferred to the females during copulation and could be responsible for switching off the female's receptivity, as occurs in some other Diptera.

Sexual activity in the majority of species, is restricted to the late afternoon or dusk period. Examples include *D. tryoni* (Tychsen & Fletcher, 1971), *D. oleae* (Moore, 1960), *D. zonatus* Saunders (Syed, Ghani & Murtaza, 1970), *Austrodacus cucumis* (French) (Bailey, 1971). Studies on *D. tryoni* (Tychsen & Fletcher, 1971) indicated that restriction of mating to the period of twilight at dusk was brought about by an interaction between light intensity and a circadian rhythm of sexual responsiveness.

The males of most species stridulate during courtship by drawing their wings rapidly backwards and forwards over a pair of 'combs' formed of large bristles situated on the third abdominal tergite, producing short bursts of high-frequency sound. The acoustical properties of stridulation have been studied to some extent in *D. tryoni* (Monro, 1953) and *D. oleae* (Feron & Andrieu, 1962) but its role in pair formation and courtship is poorly understood.

During courtship, the males of *D. tryoni* release a sex pheromone, which is secreted and stored in glands associated with the posterior ventral regions of the rectum (Fletcher, 1969). The males of all other species of Dacinae so far examined have similar glands, e.g. *D. oleae*, *D. dorsalis* and *D. cucurbitae* (Schultz & Boush, 1971), *D. oleae* (Economopoulos *et al.*, 1971), *D. kraussi* (Hardy), *D. neohumeralis* (Perkins), *D. absonifacies* (May), *D. bryoniae* (Tryon) and *Callantra aequalis* (Coq.) (Fletcher, personal communication) and *Austrodacus cucumis* (Bailey, 1971). Although in most cases there is no experimental evidence that the glands in these species produce a sex phero-mone, such a role is clearly suggested.

Economopoulos *et al.* (1971) reported that sexually mature females of *D. oleae* have an odour that differs from that of the males, and is detectable for longer periods of the day. Haniotakis (1974) has shown in olfactometer studies that *D. oleae* females attract sexually mature males during mating hours. Attraction of males to sexually mature virgin females has also been demon-strated in field experiments.

The sex pheromone producing structures of male *D. tryoni* consist of a secretory sac, and a reservoir opening directly into the rectum. Females re-spond to filter paper discs impregnated with the secretion by approaching and probing the discs with their ovipositors (Fletcher, 1969). In laboratory experi-ments only mature virgin females and some older mated females show a response. Maximum response occurred at dusk under a light intensity of 8.6 lux (Fletcher & Giannakakis, 1973).

A sensitive laboratory bioassay has recently been developed for the sex pheromone of *D. tryoni*, using enclosed glass units, and some females have been found to exhibit a probing response to concentrations as low as 10^{-4} male equivalents (Giannakakis and Fletcher, personal communication). The four major volatile constituents of the secretion stored in the reservoir have been isolated and identified as N-3-methylbutylacetamide, N-3-methyl-butylproprionamide, N-2-methylbutylproprionamide and N-3-methylbutylisobutyramide. Synthetic samples of these compounds all show some biological activity when tested in laboratory bioassays, but do not elicit as high a response as the glandular extract itself (Fletcher and Bellas, personal communication), suggesting that some of the other constituents of the secretion, which have not yet been identified, are also important.

Few observations on the sexual behaviour of the Dacinae have been made under field conditions. Bateman (1972) observed mating 'swarms' of *D. tryoni* on both host and non-host trees in an orchard, and suction trap catches (Fletcher, personal communication) indicated that around dusk mature males collected predominantly, but not exclusively, on trees bearing ripening fruit. Tychsen (1972) found that in a large field cage, males formed a swarm at dusk, occupying the leaves and branches of a small area on the upwind side of the tree. Within the swarm the males defended small territories on individual leaves where they stridulated and released their sex pheromone. Virgin females were attracted into the swarm and mating took place on the leaves or fruit when present. It is not known, however, how typical these observations of sexual behaviour are for the Dacinae in general.

The sexual behaviour of the Trypetinae

Much of the information on the sexual behaviour of the Trypetinae comes from studies on economically important species of the genera *Rhagoletis*, *Ceratitis* and *Anastrepha*.

Females of *R. pomonella* mate many times during their life (Neilson & McAllan, 1964), but they are far more receptive to the advances of males just before, during, and after oviposition than at other times (Prokopy & Bush, 1973). The fruit plays an important role in pair formation by acting as the rendezvous site for courtship and mating, and therefore it is of considerable importance in the evolution of species-specific signalling patterns (Bush, 1969*a, b*; Prokopy *et al.*, 1971). Mating occurs at any time of the day when the temperature is within a favourable range and the light intensity is sufficient to permit adequate vision (Prokopy *et al.*, 1972). Relatively few adults land on the fruit until they mature (Prokopy *et al.*, 1972). Similar observations have been made on *R. cerasi* where the males wait on the fruit for arriving females. However, mating never takes place on the fruit itself but on nearby leaves or branches, and ovipositing females on cherries are rarely intercepted by males (Bush and Boller, personal communication).

As with most *Rhagoletis* spp., both males and females of *R. pomonella* and *R. cerasi* have rather elaborate wing patterns and body markings, and at close range the visual stimulus of the female is important in eliciting the courtship approach of the male (Prokopy & Bush, 1973).

There is no evidence that auditory signals or attractant pheromones are involved in pair formation. Mature females of *R. pomonella* and *R. cerasi* do, however, deposit some kind of chemical on the surface of fruit, which acts to arrest arriving males, so that they spend more time on marked fruit than on unmarked (Prokopy & Bush, 1972). Perhaps this plays some part in pair formation because it tends to hold males in the vicinity of fruit which females are likely to visit.

The Mediterranean fruit fly, *C. capitata*, mates during the day at high light intensities when the temperature is between 21.5 and 31 °C and the relative humidity about 30 % (Myburgh, 1962). Mating in the field usually occurs on host fruits (Feron, 1962). Males mate frequently, but many females appear to be monogamous. Nakagawa *et al.* (1971) found that 40 % of females did not remate, 45 % mated an additional once or twice, and the remaining 15 % more than twice over a seven-week period. They suggested that the frequency of remating was related to the volume of sperm stored in the spermathecae.

Males of *C. capitata* vibrate their wings when they are sexually excited, and release a sex pheromone that attracts sexually mature virgin females (Feron, 1962). The wing vibration does not appear to produce sounds. It may act as a visual signal and may also enhance the evaporation of the pheromone, which is produced by paired glands situated in the last abdominal segment. Recently the two major components of this pheromone have been identified as methyl (E)-6-nonenoate and (E)-6-nonen-1-ol (Jacobson *et al.*, 1973). Both of these compounds attracted sexually excited virgin female *C. capitata* when tested on small sticky traps in the laboratory (Ohinata *et al.*, 1973). However, when tested in large field cages, the response of females was very low unless the two compounds were combined and mixed with 10 fatty acids that were also present in the male extracts. Two of these have tentatively been identified as palmitoleic and oleic acids.

In the genus *Anastrepha* the only information on sexual behaviour comes from the Mexican fruit fly, *A. ludens*, and the Caribbean fruit fly, *A. suspensa*. They appear to be quite different from one another in their behaviour. *A. ludens* (Flitters, 1964) becomes sexually active at dusk and mating is not always restricted to the host plant. The male vibrates its wings and produces sounds that attract the female. It is not known whether pheromones are also involved, or how important visual signals are at close range. Both sexes of *A. ludens* have elaborately patterned wings, but unlike most of the other species which mate during the day and have elaborate courtship displays, they mate at dusk when visual signals might be expected to be less efficient. Auditory signals have not been recorded in any Trypetinae other than *A. ludens* but they are typical of the main group of tephritids that mate at dusk, the Dacinae.

In *A. suspensa*, mating may occur at any time during a large part of the day, although it is most common in the afternoon (Nation, 1972). The mature males release a sex pheromone from glandular tissue in the pleural region of abdominal segments 3, 4 and 5. These are distended to form a small pouch on each side of the abdomen, and at the same time a thin membranous sac surrounding the anal area may be everted. The distension of the abdomen is accompanied by short bursts of slow wing vibrating or fanning. When a female approaches, the male stops fanning and initiates copulation. In the laboratory virgin females were attracted by mature males or by extracts from males 10 to 12 days old. The response of the females was greatest during the afternoon.

Conclusion

Recent research on sexual behaviour in the Tephritidae has revealed certain consistent patterns, and many interesting specialisations which differ from group to group. Attraction to the host plant is the major component of pair formation in some species, and courtship behaviour has become highly specialised, involving elaborate and ritualised visual displays in which both the male and female may take part. In other species, particularly those in which mating occurs at dusk, the males play a more active role, attracting females by auditory, olfactory or visual signals.

The mating process would seem to be a vulnerable point in tephritid biology, depending as it does, in most cases, on highly specific odours or behaviour patterns. The recent isolation and identification of the sex pheromones of several of the economically important species could prove useful for control and survey purposes. Synthetic male lures have already been used successfully to eradicate isolated populations and also to detect outbreaks and new introductions.

Population and ecological genetics

Population and ecological genetic studies on tephritid populations had not been seriously attempted until recently. The polytene chromosomes, so useful in some Diptera, are suitable for detailed genetic analysis in only a few members of this family (Krimbas, 1963; Bush, 1966*b*; Bush & Taylor, 1969). Genetic studies, therefore, have been limited to karyotype analysis in conjunction with taxonomic research (Keuneke, 1924; Emmart, 1935; Frizzi & Springhetti, 1953; Mendes, 1958; Bush, 1962, 1966*a*, *b*; Bush & Huettel, 1970) or studies on female heterogamety and other aspects of sex determination (Bush, 1966*b*).

Within the past few years the application of gel electrophoresis techniques to the study of genetic variation in enzymatic and non-enzymatic proteins has made it possible to examine the genetic structure of several important tephritid species and host races. The results of these investigations are summarised here.

The technique has thus far been applied to four general research problems: (i) the establishment of genetic markers in several species; (ii) the effects of certain insecticides on the population genetics of fruit flies; (iii) the study of genetic variation and evolution of biotypes, geographic races, and sibling species; and (iv) the establishment of mating frequencies in natural populations. It has many other applications, however, that are relevant to both laboratory and field problems. For instance, flies to be used in sterile release programmes can be monitored for genetic changes occurring during colonisation and mass rearing. Qualitative or quantitative alterations in alleles at various loci will indicate that other genes essential to the successful competition of the laboratory strain with wild flies might also be undergoing change. Genetic monitoring coupled with other tests for competitiveness can provide sensitive procedures for maintaining high-quality laboratory strains at minimum cost.

Naturally occurring genetic markers revealed by electrophoresis can also be used in migration, longevity, and sperm precedence studies and serve as chromosome markers for establishing linkage groups. Such chromosome markers can be useful in the development of strains for genetic control based on chromosome abnormalities (translocations, inversions etc.).

Gel electrophoresis – the technique

The various techniques used to carry out electrophoresis on tephritid proteins are basically similar to one another, varying principally only in the types of electrophoretic media used. Some investigators prefer acrylamide, which has the advantage of flexibility in tailoring the gel to the separation of specific proteins. Starch, the other medium of choice, has the outstanding quality that gels can be sliced into thin slabs and each stained for a specific protein thus increasing the amount of information obtainable from a single specimen. Using starch, as many as 30 proteins can be detected simultaneously in a single individual.

An IBP manual on starch gel electrophoretic techniques used to study tephritid proteins (Bush & Huettel, 1972), provides all necessary information for setting up a complete electrophoresis laboratory at minimum cost. Those wishing to use the acrylamide technique should refer to Smith (1968).

Genetic markers

One of the basic tools in many genetic studies is a suitable series of genetic markers. Such markers can be used for investigations of various aspects of sperm precedence, linkage, migration and competition. They are also useful in establishing phylogenetic relationships between species (Simon, 1969; Bush and Boller, personal communication; Bush and Prokopy, personal communication). The first priority of the IBP programme on population genetics was to establish a catalogue of such markers for use in future studies of fruit fly

Table 9. *Enzymatic and non-enzymatic proteins of Tephritidae*

Genera and Species	Stage	Protein P	ACPH+/APH	ALD	EST	LAP	AO	TO	PO	ADH	α-GPD	G-6PD	G-3PD	HAH/ODH	IDH	MDH-D	MDH-T	6-PGD	AK	FUM	HEX	PGI	PGM	GOT	Reference*
Dacinae																									
D. oleae	A	1 (2)	2 (2)	—	2 (2)	1 (2)	—	—	—	—	—	—	—	—	—	1 (8)	—	—	—	—	—	—	—	—	1,2
	L	1 (2)	3 (2)	—	2 (2)	1 (2)	—	—	—	—	—	—	—	—	—	1 (8)	—	—	—	—	—	—	—	—	1,2
D. tryoni	L	5 (2)	3 (1)	—	2 (1)	—	—	—	3 (1)	2	—	—	—	2 (1)	—	—	—	—	—	—	—	—	—	—	3
	L	5 (2)	3	—	—	—	—	—	3	2	—	—	—	1	—	—	—	—	—	—	—	—	—	—	3
D. neohumeralis	L	2 (2)	3	—	2 (1)	—	—	—	—	2 (1)	—	—	—	1	—	—	—	—	—	—	—	—	—	—	3
	A	4	2	—	2 (1)	—	—	—	—	1	—	—	—	1	—	—	—	—	—	—	—	—	—	—	3
D. cucumis	A	2	2	—	—	—	—	—	—	1	—	—	—	—	—	—	—	—	—	—	—	—	—	—	3
Trypetinae																									
R. pomonella	AL	—	1 (1)	1	3 (1)	3	1	1 (1)	—	1 (1)	—	—	1	2 (2)	1	1 (1)	—	1	—	1 (1)	2 (1)	1	1	1	4
R. mendax	A	—	1	1	3	3	1	—	—	1	—	—	1	2	1	1	—	1	—	1	2	1	1	1	4
R. zephyria	A	—	1	—	3	3	1	—	—	1	1	—	1	2	1	1	—	1	—	1 (1)	2	1	1	1	4
R. cornivora	A	—	—	—	—	—	—	1	—	1	—	—	1	—	1	1	—	1	—	—	—	—	1	1	4
R. tabellaria	A	—	—	—	—	—	—	1	—	—	—	—	1	—	1	1	—	1	—	1	—	—	1	1	4
R. basiola	A	—	—	—	—	—	—	1	—	1	—	—	1	—	1	1	—	1	—	1	—	—	1	1	4
R. cingulata	A	—	—	—	—	—	—	—	—	1	—	—	1	—	1	1	—	1	—	1	—	1	1	1	4
R. fausta	A	—	1	1	3 (2)	3	1	—	—	1	—	—	—	—	1	1	—	1	—	1	2	1	1	1	4
R. suavis	AL	—	1	1	3 (2)	3	1	—	—	1	—	—	—	2 (8)	1	1	—	1	—	1	2	1	1	1	4
R. completa	AL	—	1	1	3 (2)	3	1	—	—	1	—	—	—	2 (8)	1	1	—	1	—	1	2 (1)	1 (?)	1	1	4
R. juglandis	AL	—	1 (8)	1 (8)	3 (2)	3	1	—	—	1	—	—	—	2 (8)	1	1	—	1	—	1	2 (1)	1 (3)	1	1	4
R. boycei	AL	—	1 (8)	1 (8)	3 (2)	3	1	—	—	1	—	—	—	2 (1)	1	1	—	1	—	1	2 (1)	1 (3)	1	1	4
R. cerasi	AL	—	1	—	3	3	1	—	—	1	1 (0)	—	—	—	1	1	—	1	—	1	2 (1)	1 (3)	1	1	4
R. berberidis	AL	—	1	—	3 (2)	3	1	—	—	1	—	1	—	2 (1)	1	1	—	1	—	1	2 (1)	1	1	1	4
R. meigeni	A	—	—	—	4 (17)	1 (1)	—	—	—	2 (1)	—	1 (1)	—	1	1	1	—	1	1 (1)	1	3 (1)	1	1	—	4
Anastrepha suspensa	A	—	2 (2)	—	—	—	—	—	—	—	—	1 (1)	—	—	—	—	—	—	—	—	—	—	—	—	6
Oedaspinae																									
Procecidochares australis	A	1	1	1 (1)	4 (1)	2 (1?)	—	—	—	1 (1)	1 (?)	—	—	—	—	—	2 (?)	—	1	1 (?)	2 (2?)	1 (?)	—	—	5
Procecidochares sp. A	A	1 (?)	1 (1)	1 (1)	4 (1)	2 (1?)	—	—	—	1 (1)	1 (?)	—	—	—	—	—	2 (?)	—	1	1 (?)	2 (2?)	1 (?)	—	—	5

A = adults, L = larvae. Number of polymorphic loci in parentheses.

* 1 = Tsakas & Zouros (1969); 2 = Zouros, Tsakas & Krimbas (1968); 3 = McKechnie (1972); 4 = Bush and Boller (personal communication), Bush (personal communication); 5 = Huettel (1972); 6 = Huettel (personal communication).

† *Enzyme abbreviations:* ACPH = acid phosphatase; APH = alkaline phosphatase; ALD = aldolase; EST = esterase; LAP = leucine amino-peptidase; AO = aldehyde oxidase; TO = tetrazolium oxidase; PO = peroxidase; ADH = alcohol dehydrogenase; α-GPD = α-glycerophosphate dehydrogenase; G-6PD = glucose-6-phosphate dehydrogenase; G-3PD = glyceraldehyde-3-phosphate dehydrogenase; HDH = alcohol dehydrogenase; ODH = octanol dehydrogenase; IDH = isocitrate dehydrogenase; MDH = malate dehydrogenase; 6-PGD = 6-phosphogluconate dehydrogenase; AK = adenylate kinase; FUM = fumerase; HEX = hexokinase; PGI = phosphoglucose isomerase;¡ PGM = phosphoglucomutase; GOT = glutamate-oxaloacetate transaminase.

genetics. Thus far a diverse range of polymorphic loci have been studied in species belonging to three sub-families (Table 9). Linkage patterns have not been established in any species.

The population genetics of tephritid biotypes and geographic races

Distinct host races or biotypes have been discovered in several tephritid fruit pests (Christenson & Foote, 1960; Bush, 1966a, 1969a; Huettel & Bush, 1972) and also in species used for the biological control of weeds (Zwölfer & Harris, 1971). In the *R. cerasi* complex each host race may be further sub-divided into geographic races, some of which are almost completely reproductively isolated from one another (Boller & Bush, 1974; Bush and Boller, personal communication).

In most cases, the exact degree of genetic divergence and level of reproductive isolation between the races within a species is poorly understood or unknown. In fact the term 'host race' itself has been loosely applied to taxa ranging from populations of a polyphagous species collected on different hosts, to distinct, reproductively isolated sibling species that cannot be separated by the conventional tools of comparative morphology and cytogenetics. Here we restrict the term host race to a population of a species which exhibits a distinct, genetically based preference for one or more hosts that differ from the hosts of other populations of the same species. Races, being members of the same species, are therefore at least potentially interbreeding and have not developed complete reproductive isolation.

Exactly how host and geographic races of fruit flies evolve has generally been ignored or given only passing attention. Yet, without a precise understanding of the exact status of each race, it is difficult if not impossible to develop realistic genetic control programmes or to safely introduce beneficial species (e.g. for weed suppression).

Considerable time, effort and resources could be expended on controlling flies that are actually of no economic importance. The honeysuckle race of the European cherry fruit fly, *R. cerasi*, for instance, has recently been found to be a distinct species (Boller & Bush, 1974; Bush and Boller, personal communication), yet it cannot be distinguished morphologically or cytologically from the cherry race. Furthermore, the cherry race itself is subdivided into at least two geographic races (east and west Europe) that have developed a high degree of interrace hybrid incompatibility although very little genetic divergence can be detected. The careful evaluation of the evolutionary relationships of these species should make it easier to establish efficient control programmes.

Ecological genetics of host races

In general, there are two sources of genetic variation to be considered when studying the population genetics of natural tephritid populations. One is the variation that may exist between different geographic regions. Geographic variation at the genetic level has been shown to exist in *Procecidochares* and *Dacus* (Huettel, 1972; McKechnie, 1972), although in other genera such as *Rhagoletis* it is rare (Bush and Boller, personal communication; Bush and Prokopy, personal communication). For example, the amount of genetic variation in *P. australis* Aldrich is far greater between localities along the gulf coast of Texas than that found in populations of *R. cerasi* on *Prunus* throughout Europe.

Why variation is detectable in one species but not in another is not understood. Although only minor differences in gene frequency have been detected in the two geographic races of *R. cerasi*, these populations have already developed relatively strong reproductive isolation (Boller and Bush, personal communication; Bush and Boller, personal communication). It is clear that their close morphological similarity is being maintained not by gene flow, but by natural selection which is apparently holding the gene frequencies at their present levels.

A second complicating factor is the genetic variation that might exist between populations associated with different hosts. Often, populations which appear to be merely components of a single completely interbreeding oligophagous or polyphagous species are actually distinct host races or sibling species. Erroneous conclusions may therefore be drawn if samples from various hosts are pooled for study. Such would be the case if sympatric populations of *P. australis* infesting *Heterotheca* and the sibling species restricted to *Macnaerantnera* were analysed together (Huettel, 1972; Huettel & Bush, 1975).

Only through detailed biological studies and hybridisation experiments can genetic variation due to geographic factors and host preference be sorted out. If the true status of the populations associated with different hosts is not established, then costly errors could easily be made in biological control. At present, if any race (or undifferentiated sibling species) shows a capacity to attack an economic plant, the group as a whole has been denied use in weed control.

Genetic variation and reproductive isolation

Thus far there seems to be no clear-cut relationship between host specificity, geographic variation, and the level of genetic divergence, on the one hand, and the degree of reproductive isolation, on the other.

On the basis of emergence patterns, the honeysuckle race of *R. cerasi* is

M. A. Bateman

quite distinct from the cherry race (Boller & Bush, 1974), yet electrophoretic analysis indicates that gene frequencies are almost identical in sympatric populations (Bush and Boller, personal communication). The problem is further complicated by the fact that egg hatch may be normal in some hybrids, such as those produced by crosses between sympatric populations reared from *Lonicera* and *Prunus* in Switzerland, whereas other crosses exhibit strong incompatibility.

All that can be said at present is that genetic divergence, as revealed by electrophoresis, is no reliable indicator of the degree of reproductive isolation or host specificity that might exist in populations of geographic or host races. Because host shifts may involve alterations at only a few genetic loci (Bush, 1969a, 1974; Huettel & Bush, 1972), and may produce strong barriers to gene flow and the subsequent establishment of distinct host races and sibling species, an extensive genetic change may not be a prerequisite for speciation in tephritids.

Insecticides and gene frequency

In earlier studies, Zouros *et al.* (1968) and Zouros & Krimbas (1969, 1970b) found two polymorphic esterase loci, esterase A and esterase B in *D. oleae* segregating for a large number of alleles. In one natural population 15 alleles were revealed for esterase A and 12 for esterase B.

Only the esterase A enzyme system thus far has been implicated in insecticide susceptibility. In a laboratory study, Tsakas & Krimbas (1970) found that the organophosphate insecticide Dimethoate used to control *D. oleae* in Greece, selects against flies homozygous or heterozygous for the silent iso-allele (A_s) at the esterase A locus. This allele apparently does not synthesise an active acetylcholinesterase enzyme. Another unidentified gene apparently produces an enzyme which can be substituted in $A_s A_s$ individuals. Flies heterozygous for two different alleles at esterase B, displayed greater flight mobility. From this observation they concluded that selection for increased mobility would lead to selection for heterozygosity.

However, when the effects of organophosphate insecticides were studied in a small natural population inhabiting an olive orchard no indication of gene frequency changes due to insecticides could be found (Krimbas & Tsakas, 1971). The changes observed resulted from drift as a consequence of a reduction in population size. The same authors also estimated the mutation rate for enzyme polymorphism to be between 10^{-3} and 10^{-4}. This estimate, however, included migration as a factor mimicking mutation.

48

Mating frequencies in natural populations

The frequency with which females mate under natural conditions is essential information in designing a programme for population supression by the sterile-male release method (Zouros, 1969). As a species that mates several times in captivity may mate only once, or rarely twice, in the field, the results of laboratory tests on mating frequency are unreliable. True mating frequency can only be established by studying natural populations.

Zouros & Krimbas (1970*a*), investigated the number of fertile matings of wild females inhabiting an olive orchard of about 2000 trees, by collecting eggs from them, and analysing the resultant progeny for the polymorphic esterases A and B. Using flies of known genotype they also demonstrated that there was no sperm precedence in digamic matings in this species and that they could detect second matings if they occurred. After a series of statistical analyses of the electrophoretic results, the frequency of digamic females of *D. oleae* in the natural population was estimated to be about 17 %, and was apparently independent of population density. This method of estimating the frequency of multiple matings could undoubtedly be adapted for use on other tephritids.

4. *Myzus persicae* Sulz.
an aphid of world importance

Coordinators: M. MACKAUER & M. J. WAY

Aphids include some of the most common and destructive pests of plants. Many of the 4000 known species are widespread pests of agricultural crops in both the temperate and tropical regions except in parts of the humid tropics of the equatorial zone. Aphids can damage plants directly by feeding and indirectly as vectors of plant virus diseases.

At the suggestion of the IBP/UM (Use and Management of Natural Resources) Section Committee an international Working Group of aphidologists met in April 1967 at Silwood Park, England, to formulate the objectives of a collaborative research programme on the biological control of aphids. The meeting was attended by 31 experts from various countries.

The initial plan envisaged work to be relevant to conventional biological control. The programme was to emphasise research on aphid pathogens, parasites, and predators: their ecology, behaviour, host specificity, and impact on aphid population growth. Studies on aphid nutrition, host finding and colonisation, and related subjects, being basic to manipulation and control, had to be included. The aim, however, was to develop a model for an integrated control programme against an agricultural pest aphid. While laboratory and field studies would be combined on some or all of the above apsects, special attention would be given to a cooperative field project on the population dynamics of one particular aphid species with world-wide, or nearly world-wide, distribution.

More than 100 aphidologists throughout the world were consulted before the conference. Based on the information gathered in this way, the meeting agreed that the project should be centred around the green peach aphid, *Myzus* (*Nectarosiphon*) *persicae* (Sulz.; 1776) (Homoptera: Aphididae), and other aphids associated with it on different crops. The selection of these species was influenced by the following considerations. The green peach aphid perhaps is the most important aphid pest on a world-wide basis. It is reported as a vector of well over 100 diseases of plants in over 50 different families, including important crops such as beans, sugar beet, sugar cane, brassicas, citrus, potatoes, and tobacco (Kennedy, Day & Eastop, 1962). Attempts to prevent the spread of virus by controlling the aphid vectors so far have been

Contributors: R. L. Blackman, L. Bonnemaison, V. F. Eastop, G. N. Foster, I. Hodek, F. Leclant, H. J. B. Lowe, M. Mackauer, G. Remaudière, G. E. Russell, H. F. van Emden, M. J. Way, N. Wilding.

unsuccessful or only partially successful. Chemical control, while effective on a short-term basis, has the obvious disadvantage of producing insecticide-resistant clones when applied frequently or in large doses. Biological control generally is not considered a suitable alternative against disease vectors, mainly because biological control agents usually act more slowly than, for example, chemical and physical control measures; their effectiveness is difficult to predict; and they affect a variable, and often an inadequate, proportion of the pest population. Despite these apparent drawbacks, evidence of frequent insecticide-induced resurgences caused by the destruction of the aphids as well as of their natural enemies suggests that the role of parasites, predators, and pathogens may not be as insignificant as might have been supposed and therefore should be re-assessed. There is one area in particular where the impact of biological agents on the green peach aphid may be enhanced, namely on non-crop or alternative crop plants. More important, however, is the need for the development of an integrated control programme with the main objective of reducing the numbers of virus-carrying winged aphids.

As only limited manpower and financial support would be available in countries where *M. persicae* is not a major pest, it was agreed further that projects dealing with natural enemies of aphids in general and complementary supporting studies on other agricultural pest aphids would also be acceptable under the IBP programme.

The Silwood Park meeting recognised eight major areas that are relevant to integrated control but are inadequately studied, as follows:

(i) population dynamics of green peach aphid on different crops and/or in different climatic regions;
(ii) identification and assessment of the biological properties of the aphid species and its biotypes;
(iii) aphid/host plant relationships including host plant resistance;
(iv) aphid nutrition and rearing;
(v) role of aerial movement;
(vi) evaluation of parasite impact;
(vii) evaluation of predator impact; and
(viii) evaluation of the role of aphid pathogens.

Collaborative research on an international basis, it was hoped, would be stimulated by summarising our knowledge and understanding of aphid biology and control. To this end members of the Working Group prepared review articles on the ecology of the green peach aphid (van Emden *et al.*, 1969) and its natural enemies (Mackauer, 1968), the impact of pathogens, parasites, and predators on aphids (Hagen & van den Bosch, 1968), a text on *Aphid Technology* (van Emden, 1972*a*), and a series of mimeographed instructions for analysing the population growth of green peach aphid on potatoes (Foster and van Emden, unpublished). In addition the Working Group surveyed, by questionnaire, the nature of current activities and the manpower involved in research on aphid predators, parasites, and breeding for resistance. The formation of regional working parties was encouraged and helped in countries that had sufficiently strong and active groups of aphidologists.

The objectives of the Working Group were re-examined at a second meeting, held at Paris, France, in September 1970. The meeting was attended by 26 experts from nine countries. Following a review of the potential and main procedures of biological control of aphids, the specific objectives and achievements under IBP were discussed. As a result, one project (life cycle variations of *M. persicae*) was added and two of the original projects on nutrition and dispersal of the green peach aphid were discontinued.

In September 1972 the Entomological Society of New Zealand sponsored an International Symposium on Aphid Biology organized around the IBP project. Several members of the Working Group presented reviews on different topics in aphid biology, which were later published under the title *Perspectives in Aphid Biology* (Lowe, A. D., 1973).

A third meeting was in January 1973, again at Paris, when arrangements were made for finalising the research programmes and for publication.

This Synthesis Report summarises the work, and its results, of the different projects which were initiated and organised by the IBP Working Group on 'Biological Control of Aphids'. Most of the research was not intended to contribute materially to the solution of current problems associated with green peach aphid. As mentioned above, a well-planned, and perhaps long-range, integrated control programme would seem to offer the most effective strategy for reducing the damage caused by *M. persicae* as a disease vector. This requires, among other things, a thorough understanding of the life cycles of all organisms that form part of the disease complex and of their numerical interactions. On pp. 71–92 a method suitable for use in the field is outlined. It enables the population growth of *M. persicae* on potatoes to be estimated from samples taken at specified intervals. Obviously, the control strategy appropriate for a given situation will depend on the aphid's biotypic characteristics (p. 63) and mode of reproduction (p. 57). Some aspects of the complex field of aphid/host plant interactions are considered on pp. 67–71. On pp. 92–110 information relevant to the use and evaluation of aphidophagous insects and aphid pathogens is discussed. Two specific approaches to aphid control receive special attention on pp. 100–104 and 111–13: breeding for plant resistance; and control by genetic manipulation of aphids. The latter approach, while not as yet proven in field trials, holds considerable promise for the future. On pp. 116–19 the potential for integrated control of green peach aphid is considered and some areas where continued international cooperation would appear desirable are suggested.

General biology and population dynamics of *Myzus persicae*

General aspects

The ultimate aim of a population dynamics study is to be able to model observed fluctuations in population density over periods of several years from estimates of natality and mortality parameters obtained *independently*. Varying the values ascribed to individual parameters in the model would then make it possible to assess their relative regulatory powers and their interactions.

This ideal has not been attained in full even for species with parameters that could be adequately measured. In contrast, *M. persicae* seems innately unsuitable for such an approach. Particular problems are its overall low density on many host plants, including widely scattered uncultivated species in roadsides and wasteland, and the extensive intermixing between sub-populations by dispersal that can occur over large areas of the world. Accurate sampling is therefore difficult, even on one host plant; and it is virtually impossible to identify and study a discrete population or to quantify important parameters such as mortality during migration.

An analysis of annual density variations for the species can be attempted by taking a particular comparative 'intensity' estimate for each of a number of years, as is available from trapping records between 1942 and 1959 (Broadbent & Heathcote, 1961). Such an analysis (Fig. 4), where changes from one year to the next are plotted as logarithmic differences linked in chronological order, suggests the existence of an efficient mechanism for population stability. Whatever regulates the population appears to overcompensate for large deviations from the equilibrium position but nevertheless damps the amplitude of fluctuation. Evidence summarised by van Emden *et al.* (1969) suggests that stability of *M. persicae* depends fundamentally on intraspecific interactions. The stabilising effect of emigration lies in density-influenced production of relatively poorly fecund alatae, most of which fail to colonise suitable new food plants. The green peach aphid is remarkable because it is polyphagous but is severely restricted by the physiological condition of available food plants. Thus its host substrate selection mechanisms act for stability both in competition with other aphids (its polyphagy) and in their absence (its restriction by host condition) (van Emden & Williams, 1974).

The population ecologist is particularly interested in defining when and where in the life cycle the key regulating processes operate. This knowledge may also be important for the applied ecologist, though only if it relates to the numbers of *M. persicae* (or viruliferous *M. persicae*) that colonise and multiply on the crop at risk. Unfortunately the ill-defined limits of the region within which natural population regulation operates and other difficulties already mentioned may defy successful investigation of even single components of the dynamics of *M. persicae*. It seems unlikely, by analogy with *Aphis fabae* Scopoli, that important regulatory competitive processes operate on the pri-

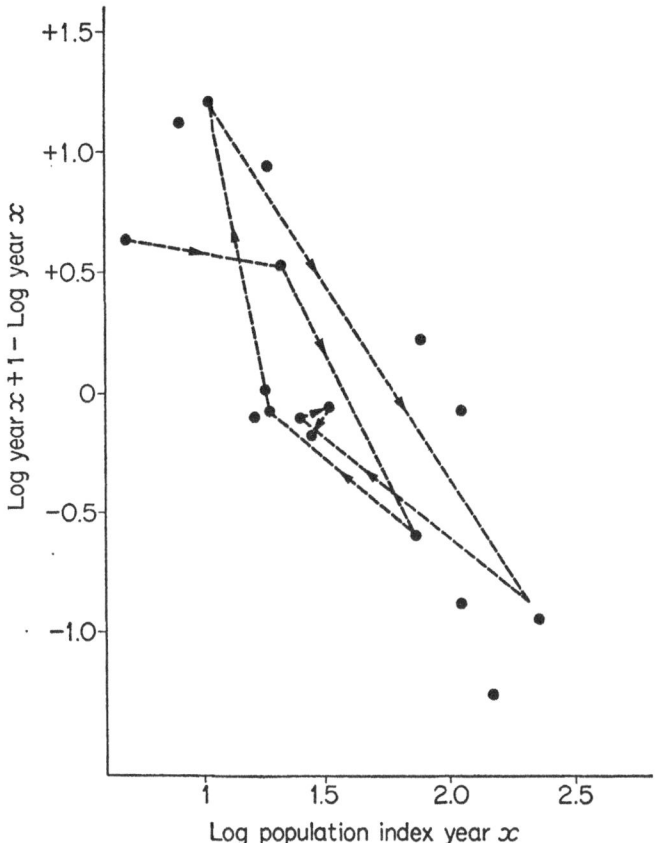

Fig. 4. Changes in *Myzus persicae* abundance from year to year plotted against abundance in the first year. The population index has been taken from trapping data, 1942–59 (Broadbent & Heathcote, 1961). Points for 1942–50 have been joined in sequence.

mary host, even in regions with very cold winters where virtually all *M. persicae* are confined to the primary host for perhaps half of the year. During the growing season, arable crops in most regions (irrigated deserts may be an exception) probably support a relatively small proportion of the overall population. Therefore, even a seemingly dramatic event in a local subpopulation, e.g. in one field, will usually have little or no impact on the overall regional population. The key regulatory processes probably operate mostly among populations on widespread weed hosts where they are extremely difficult to investigate. Only in the very simple artificial closed environment of all-the-year-round chrysanthemums in glasshouses has it been possible to demonstrate a regulatory mechanism of the kind that probably operates at critical stages of the life cycle outdoors. Thus Wyatt (1965) has shown that the density of *M. persicae* on year-round chrysanthemums can depend on intraspecific interactions which create a stable equilibrium population, the size of which depends on the host plant resistance of the particular chrysanthemum variety.

The matter is further complicated by variations in the life cycle of the species (pp. 57–63). Some races are relatively insensitive to alate-producing stimuli, but there is also a whole range of races with life cycles varying from regular holocycly to regular anholocycly. Clearly, different climatic conditions favour different types of cycle so, depending on the region, immigrants to different crops may comprise forms with entirely, or partly, separate life cycles and population dynamics. Some life cycle forms may also select different crops differentially. Thus work on inherent variation of life cycle type and other biotypic attributes such as host selection and behaviour is central to a study of population dynamics of *M. persicae*. Though it is difficult to predict how un-ravelling such complications will help us understand the population dynamics of the species less imperfectly than now, it will at least focus attention on which 'types' cause regular pest problems on which crops and on which seg-ments of the life cycle we should concentrate as a basis for integrated control.

Most past attempts at integrated control of insect pests centred on the inte-gration of chemical control with biological control by natural enemies. Out-doors, many *M. persicae* are no doubt killed by natural enemies but, being a low density pest, the aphid is often relatively unattractive to natural enemies and its restlessness and highly dispersive behaviour also makes control by natural enemies unreliable. Whilst natural enemy action must be preserved and encouraged, it is seldom likely to provide adequate natural control when integrated only with chemical insecticides. Other non-chemical methods of control must therefore be included and, in this context, there is the widespread evidence that *M. persicae* is especially sensitive to host plant condition.

If, as seems likely, host plants are involved in the population stabilising mechanisms that *M. persicae* has evolved, then it is particularly vital that we should understand the host plant interactions of this species. The interaction of host plant effects with natural enemy action may itself provide a major component in integrated control of the species. Such an interaction was postulated for green peach aphid by van Emden & Wearing (1965) and has since been demonstrated experimentally for the greenbug, *Schizaphis grami-num* Rondani, by Starks, Muniappan & Eikenbary (1972). The IBP field project was specifically designed to enable the field expression of this interac-tion to be quantified. Host plant condition, which can be manipulated by crop timing, by fertilisers, and by choice of cultivar (pp. 113–16), is not only relevant to biological control by natural enemies as indicated above but also to intra-specific regulation because it affects the threshold at which population density induces emigration. A relatively unfavourable substrate reduces the energy flow from the plant, as the producer, to the aphid, as the consumer, and is perhaps the only pest control solution which reduces the equilibrium density of an *M. persicae* population without courting instability. Intuitively this seems an appropriate strategy against a species which causes economic damage at relatively low densities through its property of transmitting virus diseases.

As mentioned earlier, dispersal enables an intermixing of a regional population across immense areas. The green peach aphid is an exceptionally dispersive species, sometimes caught at great heights and also found as large numbers colonising plants far from the source population (Dickson & Laird, 1962). Long-distance migration of large numbers may, however, be exceptional because there is abundant evidence that the vast majority of *M. persicae* migrants may land close to their source, perhaps within one kilometre (Heathcote & Cockbain, 1966). Certainly it is known for many crops that *M. persicae* alatae spread most virus to fields adjacent to their source (Broadbent, 1964). Much more, therefore, needs to be known about circumstances which affect distance migrated both in terms of effects of weather and regional topography and of interactions between the aphids themselves and between aphids and the host plant. For example, no work seems to have been done on density-induced qualitative variation in *M. persicae* alatae that might influence their migratory potential (Shaw, 1970).

The Rothamsted Survey suction trap work in Britain and parts of Northern Europe (Taylor, 1974) was initiated with IBP support. Weekly catches at a height of 12 m of different aphid species, including *M. persicae*, were recorded using traps scattered throughout the study area. These have given valuable warning of time of colonisation of crops but, so far, too few *M. persicae* have been caught at critical times of the year (e.g. at times when crops are first colonised in early summer) to demonstrate accurately the relative densities of aerial populations or subsequent levels of infestation of crops (Heathcote, Palmer & Taylor, 1969; Taylor, 1974). Such evidence further emphasises the difficulties of quantifying populations of a low-density pest such as *M. persicae*. Nevertheless, sampling the aerial population for a long time may provide the only practicable measure of the regional population density. Such trapping data are fundamentally important in indicating the potential for seasonal colonisation of different ephemerally suitable crops and weeds on which local sub-populations develop. Once established, such sub-populations may experience dramatically different events and, as already indicated, there remains the crucial questions of which sub-populations and which kinds of events are, in turn, important in terms of the dynamics of the regional population.

Life cycle variations

M. persicae exists throughout the world on a wide variety of host plants and presumably is adapted to a great diversity of environments. It must vary both within and between regions in relation to the widely differing conditions it experiences. For this reason, one might expect to find discrete races or bio-types occurring in association with particular host plants or habitats, especi-ally when one considers that *M. persicae* reproduces continually by partheno-genesis over a large part of its range. Yet *M. persicae* is a well-defined species and, although numerous subspecies, races and biotypes have been recognised according to different criteria (van Emden *et al.*, 1969), very little is known about the exact nature and significance of this intraspecific variation.

Clearly, naturally occurring clones of an aphid will show discrete differences from one another as do individuals of a sexually reproducing species, and by parthenogenetic reproduction a favoured genotype may quickly increase its relative numbers and acquire the status of a 'race' or 'biotype'. The problem of variation in an aphid species, if it is reduced to its essentials, comprises two basic questions: (i) Since aphid populations are derived by parthenogenesis and differ fundamentally from populations of sexually reproducing species, how do selective pressures operate upon them? (ii) What are the short- and long-term consequences of such selection on variation in the species as a whole? Related to the first question is the speed and magnitude of the adap-tive response of the aphid to environmental changes, whether these be natural or induced by agricultural practices, notably by pest control measures. Related to the second question is the whole problem of whether the species *M. persicae* is a single entity or a number of genetically isolated sub-units re-quiring entirely separate consideration. These questions are therefore basic to an understanding of *M. persicae* both as a species and as a pest problem.

There are several approaches to the study of variation in *M. persicae* and numerous aspects of the problem could be examined, but the life cycle dif-ferences which exist in this aphid seem to provide an exceptionally promising field. Life cycle variation, involving differences in the method of overwintering, is a significant feature of the biology of *M. persicae* in every continent through-out the world. The method of overwintering adopted, either by partheno-genetic forms on secondary host plants, or by fertilised ova on the primary host (holocycly), must have profound effects on the ecology and genetics of *M. persicae* populations. There is also evidence in the literature that life cycle differences are to some extent related to the host plant (Moericke, 1950; Waldhauer, 1957; Müller, 1958) which, if correct, could have considerable-economic significance.

The complexities of the aphid life cycle were the subject of a recent review by Lampel (1968), who took specific examples from each sub-family within the Aphidoidea and explored the numerous forms and variants of the life cycle that occur both within and between species of each group. A common inter-pretation of aphid life cycle variation is that holocyclic and anholocyclic 'races' co-exist in areas where both overwintering methods are possible (Müller, 1958). Waldhauer (1957) described distinct holocyclic and anholo-cyclic forms of *M. persicae,* but avoided the genetic implications of the term 'race'. Ossiannilsson (1959) considered that there was insufficient evidence to speak of independent races. Cognetti (1967) suggested that life cycle variation in *M. persicae* is the result of selection acting on a multiplicity of genotypes which vary in their ability to produce sexual forms. Genotypes tending towards either holocycly or anholocycly would be selected for according to which were and were not favoured by the prevailing environmental conditions. This still leaves many questions unanswered. Anholocyclic races that have lost all ability to produce sexual forms, if they occur, must be reproductively isolated from the main genetic stock of the species. If, on one hand, such races arise only infrequently and constitute rather old parthenogenetic lines, then why are they not more distinct morphologically from the holocyclic form of the species? If, on the other hand, anholocyclic races are frequently regenerated from the main genetic stock of the species, is this solely a one-way process, and how do the two forms come to co-exist in the same populations year after year?

In an attempt to solve these problems, a programme of experimental work was started at the Imperial College Field Station, Silwood Park, England, to examine the projected outcome of summer populations of parthenogenetic *M. persicae* collected on secondary host plants. Clones established in the laboratory from individual virginoparae collected randomly from a field population were subjected to a short photoperiod of 10 h of artificial light at 20 °C to trigger the production of sexual forms. Aphids were caged individually and their progeny recorded throughout life so that differences in the form of the progeny could be compared within and between clones. There were certain clearly-defined categories of response to short photoperiod. These were undoubtedly genetically determined as they were unaltered by changes of host plant or reduced photoperiods or temperatures. Holocyclic clones were terminated completely within four or five generations of transfer to short photoperiod by the production of male and female sexual forms. Production of gynoparae, males, and oviparae followed a specific sequence (Fig. 5) which was similar for all clones. Most other clones continued to reproduce parthenogenetically, but produced an increased proportion of alate virginoparae and some males under a short photoperiod (Fig. 5). The term *androcyclic* was applied to such clones, as they produced males and could not be considered to be strictly anholocyclic (Blackman, 1971*a*). Rather rarely a clone was found to have a truly intermediate character, giving females at short photoperiod that were morphological intermediates between oviparae and virginoparae and sometimes produced both sexual eggs and live young (Blackman, 1972). The only clones that were strictly anholocyclic, with no response at all to short photoperiod, were of a special biotype with distinct morphological features which on cytological examination was found to have chromosomal abnormalities (Fig. 6) (Blackman, 1971*b*).

Once the different types of photoperiodic response had been elucidated, it

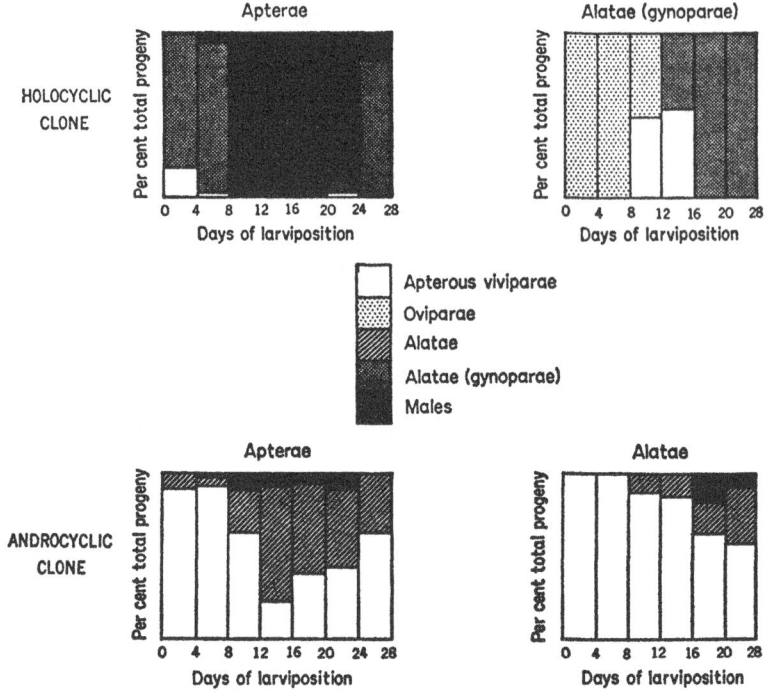

Fig. 5. The sequences of progeny produced by apterous and alate morphs of holocyclic and androcyclic clones of *Myzus persicae* after rearing on potato plants for two generations at 10 h photoperiod and 20 °C.

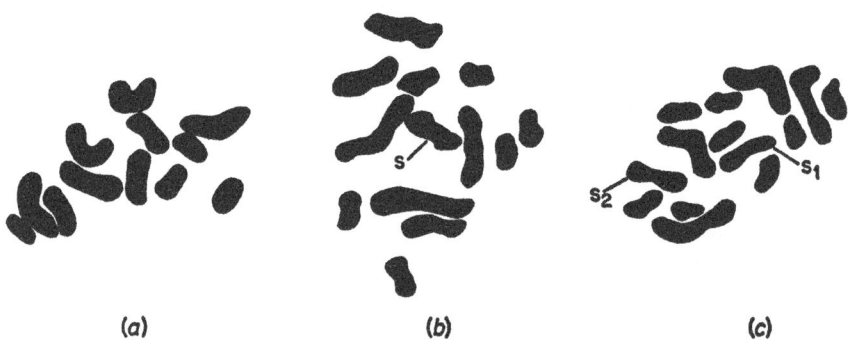

Fig. 6. Standard (*a*) and abnormal (*b* and *c*) karyotypes of *Myzus persicae*. In (*b*), the chromosome 's' is unpaired, and there are two additional small chromosomes which may have arisen by fragmentation of its homologue. In (*c*), there appear to be two unpaired elements ('s_1' and 's_2') and four additional small chromosomes, suggesting that two chromosomes of the standard karyotype have fragmented.

Table 10. *Summary of the variations in response to short photoperiod in samples from natural populations of* Myzus persicae *on secondary host plants in England and Wales, 1968–72*

Locality and date	Host plant	No. of clones tested	Holo-cyclic	Inter-mediate	Andro-cyclic	Anholo-cyclic
Silwood, Berks. July 1968	Potato	12	1	1	10	—
	Brussels sprout	12	—	—	12	—
Trumpington, Cambridge July 1969	Potato	23	16	—	6	1
	Kale	20	2	—	18	—
	Sugar beet	14	5	5	3	1
Silwood, Berks. July 1970	Potato	14	10	—	3	1
	Mustard	18	1	1	16	—
Brabourne, Kent July 1970	Potato	6	6	—	—	—
Newcastle July 1971	Potato	29	6	1	22	—
Anglesey July 1972	Potato	41	—	3	38	—

was possible to categorise clones set up from sampled populations on a basis of their life cycle characteristics. In five years a total of 189 clones from 10 populations of *M. persicae* on five different crop host plants in England and Wales were examined (Table 10).

Some striking differences were revealed in the proportions of holocyclic and androcyclic forms in populations on different crops, in different localities, and in different years at the same locality. The proportion of holocyclic forms was far greater (54 %) on potatoes than on adjacent brassicas (6 %). This substantiates the conclusion that there is a significant degree of host plant specificity within the species for important crops, which is directly related to life cycle differences. The results also indicate that the holocycle may often be more important in southern England than has previously been realised, although in the North and in north Wales androcyclic forms may predominate. The relative proportions of the two forms on the same crop and in the same locality may vary from year to year, perhaps in relation to the relative success of the two methods of overwintering in contributing to the initial colonisation of, or later build-up on, secondary hosts. Wide fluctuations from year to year in the proportions and selective advantages of *M. persicae* which overwinter sexually and parthenogenetically may have far-reaching effects on the population dynamics and population genetics of the species, about which little is known.

The production of males by androcyclic clones made it possible to begin a study of the genetic basis of the life cycle variability, by crossing males from

M. Mackauer & M. J. Way

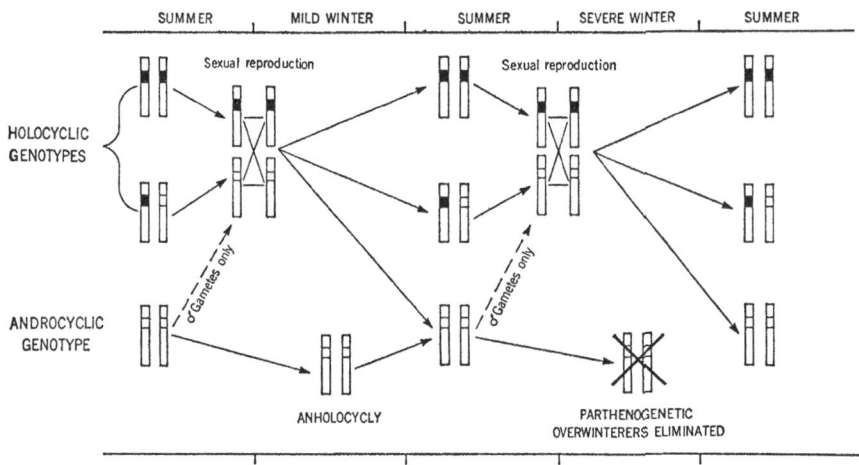

Fig. 7. Maintenance of life-cycle variation in *Myzus persicae* through three seasons. Androcyclic clones are generated afresh by the breeding system each year, even after a severe winter when parthenogenetic morphs outdoors are eliminated.

an androcyclic clone with sexual females from a holocyclic clone. Fourteen F_1 lines were obtained from the initial cross and subsequent studies of 314 clones over three sexual generations have indicated that androcycly may be inherited as a simple recessive character (Blackman, 1972). Rather than the polygenic system proposed by Cognetti (1967), the main features of the life cycle variation of *M. persicae* may be explicable in terms of a kind of genetic polymorphism (Ford, 1964), whereby the occurrence and relative proportions of holocyclic and androcyclic phenotypes are largely dependent on the frequencies of the alternative alleles of a major gene. These would in turn, of course, be determined largely by the environment, especially the severity of the winter climate. Severe winters would greatly reduce or even eliminate the androcyclic phenotypes, but the following summer the allele for androcycly would still be present in heterozygous condition in a relatively large proportion of parthenogenetic lines. At the same time the breeding system would generate afresh each year some genotypes homozygous for the recessive, androcyclic character, so that forms with the potential to overwinter parthenogenetically would always be present (Fig. 7).

If this is the explanation of how holocyclic and anholocyclic methods of overwintering co-exist in Britain, does it also fit the overall pattern of life cycle variation in *M. persicae* throughout the world? More information is needed, from many different parts of the world. In 1971, 180 specialists throughout the world were sent a questionnaire seeking information on the life cycle of *M. persicae* in their regions and a circular proposing an international collaborative study to try to place the subject on a more quantitative basis. Forty-nine questionnaires were returned. The proposal for a collaborative study was not adequately supported and very little quantitative data is yet available for assessment of the relative occurrence of holocyclic, androcyclic, and anholocyclic forms in parts of the world other than England and Wales. Many of the answers given in the questionnaire were necessarily based on casual or limited observations of the life cycle of *M. persicae*, and to analyse them in detail would probably give rise to misleading statements. Nevertheless, it seems worthwhile to summarise in Fig. 8 what is known of life cycle variation throughout the world, including both questionnaire returns and published literature, in the hope of prompting studies in areas where information is most needed.

Biotypes

Most of the intensively-studied species of aphids consist of populations with differing biological properties. These populations are often known as biotypes. A biotype has been defined as consisting of all individuals of equal genotype. In practice biotypes are recognised by biological function rather than karyotype or morphology and consist of those individuals that behave similarly as far as immediate interests are concerned. Biotypes are frequently recorded for aphids because the host plant range, the character that interests many observers, is also an important evolutionary isolating factor. Shaposhnikov (1971) summarised and gave references to recent literature on the evolution of aphids. An account of the history, origins and problems posed by biotypes was given by Eastop (1973). Most previous studies of aphid biotypes have concerned differences in biology, host plant range, virus transmitting ability, reaction to stimuli, particularly light and colour, and insecticide resistance. Van Emden *et al.* (1969) listed many references to the biotypes of *M. persicae*. Sudderuddin (1973) discussed work on insecticide resistance in *M. persicae*. Halimie & Ford (1972) and Parry & Ford (1969) recognised biotypes by their different uptake of food.

The recognition of biotypes in the field is more difficult. H. J. B. Lowe (1973a) showed that previous and parental host plant experience can produce differences in behaviour resembling those between biotypes. The green peach aphid was not important on tobacco in Florida until 1946 (Wilson *et al.*, 1948). It is not clear whether the aphid's sudden importance on tobacco was a North American mutation or because of an introduction from elsewhere. The outbreaks of sorghum greenbug (Harvey & Hackerott, 1969) and spotted alfalfa aphid (Nielson *et al.*, 1970) in North America probably were results of introducing new genetical material from the Old World. Some biotypes of *S. graminum* lack a pair of chromosomes in some cells (Mayo & Starks, 1972) but differences between biotypes need not be primarily genetic. Also different symbiotic organisms in aphids of similar genotype could produce apparent biotypic differences.

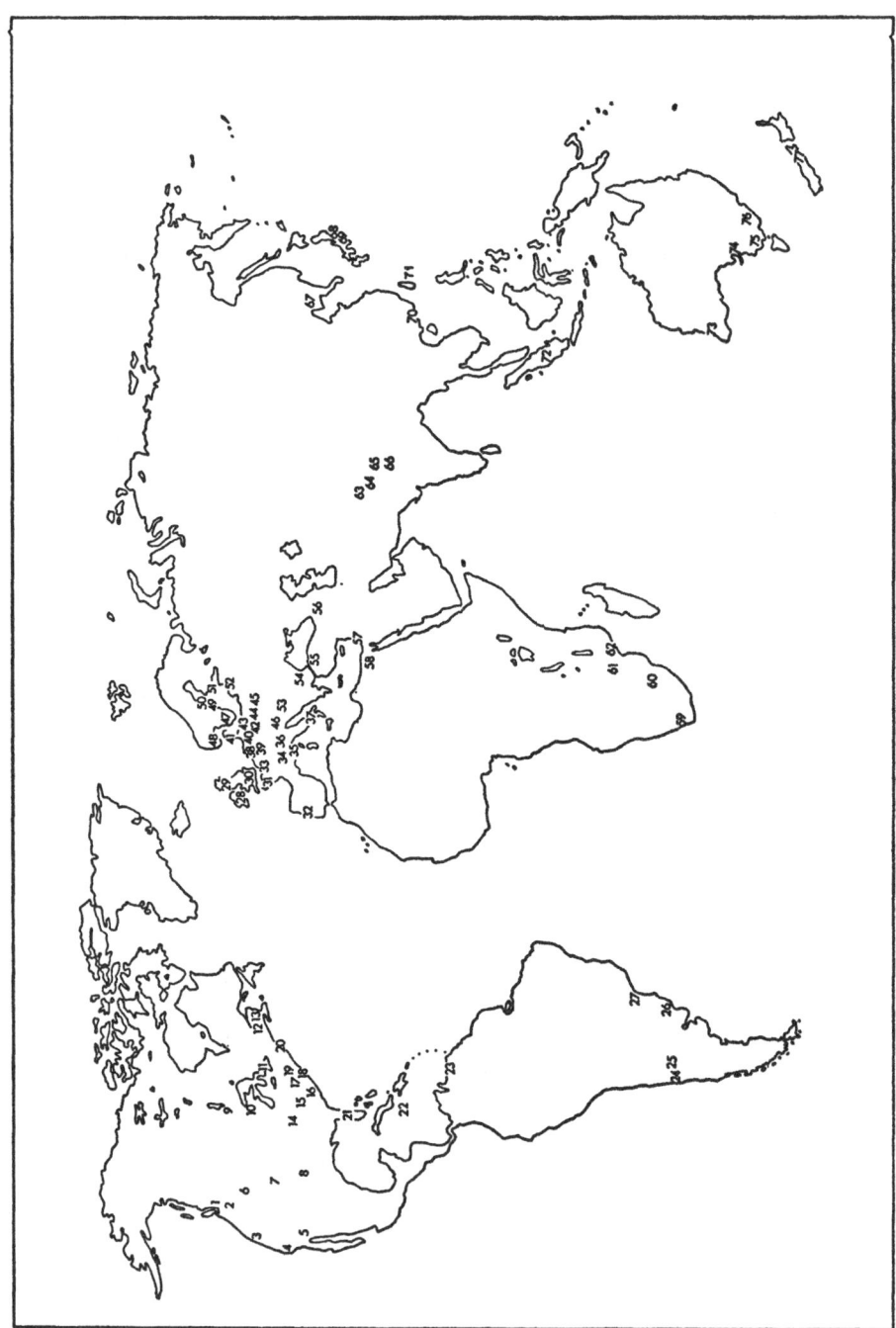

Fig. 8.

Myzus persicae *Sulz.*, *an aphid of world importance*

Fig. 8. Occurrence of holocycly and anholocycly in populations of *Myzus persicae* in different parts of the world.

North America: (1) Fraser River delta of British Columbia, Canada: anholocycly in sheltered situations (Wright, MacCarthy & Forbes, 1970); holocycly in colder interior of British Columbia (Forbes, pers. comm.). (2) Eastern Washington, USA: mainly holocycly but some anholocycly also (Davis & Landis, 1951; Tamaki, pers. comm.); sexuales and eggs not yet found in western Washington (Landis, pers. comm.). (3) North and central California, USA: both anholocycly and holocycly, the latter becoming prevalent farther north (Lange, pers. comm.). (4) Southern California, USA: anholocyclic overwintering prevalent, oviparae rarely found, small numbers of males in late winter (Dickson, pers. comm.; Essig, 1948; Walker, pers. comm.). (5) Southern Arizona, USA: anholocycly; all-year-round flight activity (Coudriet & Tuttle, 1963). (6) Idaho, USA: holocyclic overwintering (Bishop, pers. comm.). (7) Colorado, USA: holocyclic overwintering (Berry & Simpson, 1967; Newton, Palmer & List, 1953). (8) Oklahoma, USA: anholocycly (Walton, 1954). (9) Manitoba, Canada: no outdoors overwintering, possibly immigrates from further south (Robinson, pers. comm.). (10) Minnesota, USA: probably little outdoors overwintering, possibly immigrates from further south (Radcliffe, pers. comm.). (11) South-west Ontario, Canada: mainly holocyclic (Elliott, 1968, and pers. comm.). (12) North-east Maine, USA: mainly holocyclic, utilising *Prunus nigra* as a primary host (Shands & Simpson, 1969). (13) New Brunswick, Canada: holocycly now rare, probably little outdoors overwintering (MacGillivray, 1972, and pers. comm.). (14) Missouri, USA: holocycly and anholocycly (Taylor, 1908). (15) Kentucky, USA: mainly anholocycly but also holocycly (Fusco & Thurston, 1968; Thurston, pers. comm.). (16) North Carolina, South Carolina, and Virginia, USA: mainly anholocycly (Lawson & Chamberlin, 1957). (17) Western Maryland, USA: mainly holocyclic (Smith, pers. comm.). (18) Eastern Maryland, USA: anholocycly (Harrison, pers. comm.; Smith, pers. comm.). (19) Pennsylvania, USA: mainly holocyclic (Horsfall, 1924). (20) Connecticut River valley, USA: holocyclic only (Lawson & Chamberlin, 1957). (21) Florida, USA: anholocycly or possibly immigration from other regions (Mason, 1922; Wolfenbarger, pers. comm.).

Central and South America: (22) Puerto Rico: anholocycly (Smith, Martorelle & Pérez-Escolar, 1958). (23) Venezuela: anholocycly Cermeli, pers. comm.). (24) Central Chile: mainly holocyclic (Zuñiga, 1966, and pers. comm.). (25) Mendoza, Argentina: holocyclic and anholocyclic overwintering (Espul & Mansur, 1968). (26) Rio Grande, Brazil: anholocycly (Link, pers. comm.). (27) São Paulo, Brazil: anholocycly, but males collected in coldest months of year (Costa, 1969, and pers. comm.).

Europe: (28) Ireland: mainly anholocycly (Dunne, 1971, and pers. comm.). (29) Scotland: holocycly and anholocycly, predominantly the latter (Fisken, 1959; Shaw, 1957). (30) England and Wales: holocycly and anholocycly, predominantly the latter (Blackman, 1971*a*; Broadbent, 1949; Broadbent & Heathcote, 1955; Heathcote, pers. comm.; Jacob, 1941). (31) Bretagne, France: mainly anholocyclic overwintering (Poisson, 1940; Robert, pers. comm.). (32) Portugal: holocycly and anholocycly, predominantly the latter (Ilharco, pers. comm.). (33) Versailles, France: Bonnemaison, pers. comm.). (34) Valence and Rhône valley, France: holocycly and anholocycly (Leclant, pers. comm.). (35) South-eastern France: holocycly and anholocycly (Iperti, pers. comm.). (36) Switzerland: holocycly, with immigration into higher regions (Fenjves, 1945; Meier, pers. comm.). (37) Italy: holocycly and anholocycly (Cognetti, 1967). (38) Netherlands: mainly holocyclic, anholocycly outdoors occurs rarely if at all (Hille Ris Lambers, 1955, and pers. comm.). (39) Bonn, West Germany: mainly holocyclic, some anholocycly (Haine, 1950; Moericke, 1950). (40) Elsdorf, West Germany: holocycly and anholocycly (Steudel, 1952). (41) Denmark: mainly holocyclic (Heie, 1954). (42) Niedersachsen, West Germany: holocyclic (Gersdorf, 1955, and pers. comm.). (43) Rostock, East Germany: mainly holocyclic, some anholocycly (Müller, 1958, and pers. comm.). (44) Berlin: holocyclic (Heinze, 1948). (45) Poznán, Poland: holocyclic (Stacherska, pers. comm.). (46) Austria: holocyclic (Böhm, 1962). (47) Southern Sweden: holocyclic (Ossiannilsson, 1952, and pers. comm.). (48) Norway: no overwintering outdoors (Stenseth, pers. comm.; Tambs-Lyche, 1950). (49) Uppsala, Sweden: no overwintering outdoors (Ossiannilsson, 1952, and pers. comm.). (50) Hudiksvall, Sweden: no overwintering outdoors (Pettersson, pers. comm.). (51) Southern Finland: no over-

M. *Mackauer & M. J. Way*

Biotypes and plant breeders

Cereal breeders and parasitic fungi exist in a state of symbiosis. New and fungus-resistant varieties of cereals succumb after a few years to what are interpreted as new strains of fungi. A similar situation seems likely to develop with aphids if serious attempts are made to breed aphid-resistant varieties of major crops. Eastop & Russell (1967) emphasised that 'biotypes' are only homogeneous for certain characters and are not likely to be permanent taxa. Anholocyclic populations could contain more or less permanent biotypes, but biotypes of holocyclic populations will usually end with the sexual generation in the autumn. Müller (1971) described isolating mechanisms that could perpetuate biotypes in holocyclic species. Frazer (1972) considered the significance to plant breeders of biotypes, particularly in view of their transitory nature. Markkula & Roukka (1970) gave an account of the fecundity of biotypes of pea aphid, *Acyrthosiphon pisum* Harris, on different legumes and of the variation of amount of conditioning by previous host plants. Dunn & Kempton (1972) found biotypes of the cabbage aphid, *Brevicoryne brassicae* L., able to colonise most of the clones of Brussels sprouts previously selected for resistance to the species. Mackauer (1973*a*) reported that the demographic and biological characteristics of biotype R1 of pea aphid remained unchanged over a 10 year period, when maintained under constant laboratory conditions. He suggested that a biotypic characterisation is useful for describing the bio-

wintering outdoors (Heikinheimo, 1959, and pers. comm.). (52) Latvian S.S.R.: no overwintering outdoors (Zirnitis, 1944). (53) Budapest, Hungary: holocyclic (Szalay-Marzsó, pers. comm.). (54) Bulgaria: holocyclic (Kovachevski, 1942).
Asia and Africa: (55) Ankara, Turkey: holocycly and anholocycly (Düsgünes, pers. comm.). (56) Ararat plain, Armenian S.S.R.: holocyclic (G. M. Marjanian, pers. comm.). (57) Israel: anholocycly (Harpaz, pers. comm.; Swirski, pers. comm.); possibly an abortive holocycle on apple in winter (Zimmermann-Gries & Swirski, 1956). (58) Cairo, Egypt: anholocycly, but an abortive holocycle occurs in later winter/spring (Willcocks & Bahgat, 1937). (59) Stellenbosch, South Africa: anholocycly, but the holocycle occurs inland at Ceres (Durr, pers. comm.). (60) Pretoria, South Africa: holocycly and anholocycly, mainly the latter (Daiber, pers. comm.; Daiber & Schöll, 1959; Schöll & Daiber, 1958). (61) Rhodesia: anholocycly (Shaw, pers. comm.). (62) Mozambique: anholocycly (Ilharco, pers. comm.). (63) Lahore, Pakistan: anholocycly, but with abortive holocycle in late winter/spring (Das, 1918). (64) North-western India: holocycly and anholocycly (Batra, 1953). (65) Simla, India: anholocycly (Verma, pers. comm.). (66) New Delhi, India: anholocycly, but with abortive holocycle in late winter/spring (Verma, pers. comm.). (67) Manchuria: holocycly (Kawasaki, 1940). (68) Utsunomiya, Japan: holocycly and anholocycly (Tanaka, pers. comm.). (69) Hatano, Japan: holocycly and anholocycly (Takaoka, 1960, and pers. comm.). (70) Canton, China: anholocycly (Hoffmann, 1937). (71) Taiwan: anholocycly (Tao, pers. comm.).
Australasia: (72) Sumatra: anholocycly (de Jong, 1929). (73) Western Australia: anholocycly (Norris, 1943). (74) South Australia: holocycly and anholocycly (Fowler, 1934). (75) Victoria, Australia: holocycly and anholocycly (Ward, 1934). (76) Canberra, ACT, Australia: holocycly and anholocycly (Anonymous, 1944). (77) New Zealand: holocycly and anholocycly (Lowe, 1962, and pers. comm.).
(*Above:* pers. comm. = personal communication.)

logical potential of a reproductively isolated aphid clone under a known, constant, and reproducible set of conditions.

The recent large populations of *Nasonovia ribisnigri* Mosley on *Petunia* and *Nicotiana* in Western Europe are an example of an aphid acquiring a host plant protected by sticky hairs, presumably helped because its usual unrelated hosts are protected by similar sticky hairs. The paradox is that most of the 4000 species of aphids are highly host specific or, conversely, most of the 230000 species of vascular plants apparently are resistant to more than 99 % of the species of aphids, and yet many individual aphids can be reared on plants on which they are not or only seldom found in nature. It is thought that this situation is the result of the evolution of a biology that depends on environmentally induced morphs, parthenogenesis, and vagility (Eastop, 1973). While plant breeding may be successful against some pests and helpful against most, it is unlikely by itself to produce aphid-free crops.

Conclusions

Biotypes are an expression of the way by which living organisms may combat their principal mortality factors. When a species is unable to meet changing circumstances it is doomed by them, or, in the case of a pest, controlled. Biotypes are inherent in the aphids' way of life and can be countered by diversifying the causes of mortality, that is, by integrated control.

Host plant interactions

M. persicae is an aphid that is particularly sensitive to the condition of its host plant. It has been suggested earlier (pp. 54–7) that the interaction of host plant substrate with natural enemy action may be a key to the biological component of integrated control of this species.

The IBP contribution to the research on host plant interaction of *M. persicae* has centred on the Department of Horticulture at the University of Reading, England, where techniques for studying the interaction for *B. brassicae* were brought to bear also on *M. persicae*. However, because of ecological differences between the two species (van Emden *et al.*, 1969), they have continued to be studied as a contrasting pair. This led to emphasis on brassica plants instead of on potatoes.

Efforts to progress beyond the general correlation (van Emden *et al.*, 1969) which has been established for many years between the total pool of soluble nitrogen and aphid performance began with an analysis of aphid performance in relation to leaf age (van Emden & Bashford, 1969). No evidence could be found to support the general assumption, which stems from Mittler's (1958) work on willow aphids, that old leaves are favourable for aphids because such leaves are likely to be rich in translocated nutrients, particularly nitrogen.

Mittler (1958) analysed sap from trees, but of course trees show a sudden 'synchronised' senescence in all leaves which is very different from the continual transfer of hydrolysed materials from senescing leaves to the apex in herbaceous plants. Analyses of old leaves of Brussels sprouts and of their petioles (van Emden & Bashford, 1969; and unpublished) at various stages of senescence showed a low soluble nitrogen content when compared with mature leaves; presumably senescence in the old leaves of herbaceous plants is a gradual process and translocation of hydrolysed materials out of the leaf is rapid.

The performance (fecundity) of both aphids per unit total soluble nitrogen concentration was in fact better on old than young leaves, though absolutely *B. brassicae* did less well on the former in contrast to *M. persicae*. Later work (van Emden, 1972*b*) indicated that this was largely because of high non-limiting levels of amide in the young leaves, accounting for a larger proportion of the high total soluble nitrogen content of young leaves without greatly contributing to aphid performance. For *M. persicae*, the lower levels of sinigrin and 'unfavourable' amino acids (see later) in old leaves correlated with the improved performance of the aphid. This was connected with the suggestion (van Emden, 1973) that the normal preference by *B. brassicae* and *M. persicae* for young and old leaves, respectively, was at least partly related to the aphids' different response to mustard oils and to a different balance of amino acids in the substrate.

Analagous results were obtained in experiments with excised leaf material (see later).

Experiments on the growth responses of the two aphids to plant age enabled the first tentative correlations between aphid performance and individual amino acids to be made (van Emden & Bashford, 1971). There was a tendency for susceptibility of brassica plants to *B. brassicae* and *M. persicae* to decline progressively with time from an early peak at six weeks. Aphid performance was poorly correlated with total soluble nitrogen in the leaves, especially in plants less than nine weeks old. Aphid performance could, however, be correlated with the concentration of amide and some individual leaf amino acids, though the amino acids differed for the two aphid species. Increases of amide and threonine for *B. brassicae*, and of amide and methionine for *M. persicae*, were correlated with increased aphid performance. Phenylalanine and γ-aminobutyric acid were inversely correlated with performance of the two aphids, respectively, a result which was initially somewhat surprising.

More information on the relative value of mustard oil and amino acid concentrations for predicting aphid performance on brassicas resulted from an experiment (van Emden, 1972*b*) in which two crucifer species (turnip and *Sisymbrium* sp.) were each grown under six 'physiological' treatments (combinations of two soil types and three plant ages). Multiple regression of aphid performance with leaf analyses identified further amino acids (Table 11)

Table 11. *Amino acids provisionally correlated with performance of* B. brassicae *and* M. persicae (*van Emden*, 1972*b*)

Species	Positively correlated	Negatively correlated
B. brassicae	Amide Threonine Glutamic acid	Phenylalanine Glycine
M. persicae	Amide Methionine Leucine	γ-Aminobutyric acid Tyrosine Proline

which correlated with aphid growth rate, used throughout later experiments to quantify 'performance'.

A simple three-dimensional model was devised of the correlation of (i) total allylisothiocyanate concentration (mustard oil), and (ii) balance of amino acid 'favourability' as determined from the aphid growth rate observed in the experiment. This model accounted for 75 % (*B. brassicae*) and 81 % (*M. persicae*) of the aphid growth rate variation in the treatment and plant species combinations of the experiment. The model assumed a positive response to mustard oils for *B. brassicae* and a negative response for *M. persicae*. Physiological condition of the plants contributed considerably to the variability of mustard oil concentration; similarly there was an effect of plant species on amino acid concentrations.

The complete model was used to test predictions for a number of Brussels sprout varieties chosen at random from a seedsman's catalogue. Two extreme varieties (on the basis of leaf analyses) were compared for two years in a field trial with *B. brassicae*, and the chemical prediction was confirmed. Attempts to translate the amino acid part of the model to a range of chrysanthemum varieties and *M. persicae* were successful for three of the four varieties compared, and this work is continuing.

The model was further tested in an experiment combining two treatments known to influence the nutritional quality of the plant substrate for the two aphid species studied (van Emden *et al.*, 1969), fertiliser treatment, and excision of leaf discs. Combinations of nitrogen and potassium fertilisation (van Emden, 1966) were chosen to provide Brussels sprout plants designated 'resistant' (low nitrogen, high potassium) and 'susceptible' (high nitrogen, low potassium). Analyses of discs and attached leaves of such plants showed that the total amino acid concentration in leaf discs rises dramatically, presumably because hydrolysis of proteins induced by discing is not followed by export of nutrients as it is in attached senescing leaves. Discing was found to influence the amino acids used in the model as follows: although amide is greatly increased, so to a lesser extent are the amino acids which correlate with poor aphid performance, and methionine ('favourable' for *M. persicae*) is decreased. On discs from 'resistant' leaves (where attached leaves have low amide) the aphids appeared able to benefit from the increased amide. In 'susceptible' leaves amide is already high and the growth response of the aphids (particularly *M. persicae*) appeared more related to the deleterious effects of discing. The experiment was designed to test the prediction that discing would be likely to lower the nutritional differential between 'resistant' and 'susceptible' plants created by fertiliser treatment, and the prediction was indeed confirmed.

The following conclusions seem consistent at present with the results of several experiments on the two aphid species involving a variety of host material (van Emden, 1973):

(i) The general correlation between total soluble nitrogen in plants and aphid performance (van Emden *et al.*, 1969) appears to reflect the level of amide in plants. Amide correlates with the growth rate of both aphid species, and very frequently is the quantitatively dominant fraction in the amino compound composition of the total soluble nitrogen in a leaf.

(ii) Secondary substances (e.g. mustard oils) are probably both phago-stimulatory and deterrent to any single aphid species at different concentrations. The exact concentration at which either effect becomes pronounced varies with the levels of other gustatable compounds (e.g. sugars) in the plant. In cultivated crucifers, mustard oils are usually in the phagostimulatory range for *B. brassicae*, though they rise to deterrent levels in some wild species. *M. persicae* is 'deterred' by levels which are still phagostimulatory to *B. brassicae*, and a high growth rate is normally found only on tissues with a low mustard oil content.

(iii) More *M. persicae* than *B. brassicae* variability can be included in regressions of aphid growth rate on amino acids. A number of speculative explanations can be suggested: different amino acids correlate with the performance of the two species, and those that correlate with *M. persicae* indeed show a greater variation in concentration with treatment than those that correlate with *B. brassicae*. Moreover, dense aggregates of a *B. brassicae* infestation act as a 'physiological sink' to mobilise nutrients from distant leaves of the plant (Way, 1973*a*) to a degree which is not likely to occur with the sparser populations of *M. persicae*. There may also be a difference between the two species in the diversity of internal symbionts and their ability to synthesise amino acids and partially to compensate for nutritional deficiencies in the plant sap.

(iv) Certain amino acids have been identified (Table 11) as correlated with aphid performance. Some correlate positively, others negatively. It must be stressed that the particular amino acids listed are implicated solely by correlation and that any direct effects of most of the compounds listed have yet to be demonstrated. However, amide and methionine have been identified as important to *M. persicae* in chemically-defined diets (Dadd & Krieger, 1968; Mittler, 1970). Amino acids have usually been regarded in the above light, i.e.

as nutrients likely to benefit aphids. It is therefore of interest that some amino acids correlated negatively with aphid performance. At this stage any attempt to put forward a causal relationship must be entirely speculative, but the work of Miles (1968) would suggest that some amino acids (e.g. tyrosine, phenylalanine and γ-aminobutyric acid) might affect the aphids via a chemical response *in the plant* to wounding, which involves the production of phenols through a pathway involving certain amino acids. Obviously, such amino acids would not similarly influence the aphid when it is reared on an artificial diet.

Additional to these studies on aphid performance, some attempt was made to examine the effects of *M. persicae* and *B. brassicae* infestation on the growth of the host plant, Brussels sprouts. Various growth responses of the plant to infestation were identified, particularly a general reduction in dry and fresh weight, height, internode lengths, leaf size, and number. A most striking reduction in root dry weight was observed, which is only partly because of removal by the aphid of assimilate otherwise available for storage. Plants which were subjected to probing and salivation by aphids, and on which continued feeding by the aphids was prevented, showed similar reductions accompanied by increased respiration of the plant. It therefore appears that the increased respiration of the plant, a response to wounding and/or aphid saliva, contributes considerably to reduction of dry weight. In spite of the apparently greater visible damage to plants normally associated with *B. brassicae* in comparison with *M. persicae*, the latter species caused equivalent growth reductions of the plant. Moreover, the slight stimulation of plant growth observed in comparison with un-infested plants when low densities of *B. brassicae* were present could not be observed at any density with *M. persicae*.

The green peach aphid is generally considered an important pest largely because of its property as a virus vector. The latter experiments, however, suggest that even sparse populations (which usually develop on its secondary host plants, particularly in Europe) may reduce plant growth and yield rather more than the lack of visual symptoms would suggest.

Population dynamics

Methodology

Introduction

It is clear from pages 54–7 that a study of *M. persicae* on its secondary host plant(s) is only one component of the aphid's population dynamics and perhaps an ecologically insignificant one. Nevertheless, the crop cycle was considered to be the only part of the aphid's life history that was practicable for a field project to be offered for world-wide participation, especially as crops serving as secondary host plants can be grown over a far wider area in the world than can the primary host, peach. Moreover, most human intervention aimed at controlling the species takes place on potatoes, brassicas, and other annual crops. The project was therefore based on *M. persicae* on potatoes. The purpose was to quantify the relative control potential on the crop population of naturally occurring predators, parasites, and pathogens in relation to variations in the breeding rate of the aphid.

The project was based on Hughes' (1963, 1972) time-specific analysis of the age-structure of aphid populations. Such an analysis requires stable instar distributions; the great majority of data sets confirmed the presence of such stable distributions in the field, even fairly early in the season. Only six of 34 data sets contained unstable instar distributions. A subsidiary aim of the field experiment was to test Hughes' method, devised in Australia for *B. brassicae*, on *M. persicae* over a wide range of conditions; the method was clearly shown to be applicable.

A detailed methodology was issued to participants in attempts to standardise the routine assessment of population density and the collection of data for the time-specific age analysis. Most participants obtained satisfactory results using the method outlined below.

Methods

Crop. A site of 0.5 ha optimal crop area of a variety of potato normally grown in the region should be selected as the study area. Non-persistent insecticides could be used if necessary, but their use must be recorded; haulm destruction should be delayed until after the aphid population had crashed.

Identification of aphids. Advice was given on the identification of aphids that infest potato. A key to the instars of *M. persicae*, identification of which is required for the age-specific analysis, was provided.

Meteorological observations. Mean daily temperatures and rainfall and any unusual or particularly extreme weather conditions should be recorded.

Sampling. This was to be carried out so that all data could be standardised to numbers per hill (hill = the aerial product of a seed tuber) or numbers per haulm (haulm = the aerial product of one seed tuber bud).

Regular sampling: Population samples were to be taken from the whole crop at intervals of not more than 100 day-degrees (in °C) apart, to determine the phenology of the aphid population and its predators and/or parasites, and to identify the most suitable times for age-specific analyses. Advice on sampling methods was given, but participants were free to use their own sampling techniques if long-term studies of *M. persicae* had been undertaken before the beginning of the IBP project.

Twin-samples: These formed the basis of the age-specific analysis. It was recommended that such samples should be taken three times during the season (Fig. 9): (i) during the period of exponential growth (about 300 day-

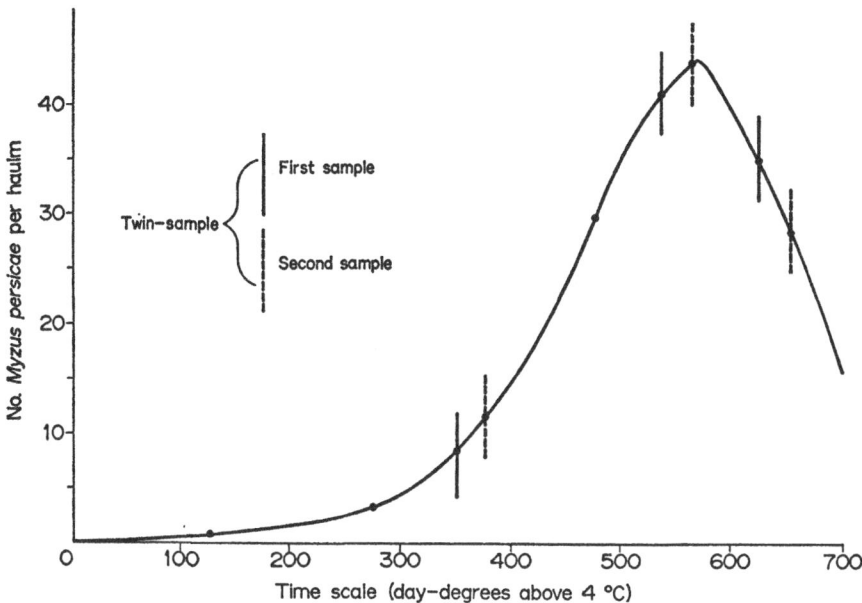

Fig. 9. A population curve of *Myzus persicae* plotted on a physiological time-scale to illustrate the ideal timing of the three twin-samples suggested in the IBP field project.

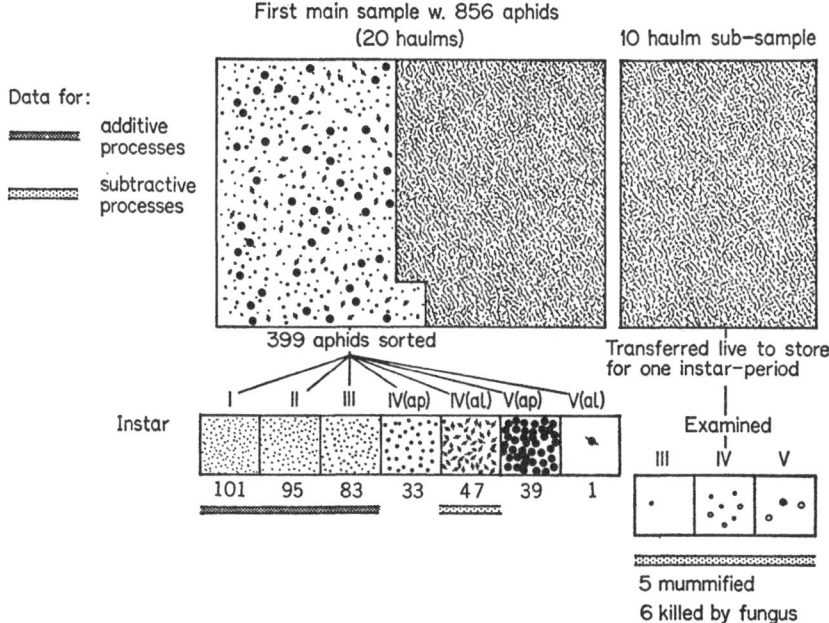

Fig. 10. Diagram illustrating the processing of twin-sample material to obtain the data for a time-specific age analysis: w, washed; ap, apterous; al, alate.

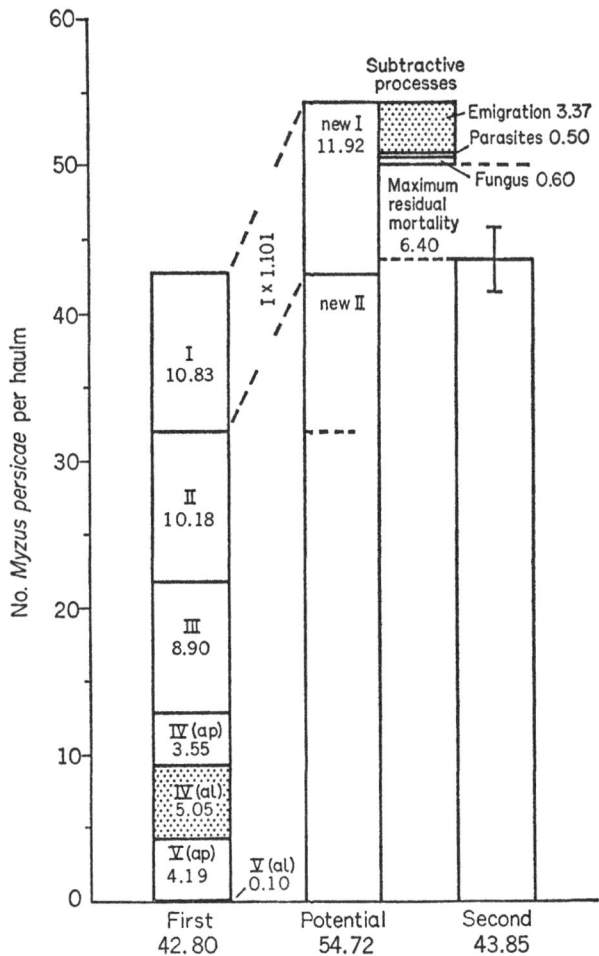

Fig. 11. Example of a time-specific age analysis from twin-sample data (see text for details): ap, apterous; al, alate.

degrees (°C) after *M. persicae* was first detected on the crop); (ii) when the population was nearing its expected peak density; and (iii) within a week of the first time the population appeared to be declining irreversibly.

At least 20 haulms were to be taken for the first main sample. Aphids were removed from the haulms by washing. The method enabled the total number of insects in each sample, or part of it, as well as the relative proportions of aphid instars and morphs (Fig. 10) to be counted efficiently. At the same time a second sample (sub-sample) was to be taken, and eggs, predators, and mummified or diseased aphids removed from it. This sub-sample was to be kept for one instar-period (determined by temperature) and then examined for mummified aphids and for aphids infested with fungus disease.

A second main sample was taken in the field one instar-period after the first.

Time-specific age analysis of twin-samples (Fig. 11). The technique is based on two basic assumptions: (i) that all instars are of equal duration (this assumption can be presumed valid if the numbers of the different instars frequently form a geometric series in nature); and (ii) that there is a linear relationship between the temperature (in °C) and the speed of development above a (experimentally obtained) temperature threshold for development.

Where stable instar distributions, i.e. a geometric series of numbers in the first three instars, are identified, it is possible to predict the size of the aphid population one instar-period after the *first* main sample from the ratio of the first three instars. This enables the *potential* population after one instar-period to be predicted on a physiological time-scale. The actual population at that time is known from the second sample. This is usually lower than the potential population, and some of the *subtractive processes* contributing to the 'gap' can be quantified: (i) emigration during the instar-period is measured as a function of the number of alatiform fourth-instar nymphs present in the first sample; (ii) the number of aphids killed by parasites (= mummification) is determined from the sub-sample; and similarly (iii) the number of aphids killed by fungal pathogens; (iv) the remaining 'gap' represents the 'maximum residual mortality' which equals the maximum mortality that can be ascribed to predation.

Population dynamics on potatoes

The international field experiments on potatoes

The field experiment was taken up by 16 workers in 10 countries. Thirteen sets of twin-sample data have been received, representing the data of eight participants in five countries and covering the northern hemisphere from 37° to 56° North.

The great majority of the data sets received confirm the presence of stable instar distributions in the field, even fairly early in the season. This is a prerequisite for the analysis of population change devised by Hughes (1963).

The following workers contributed data:

S. Barbagallo (Project Leader) and R. Inserra, Istituto di Entomologia Agraria dell'Università, Catania, Sicily, Italy; Location: 37° 18′ N, 2° 30′ E (Syracuse, Italy); Data: Three twin-samples were taken in 1970; complete set of data available for 1970.

T. M. W. Davidson (Project Leader) and S. McLean, Scottish Plant Breeding Station, Pentlandfield, Roslin, Midlothian, Scotland, Britain; Location: 55° 52′ N, 3° 11′ W (Edinburgh, Britain); Data: One twin-sample was taken in 1970; complete set of data available for 1970.

J. H. Fidler (Project Leader), Ministry of Agriculture, Fisheries & Food, Lawnswood, Leeds, Britain; Data: Aphid extraction techniques were tested in 1969.

B. Gałecka and A. Kajak (Project Leaders), Polish Academy of Sciences, Institute of Ecology, Dziekanow k. Warszawy, Poland; Location: 52° 30′ N, 21° 30′ E (Warsaw, Poland); Data: Full sets of data available for 1970 and 1971 but aphid numbers too low for twin-sampling.

S. H. Hodjat (Project Leader), Mostafavi and F. Hashemi, Ministry of Agriculture, Plant Pest & Diseases Research Institute, Tehran, Iran; Location: 35° 48′ N, 50° 58′ E (Karaj, Iran); Data: Population data available for 1970 but aphid numbers too low for twin-sampling.

Dionisio Link (Project Leader), Universidade Federal de Santa Maria, Departamento de Fitotecnia, Centro de Ciencias Rurais, Santa Maria, Brazil; Location: 29° 40′ S, 53° 42′ W (Santa Maria, Brazil); Data: A full set of data available for 1971 but aphid numbers were too low for twin-sampling.

M. E. MacGillivray (Project Leader) and D. P. MacLaggan, Canada Department of Agriculture, Research Station, Fredericton, NB, Canada; Location: 45° 55′ N, 66° 37′ W (Fredericton, NB); Data: Full sets of data available for 1970 and 1971 but aphid numbers too low for twin-sampling.

F. L. McEwen (Project Leader) and Gwen Ritcey, University of Guelph, Department of Environmental Biology, Guelph, Ontario, Canada; Location: 43° 32′ N, 80° 38′ W (Preston, Ontario); Data: Six twin-samples were taken, three in 1970 and three in 1971; complete sets of data also available, no information on predators in 1970.

W. Meier (Project Leader), Eidg. Forschungsanstalt für Landwirtschaftlichen Pflanzenbau, Zürich-Reckenholz, Switzerland; Location: 47° 23′ N, 8° 33′ E (Zurich, Switzerland); Data: Three twin-samples were taken in 1970; complete set of data available for 1970.

A. C. Milne (Project Leader) and G. N. Foster, University of Newcastle upon Tyne, Department of Agricultural Zoology, School of Agriculture, Newcastle upon Tyne, Britain; Location: 55° 00′ N, 1° 46′ W (Newcastle upon Tyne, Britain); Data: Three twin-samples were taken in 1969, four in 1970 and four in 1971; complete sets of data available for 1969, 1970 and 1971.

C. E. Taylor (Project Leader), S. Gordon and A. Dixon, Scottish Horticultural Research Institute, Invergowrie, Dundee, Scotland, Britain; Location: 56° 19′ N, 2° 57′ W (Dundee, Britain); Data: One twin-sample was taken in 1969, one in 1970 and three in 1971; complete sets of data available for 1970 and 1971.

R. Thurston (Project Leader) and L. Wu, University of Kentucky, Department of Entomology, College of Agriculture, Lexington, Ky., USA; Location: 38° 02′ N, 84° 30′ W (Lexington, Ky.); Data: Five twin-samples were taken in 1971; set of data available for 1971; no information on other aphids or predators.

J. A. T. Woodford (Project Leader), University of Cambridge, Department of Applied Biology, Cambridge, Britain; Location: 52° 15′ N, 0° 5′ E (Cambridge, Britain); Data: Two twin-samples were taken in 1971; complete set of data available for 1971.

Table 12. *Type of summary data available for a twin-sample*

Location	Newcastle
Variety	Majestic
Twin-sample dates	16–18 August 1969
Population density at beginning of twin-sample (in aphids/haulm)	16.90
Population density at end of twin-sample (in aphids/haulm)	8.90
Potential increase rate (of numbers in first instar)	0.95
Overall potential increase rate of population	1.23
Potential population density (in aphids/haulm) at end of instar-period	20.77
Average T (in °C) of week preceding twin-sample	17
Aphids/haulm emigrating during instar-period	0.62
Emigrants as percentage of potential population density	3.00
Aphids/haulm parasitised during instar-period	0.50
Parasitised aphids as percentage of potential population density	2.40
Aphids/haulm killed by fungus during instar-period	1.00
Aphids killed by fungus as percentage of potential population density	4.80
Maximum residual mortality (in aphids/haulm)	9.74
Maximum residual mortality as percentage of potential population density	46.90

Results

The results are based on calculations from data sent in by the various participants. The numbers quoted below are extracted from a large summary table
(Table 12) of identical calculations comparing the various sets of data; the full
data will be published elsewhere.

Potato as a host plant for M. persicae. Seven varieties of potato (Bintje, Katahdin, Kennebec, Kerr's Pink, King Edward, Majestic, and Pentland Crown)
were twin-sampled. On most of these varieties reproductive rates of the aphid
were below the optima obtained in the laboratory under low-light conditions
or on potato leaf discs. In the laboratory, population increase per instar-
period can be as high as 1.7 at 15 °C or 1.5 at 20 °C (the latter representing a
greater daily increase as instar-periods are shorter at higher temperatures).
Unusually high rates obtained in the field were 1.99 at 14 °C on Kerr's Pink
at Dundee in 1971 and 1.60 at 24 °C on Katahdin at Lexington in 1971.
Unusually low increase rates were found on King Edward at Cambridge
in 1971.

Some of the highest rates of increase were found at peak density; it would
appear that subsequent reductions in increase rate were not a direct consequence of high aphid population density.

Peak populations were very variable and ranged from six to 340 aphids per
hill; *M. persicae* was usually outnumbered by other species, particularly by
the green potato aphid, *M. euphorbiae*. There was no apparent relationship
between peak numbers and temperature, altitude, or longitude.

Thus rates of increase in the field were low compared with the potential for
increases on potato, as measured in the laboratory. They were very low com-

pared with the laboratory potential for increases on tobacco. Barlow's (1962) studies indicated potential increase rates on tobacco exceeding 2.0 (e.g. 2.6 at 15 °C and 2.4 at 20 °C). This suggests that potato is a sub-optimal host for *M. persicae*; the very large populations that might be predicted from laboratory studies are unlikely to occur in the field.

Most populations had either one or two distinct peaks. The first occurred between 280 and 550 day-degrees (°C) after aphids were first detected in the crop except at Preston, Ontario, in 1970 when population growth was slow and a major decline did not occur until after 760 day-degrees (°C).

Realisation of potential increase rate as measured in the field. As pointed out above, potential increase rates of the field populations appeared to be limited by host plant substrate. The potential increase rate (taking into account emigration) in half the twin-samples was largely realised, in that more than 75 % of the predicted population was subsequently recorded at the appropriate time in the field. Nevertheless, potential populations were more than halved by natural enemies in a quarter of the twin-samples.

A detailed study at Newcastle upon Tyne (Foster, 1972) revealed a decline in potential increase rate associated with ageing of the variety Majestic, followed by a recovery at senescence. This trend was not obvious in other surveys, but could be detected as a high–low–high sequence of values for potential increase rate at Preston in 1970, at Zurich, Switzerland, in 1970, and at Lexington, USA, in 1971.

The potential increase rate under-predicted the final population size in six of the 34 samples. In two of these cases, the population structure was unstable and the estimates of potential increase rate unreliable. In a further two, aphid density was very low and sampling error proportionally large; the 'potential' density underpredicted the observed density by 15.3 and 27.5 %. In the two other cases, where data were reliable, the prediction errors were 2.6 and 6.1 % of the observed densities. It is impossible to test these predictions for significance, but it is assumed that they are within the limits of sampling error associated with the distribution of aphid counts.

Reduction of the population by emigration of alatae. Emigration plays a large part in reducing populations of aphids which aggregate (van Emden *et al.*, 1969). The green peach aphid is not an aggregating species; few alate aphids were produced, even after the population peak. Emigration comprised a steady departure of relatively few individuals from just before peak density until the end of the season. Five twin-samples taken about ten days after the peak showed emigration losses at Newcastle upon Tyne as 3 % (1969), 1.9 % (1970) and 2.4 % (1971) of potential density, at Cambridge as 0.6 % in 1971, and at Preston as 0.1 % in 1970. Twin-samples more than 20 days after peak showed losses of 6.9 % in Newcastle upon Tyne in 1969 and 0.3 % in Preston in 1970.

Emigration was the most variable factor found during this survey. There was no significant correlation between density and rate of emigration. This was probably partly because the data for each sampling occasion were pooled and therefore local variations in alate production with density in any one field were not identified, but certainly there was no indication that most alatae were produced where overall peak populations were highest. Thus the highest emigration rate (8.7 % at Newcastle upon Tyne in 1970) occurred in a population that reached a peak density of 34.7 aphids per haulm; a much lower rate (0.8 %) occurred in a peak population of 55.2 aphids per haulm, at Preston, in 1971. The most consistent feature of emigration was the very low pre-peak rate of alate production, which did not exceed 0.6 % of potential density except at Newcastle upon Tyne in 1970.

Reduction of the population by parasitism. Parasitism was measured by the appearance of mummies in a sub-sample of adult aphids and fully grown nymphs kept in the laboratory for one instar-period. This probably underestimates the population effect of parasites, as disturbance of individuals by parasite activity may cause a decrease in aphid multiplication, and parasitised aphids may fall to the ground and die without forming mummies. (Such mortality would be ascribed to predation in these experiments.)

When the levels of parasitism found in the experiment are considered, however, it is unlikely that the additional losses referred to above would have contributed greatly to overall aphid mortality. Percentage parasitism was low in all twin-samples. Less than 1 % parasitism was recorded in 20 samples, 1–2 % in five, 2–3 % in four, and the maximum rate recorded in the remaining five twin-samples was 6.2 %.

There was no suggestion in the data that high aphid densities would be likely to be accompanied by an increased rate of parasitisation. Indeed, the higher rates of parasitisation (above 2 %) tended to be found in populations with a low increase rate, and this was particularly true with rapidly breeding populations early in the season when control would be most desirable. Although there was little evidence of increase in parasitism during any one season, pre-peak values were invariably the lowest in any series. Thus it seems unlikely that the early generations of primary parasites were subject to a high rate of hyperparasite attack as has often been reported, e.g. for *B. brassicae* (Hafez, 1961). The highest overall rate was found at Lexington, Kentucky, where aphid density on potato was low though neighbouring tobacco fields were heavily infested.

Reduction of the population by entomogenous fungi. This was assessed from a sub-sample kept in the laboratory as for parasitism; to some extent the same problems of aphids falling from the plants apply. No populations suffered greatly from fungus attack, however, although one loss of 6.7 % was recorded.

M. Mackauer & M. J. Way

In general, values were much lower than this (0 % in 14 twin-samples, 0–1 % in seven, 1–2 % in six, 2–3 % in two and 3–6, 7 % in five). Values exceeding 1 % were recorded only after the population peak. Even heavy rainfall (e.g. weeks with more than 40 mm rainfall) did not appear to promote losses from entomogenous fungi.

Reduction of the population by other factors. The mortality that remains when the losses due to emigration, parasitism, and disease have been subtracted from 'potential' population density is referred to as the '(maximum) residual mortality'. Very high levels of residual mortality were observed in post-peak twin-samples. High mortality often occurred well before the aphid population peak and was therefore of control value. The residual mortality includes young not born because of the loss of adults, and is therefore expressed as percentage loss of the potential population at the end of the instar-period rather than as percentage loss of the population recorded on the crop.

Hughes (1963) regarded residual mortality as the *maximum* mortality ascribable to predation, for it includes other mortalities (e.g. rainfall, losses by dislodgement of parasitised aphids or aphids falling with cut foliage and flowers), as well as sampling errors. Where there is little residual mortality, however, it is suggested that predators would be having little effect. The decision whether or not a large residual mortality indicates a large predator impact is of course more difficult to make. There is little to suggest that rain is important as a direct mortality agent for *M. persicae*. Residual mortalities could not be related to the rainfall during an instar-period. For example, 18.9 mm of rain at Newcastle upon Tyne in 1970 was associated with a residual mortality of only 3.5 % of potential population. There is no easy way of estimating the errors of the various estimates contributing to Hughes' analysis. Certainly, if they 'swamped' the partitioned mortality, one would expect *negative* residual mortalities with greater frequency than they occurred (17 % of twin-samples). Residual mortalities therefore include a large predator component, but this may often be seriously over- or underestimated owing to sampling error. Residual mortalities can be discussed only insofar as they correlate with records of predator occurrence and abundance on the crop and insofar as they show a consistent pattern. Residual mortalities in the survey are therefore provisionally and cautiously ascribed to predator activity. It would be complicated and space-consuming to attach provisos to all the comments that follow, but the limitations must be borne in mind below where that half of the twin-samples in which residual mortality was a large component are examined in more detail.

Predation before peak population was highest at Preston in 1970 (74 %), Newcastle upon Tyne in 1970 (51.7 %), and Cambridge in 1971 (48.6 %). At Newcastle, high early mortalities were ascribed to non-specific entomophagous arthropods of the ground fauna, e.g. linyphiid spiders, anystid mites, and

80

staphylinid beetles. An early build-up of anthocorid bugs was responsible for high mortality at Preston.

Predation after the peak was large in all cases, the period immediately after each main peak being associated with losses exceeding 25 % (except 18 % at Syracuse in 1970). These losses were caused mainly by syrphids and coccinellids which at most sites did not become abundant until after the aphid peak. An exception occurred at Newcastle upon Tyne in 1971 when an early invasion by *Coccinella undecimpunctata* L. prevented the population of *M. persicae* from rising above four aphids per haulm. This impact on green peach aphid was associated with a large, uncontrolled build-up of *Macrosiphum euphorbiae*. At least three other low-density populations were associated with early and/or large coccinellid populations: at Karaj, Iran, in 1970, at Fredericton, Canada, in 1971, and at Santa Maria, Brazil, in 1971. By contrast, the early attack by anthocorids at Preston in 1971 failed to prevent exponential aphid growth.

Summary of the sites

Brazil (Santa Maria). The lowest peak population recorded was in 1971. Total aphid density did not exceed two per hill and *M. persicae* density was not more than 0.5 aphids per hill. Coccinellids were the only predators recorded; they were found from the first day of sampling onwards, maintaining an aphid/predator ratio between 2:1 and 11:1 and causing extinction of the aphid population about 800 day-degrees (°C) after invasion.

Britain (Cambridge). A moderate infestation developed in 1971. The population increased smoothly and exponentially but, after 400 day-degrees (°C) crashed to extinction at a rate slightly faster than that at which it had grown. A twin-sample taken when increase was fastest indicated an unusually low potential increase rate (0.92) with a negative residual mortality. A twin-sample taken in the fastest part of the decline phase indicated an even lower potential increase rate coupled with 49 % predation, 6 % disease, 3 % parasitism, and less than 1 % emigration. A wide range of predators occurred at the peak but coccinellid larvae were the most numerous and most effective aphidophages.

Britain (Dundee). Full surveys were conducted in 1970 and in 1971. Large populations developed in 1969 and 1971, but peak density was low in 1970. Twin-samples taken at the 1969 and 1970 peaks indicated high rates of emigration and predation. The chief predators were anthocorids and syrphid larvae. Peak potential increase rates were low compared with 1971 when an unusual situation developed. Twin-samples which were taken immediately before and after the main peak indicated identical potential increase rates, similar but declining rates of emigration and parasitism, and a high initial kill by fungus disease. The only major difference was in the residual loss, from 14 % (12 aphids per haulm per instar-period) before the peak to 45 % (63 aphids per haulm per instar-period) after the peak. Predation alone was responsible for population decline, although most surveys indicated that other factors, in particular a reduction in potential increase rate, were associated with population crashes. Significantly perhaps, when potential increase rate rose sharply (from 1.6 to 2.1) 170 day-degrees (°C) after the main peak, population density also rose despite sustained high residual mortality. The predators alone (in this case, syrphid larvae and coccinellids) were insufficient to maintain control against a population that did not lose its growth potential.

Britain (Edinburgh). This site was in a seed-growing area. In 1970, the population that developed was small and late, maximum numbers occurring at the end of August. The season was cool (the average daily temperature did not exceed 18 °C) and wet. A twin-sample was taken as the population declined. The potential increase rate was low; the instar distribution significantly departed from a geometric progression. Nevertheless, only a large residual mortality could account for the rapid decline; the estimated losses (33 % predation, 7 % disease, 5 % parasitism, and 6 % emigration) are thought to be a good approximation of reality. No coccinellids were found but anthocorids and syrphid larvae were present in August.

Britain (Newcastle upon Tyne). In 1969, the aphid population maintained a high density for a very short period. High residual mortalities at and just after the peak (42 and 47 %) were attributed to predation by syrphid larvae. Neither parasites nor fungal diseases were recorded at the peak, but losses due to these factors were high in the decline phase. In 1970, population growth was interrupted at an early stage when 52 % residual mortality was recorded; this was at a time when there were few specifically aphidophagous insects – predation was here ascribed to the general predator component of the ground fauna, in particular staphylinid larvae, linyphiid spiders, and anystid mites. The aphid population recovered and maintained a high density for 150 day-degrees (°C) during which predation was minimal, the main regulating factors being emigration and a decline in birth rate. The appearance of aphidophagous insects, mainly syrphid larvae, upset the balance caused by intrinsic factors and caused a population collapse as rapid as that of 1969. The 1971 population was minute, never exceeding 3.5 aphids per haulm, although potential increase rate was unusually high throughout the season. The first two twin-samples indicated high predation (18 and 55 % residual mortalities) associated with a build-up of *C. undecimpunctata*, a coccinellid not seen in either 1969 or 1970.

Myzus persicae *Sulz.*, *an aphid of world importance*

Canada (Fredericton, NB). Small populations developed in 1970 and in 1971. A very slow build-up to 16 *Myzus persicae* per hill after 800 day-degrees (°C) in 1971 could not be twin-sampled because the crop was killed by potato blight. Coccinellids were the only predators recorded in 1970; no information on predators was available for 1971.

Canada (Preston, Ontario). A large population developed slowly in 1970 and levelled off at about 50 aphids per haulm for 200 day-degrees (°C). Rate of aphid increase was low early in the season, perhaps indicating resistance in the variety Kennebec. Three post-peak twin-samples indicated heavy predation (49–54 %), low and declining parasitism, and low fungal infection. Emigration was less than 0.5 % of the total loss. A much larger population developed in 1971. An early setback in population growth 250 day-degrees (°C) after invasion of the crop was due apparently to the build-up of anthocorid bugs. An early twin-sample indicated 74 % predation. The bugs declined in number 100 day-degrees (°C) later. A twin-sample taken during exponential growth showed a minute negative residual mortality, indicating a sudden relaxation in predator attack intensity. The potential increase rate declined from 1.6 (during the exponential phase) to 1.4 at the peak, and 25 % predation was sufficient to start a population crash (85 % reduction in aphid density in 150 day-degrees (°C)). Emigration, parasitism, and disease all made negligible contributions to aphid losses. The crash was associated with a reduction in aphid increase rate and increasing populations of anthocorids, coccinellids, and syrphids.

Iran (Karaj). The population oscillated at low density from May to early August in 1970. No twin-samples were taken. It appears that the population was attacked by coccinellids and anthocorids.

Italy (Syracuse). A large population developed in 1970, with a smooth curve up to and down from the peak. In the exponential phase, two twin-samples showed high potential increase rates, 6–7 % predation and no emigration. Predators were quite abundant, but could hardly deviate the population from its exponential increase rate. Directly after the peak, potential increase rate dropped to 1.1 and then dropped further to 1.02, predation increased to 18 %, and emigration to 8 %, expressed as loss from potential. This combination of factors caused a crash in the form of a 99.2 % reduction in aphid density at 225 day-degrees (°C). The commonest predators were coccinellids, lacewing larvae, anthocorids, and syrphid larvae.

Poland (Warsaw). Very low populations occurred in 1970 and 1971. Twin-samples were attempted but were insufficient in size to permit evaluation. *M. persicae* never comprised more than 0.5 % of the total aphid infestation in

1970 and 7 % in 1971. In both years coccinellids were present from early in the season and were commonest shortly after the start of the decline of the main aphid infestation. Lacewing and syrphid larvae were present in small numbers in 1971.

Switzerland (Zurich). A small population developed in 1970. The population density of the first twin-sample was too small to permit accurate estimates of growth and mortality factors. Similarly insufficient aphids were sampled at the second and third twin-samples to give reliable estimates of potential increase rate; on both occasions estimates were unusually low and the instar distributions showed a statistically significant departure from geometric progressions. Consequently, all three twin-samples gave negative residual mortalities. Nevertheless it was clear that good aphid control was caused by a very early build-up of populations of coccinellids and lacewing larvae. Anthocorids and syrphid larvae came later and were less important.

USA (Lexington, Kentucky). The population was small, never exceeding five aphids per haulm. The variety Katahdin grown at Lexington has been used as a standard 'susceptible' variety (Adams, 1946), and it is possible that day temperatures may have been approaching the upper threshold for aphid development; an average temperature of 23 °C was recorded for most of June. Routine sampling indicated two peaks, one 280 day-degrees (°C) after colonisation and one 170 day-degrees (°C) later. Five twin-samples were attempted but only three were taken at times when aphids were at all numerous – at the two peaks and in the dip between them. A high–low–high trend in potential increase rate was associated with a low–high–low trend in residual loss (maximum at 23 % of potential population between peaks). No information is available concerning predators. Parasites were recorded at each twin-sample, the maximum loss being 6.2 % of the potential population. Fourth instar alatae and fungus disease were not recorded. As in Iran, low aphid density was associated with high temperatures.

General conclusions

All the data collected suggest that economically acceptable biological control of *M. persicae* as a virus vector by promoting the existing natural enemies is out of the question after the aphid has reached a potato crop. However, peak numbers can be limited by predators on occasion, especially where the aphid's increase rate is low. This perhaps highlights a potential for biological control outside the crop to reduce the numbers of migrants that reach the crop. Moreover, the project has illustrated how a collaborative international project could be brought to bear on other target aphid species that perhaps are better suited to biological control measures.

It would seem that the indigenous parasite 'races', even without reduction in numbers by hyperparasites, produce rather insignificant mortality in *M. persicae* populations. Work on parasites might therefore concentrate on the potential of using 'foreign' races and/or inundation rather than on the preservation of biological control agents.

Fungal attack also appears too sporadic to hold out much hope for biological control, though the development of new strains or of new techniques for disseminating artificially introduced fungi might modify this conclusion. There were no signs of field mortality from fungal attack early in the season, even under damp conditions, and in general *M. persicae* populations were too sparse to make satisfactory targets.

Predators, on the other hand, appeared surprisingly important in several diverse areas. There seems a real potential here for a control of the size of the aphid peak, though predator abundance was not evaluated critically in the survey and it is therefore difficult to attribute aphid losses to particular groups of predators. Coccinellids would appear potentially the most valuable predators, as they can appear in quite large numbers fairly early in the season, though they are unlikely to prevent a build-up of *M. persicae* to a distinct peak. The predators of aphids infesting potato are diverse, ranging from exclusively aphidophagous insects whose numbers are governed by aphid density to non-specific entomophages which can build up in numbers independently of aphids. The latter group may exert control early in the season when aphid numbers are low, or later by chance inundations. There is a real need for more work to be done on the lesser known predators, in particular the aphidophagous cecidomyiids, hemerobiid lacewings, spiders, predatory mites, and beetle larvae.

The project data further indicate that predators usually only induce an actual decline in *M. persicae* population size when the aphids are increasing at well below their maximum rate, even on a 'poor' host such as potato. Thus, although 17.6 % predation accompanied an 8.6 % decline of the population in June 1970 in Syracuse (when the overall potential increase rate was only 1.24), a 43.7 % predation at Dundee in August 1971 (when the increase rate was 1.99) resulted in a population *increase* of 8.8 %. Figure 12 shows the distributions of increase rates and per cent predation aligned to keep the population constant. Although quite active predation is frequent, its distribution in relation to aphid increase rate is poor and many populations escape control. The choice of potato for the project seems to have been fortuitously useful, for it has given examples of increase rates adequately low for predator impact before the aphid population peak. No such effect might have appeared if tobacco, for example, had been chosen. Studies on plant resistance and analogous changes in the plant brought about by age etc. would therefore appear essential to any effective biological control relying on the existing predators. Although predators often greatly accelerated the decline of the aphid population once the rate of aphid increase was reduced late in the

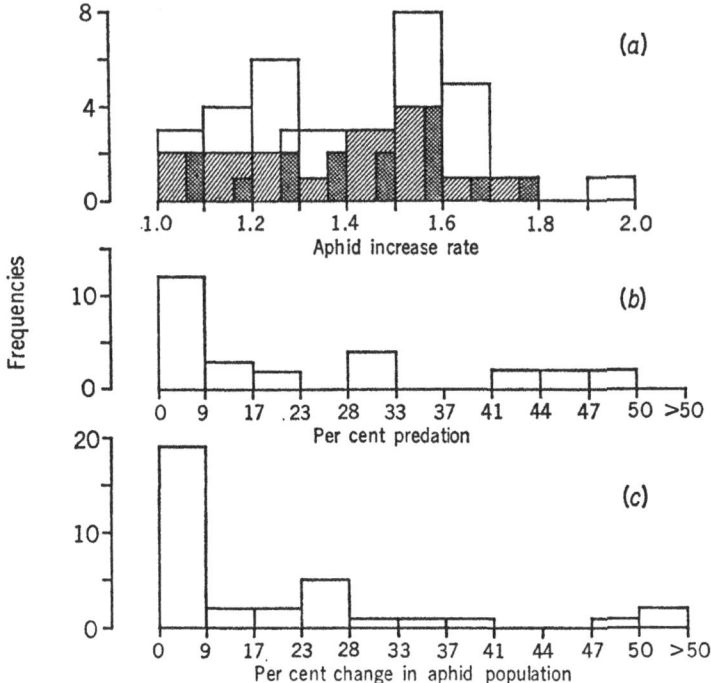

Fig. 12. Frequency distributions of the field data relating to the impact of predators on aphid populations. (*a*) Frequency distribution of aphid increase rates (total columns), the coincidence in the field of enough predation theoretically able to prevent an increase in the population (stippled), and the occurrence of such control (diagonal shading); (*b*) frequency distribution of per cent predation on a scale which theoretically prevents populations with an increase rate on scale (*a*) above from rising; and (*c*) frequency distribution of per cent reduction in aphid populations between the two samples of a twin-sample: the 0–9 category includes all samples showing a reduction of less than 9 % and therefore all increases.

season, and this is probably valuable in influencing the size of overwintering populations, the decline is followed by too many stages in the life cycle to make such predation of direct and predictable value for the following year. A surprise was the low and steady rate of emigration revealed by the project, which means that predators are removing potential migrants even well after the population peak.

Population dynamics on brassicas

Several species of Cruciferae are secondary host plants of *M. persicae*. In greenhouses the aphid develops particularly well on rapidly growing plants such as radish (*Raphanus sativus*), turnip (*B. campestris rapa*), mustard (*B. alba*), and Chinese cabbage (*B. pekinensis*); the aphid's fecundity is reduced on cauliflower (*B. oleracea* var. *botrydis*), Brussels sprouts (*B. oleracea* var.

gemmifera), colza (*B. napus napobrassica*), and common cabbage (*B. oleracea capitata*).

Anholocyclic overwintering of *M. persicae* has been observed on winter cabbage, colza, and spinach. In Denmark, apterous individuals can survive anholocyclically if the mean temperature during the three coldest months is above 4 °C (Heie & Petersen, 1961). In the Federal Republic of Germany, anholocyclic overwintering occurs if the maximum monthly temperature remains above 10 °C (Heinze, 1939). In France, under certain conditions, some apterous individuals survive if the minimum temperature does not fall below −18 °C (Bonnemaison, 1950). The small populations that survive when winters are mild start reproducing with the beginning of spring. These insects produce winged virginoparous females which disperse to secondary host plants together with the winged fundatrigeniae.

At Versailles, potatoes (var. Bintje), sugar beets, and winter cabbage (var. Norway) or summer cabbage (Milan cabbage, var. Marcellin) were planted in a Latin square design to determine their relative attractiveness to *M. persicae* and to other aphids. The aphids' population growth and the impact of predators and parasites was also measured. Beginning with the arrival of the first winged aphids, i.e. from 10 to 15 May onwards, aphid populations were counted on one leaf each of designated plants of sugar beet and cabbage and on one upper, one middle, and one lower leaf each of potato plants. The observations were continued for three years, from 1967 to 1969. It was found that, early in the season (15 and 16 May), summer cabbage and potatoes attracted similar numbers of winged *M. persicae*, i.e. an average of 35 and 38 individuals, respectively, per 100 leaves; no aphids of anholocyclic origin were found on sugar beets.

Figure 13 gives the mean total numbers of *M. persicae* per 100 marked leaves of cabbage, potato, sugar beet, and winter cabbage. It should be noted that the aphid's rate of reproduction is higher on mature leaves of cabbage than on those of potato and in particular of sugar beet. Populations of *M. persicae* decline sharply during the summer. This decline is caused by the action of predators and parasites and also by changes in the physiological condition of the cabbage leaves. Laboratory experiments have already indicated that the aphid's fecundity on cut leaves of winter or summer cabbage reaches a minimum in July and a maximum during the period from December to February (Bonnemaison, 1971). The waxy layer of the cabbage leaves also contributes to the reduction in fecundity of *M. persicae*, as was reported by Way & Murdie (1965) working with Brussels sprouts; the spraying of wetting agents on cauliflower and turnip leaves causes the aphid's rate of reproduction to increase again (Heathcote & Ward, 1958).

Under conditions of high temperatures and dryness the turgescence of old leaves decreases and, as a consequence, the fecundity of *M. persicae* declines. High temperatures affect aphids on different host plants in a different way. In

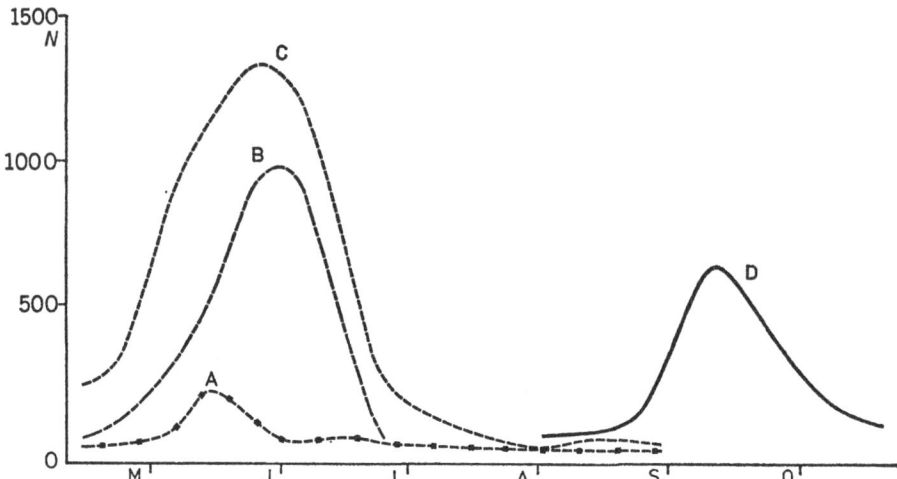

Fig. 13. Population changes of *Myzus persicae* on sugar beet (A), potato (B), summer and winter cabbage (C and D) at Versailles, France. (Mean number of aphids, *N*, on 100 leaves during 1967–9.)

1967, the maximum temperature in the shade was 23.9 °C on 30 June, 29.4 °C on 1 July, and 25.9 °C on 2 July; on 4 July populations of *M. persicae* on cabbage were only 89 %, on sugar beet only 24 %, and on potato only 40 % of what was observed one week earlier. In 1969, the maximum temperature was 27.7 °C on 27 July and 32.7 °C on 28 July; on 1 August the aphid population on cabbage had declined to 24 % of that observed on 23 July (Bonnemaison, 1971).

The relative impact of natural enemies on *M. persicae*, as influenced by different host plants, is difficult to measure. This is because additional aphid species are present, e.g. mainly *B. brassicae* and *Macrosiphum euphorbiae* on cabbage; *Aphis fabae* and *Aulacorthum solani* (Kalt.) on sugar beet; and *M. euphorbiae*, *A. solani* and *Aphis nasturtii* Kalt. on potato (Fig. 14).

Adult coccinellids are present on cabbage in May and June but disappear towards the end of June when their eggs become rare. Coccinellids are more numerous and remain longer on sugar beet, and particularly on potato, than on cabbage. In July, larval and adult coccinellids are the most common predators on potato. On sugar beet, they are as numerous as the eggs and larvae of syrphids. Coccinellids are rare on cabbage and sugar beet in the fall.

Lacewings are regularly observed on cabbage and on sugar beet from May to August and on potato from May to July.

Syrphid larvae are the most common predators on summer cabbage from June to October and on winter cabbage from September to November. They are usually present, but only in small numbers, during June and July on potato and from June to September on sugar beet.

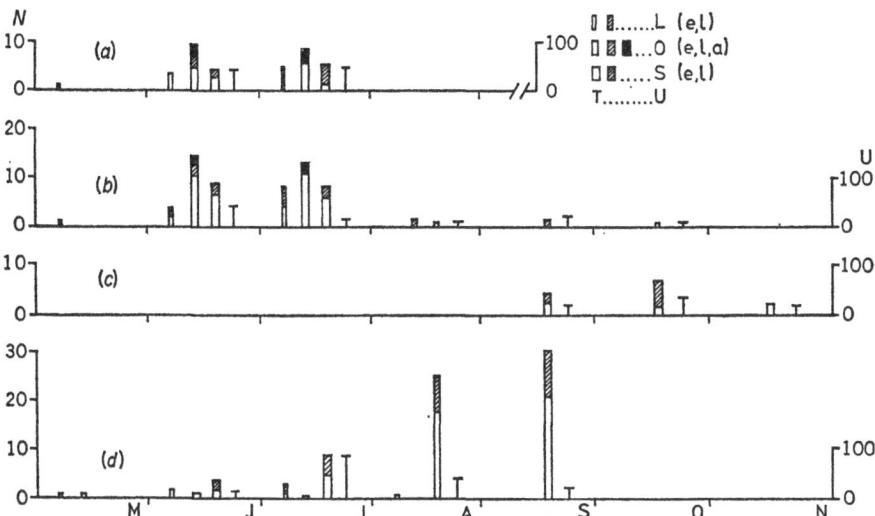

Fig. 14. Numbers (*N*) of lacewings (L), coccinellids (O), syrphids (S), and parasitised aphids (U) in samples of 100 leaves of potato (*a*), sugar beet (*b*), winter and summer cabbage (*c* and *d*) during the years 1967 to 1969: e, eggs; l, larvae; a, adults.

The distribution of parasites, mainly *Ephedrus persicae* Froggatt, *E. plagiator* (Nees), *Diaeretiella rapae* (M'Intosh), and *Aphidius matricariae* Haliday, appears not be influenced by the different aphid host plants.

Population dynamics on peach

Whereas the population dynamics of *Myzus persicae* on potato was studied on a worldwide basis, only some research was carried out on the aphid's dynamics on the primary host plant. This was done at several localities in France and, later, in the Netherlands, Switzerland, and Australia. The studies in France were concerned with an assessment of integrated control procedures for aphids in peach orchards. In order to define tolerance levels that are practicable, information was required on (i) the population growth of aphids on a certain number of reference trees; (ii) the impact of natural enemies; (iii) the influence of climatic conditions; and (iv) the spatial distribution of the pest in orchards.

Preliminary observations indicated two things. First, peach trees can tolerate a fairly high number of *M. persicae* without measurable economic damage; and second, the aphid infestation is fairly heterogeneous in orchards. These two aspects were considered in the development of a sampling method which classifies trees according to their degree of infestation based on the powers of 5. One whole tree is considered a sampling unit. For example, a second-degree infestation means that the particular tree is infested with between 5^1 and 5^2 aphids, i.e. with between six and 25 aphids; and a degree 5 infestation means that the particular tree is infested with between 5^4 and 5^5 aphids, i.e. with between 626 and 3125 aphids.

A simplified method based on the previous procedure consists of estimating the total number of infested twigs per tree. For each of 50 reference trees in one orchard the total number of infested one-year-old twigs is counted disregarding the actual number of aphids on the branches. The degree of infestation is then determined as follows. A tree with between one and five infested twigs is assigned to class 1; one with between six and 25 infested twigs is assigned to class 2; one with between 26 and 625 infested twigs to class 3, etc. The number of trees in each class is multiplied by the class mark, and the products are added up. This value accurately estimates the intensity of infestation in an orchard.

The development of *M. persicae* populations on peach varies considerably from one region to another. In a Mediterraenean-type climate oviposition takes place in December and eggs hatch very early, namely in mid-January. The emigration begins in early April with the winged individuals of the third generation and continues for at least three months. Emigration is influenced by the physiological condition of the host plant. Under irrigation, peach continues to grow until late in the season and thns produces young shoots that are suitable for the aphid's reproduction. High temperatures in May and June accelerate plant growth, at the same time enabling the spread of the aphid (particularly of its nymphal stages) along twigs and branches; this leads to the colonisation of new growth. Aggregation – which tends to reduce the aphid's reproductive achievement and, in addition, leads to the production of winged individuals and consequently to emigration – is thus reduced. Such a situation is observed in southern France (Provence, Languedoc, Roussillon) and in Australia (Fig. 15).

In colder climates there is a considerable time lag between oviposition in November and egg hatch, which begins in March. Where trees are neither pruned nor irrigated, as happens in the Paris area, emigration starts already in the second generation, i.e. in mid-May. Nearly all individuals of the third generation are winged and the infestation of peach trees by aphids disappears at the latest in early June (Fig. 15).

In regions where both holocyclic and anholocyclic overwintering of *M. persicae* occurs, the development of the spring populations on the primary host is completely independent of that observed at the same time on secondary hosts. It has been experimentally shown that winged viviparous females are unable to colonise young peach twigs if they have been obtained from populations on secondary host plants. The same applies to winged fundatrigeniae which originate on peach but cannot reproduce on it. These two facts are confirmed by the development of the infestation in a young orchard where trees are still spaced out. Here the greatest number of infested trees is observed

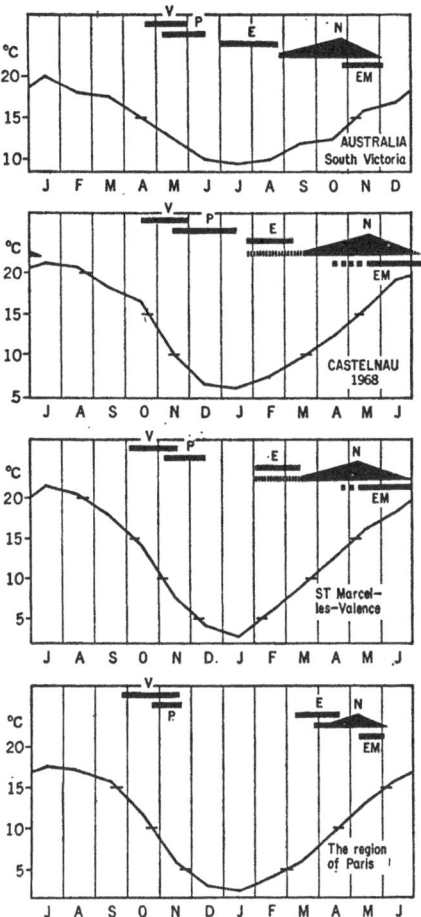

Fig. 15. Average monthly temperatures and comparative development of *Myzus persicae* on peach in Australia (South Victoria, latitude 36° S) and in three regions of France: Castelnau (Mediterranean coast, latitude 43° 30′ N), St Marcel-les-Valence (central Rhône valley, latitude 45° N), and the region of Paris (latitude 49° N). There is a close correlation between the number of months with a temperature below +5 °C and the length, in months, of the egg stage of *M. persicae*. E, period of egg hatch; EM, migration; N, period of economic damage; P, oviposition period; V, return flight.

when the fundatrices appear; later the number continually decreases and typical damage by fundatrices can be noted on each tree. If winged individuals could reproduce on peach, the number of infected trees would increase dramatically – which does not happen.

The main mortality factors that affect *M. persicae* on peach vary considerably according to the season. In southern France, where eggs hatch early, cold rainy weather kills many young nymphs of the fundatrix generation if these

91

M. Mackauer & M. J. Way

are not yet protected in young leaves and flowers. With the beginning of the second and especially with the third generation, the aphids are concentrated in the deformed leaves that appear to be particularly attractive to syrphids. As a consequence, an appreciable number of primary colonies (each one generated by one fundatrix) is completely destroyed by syrphid larvae, most frequently by *Epistrophe balteata* De G. Beginning with higher temperatures at the end of spring, the impact of predators increases; however, infestations on heavily attacked peach trees continue undiminished as long as the physiological condition of the plant enables a high rate of aphid reproduction. It would seem that predators are potentially capable of destroying the aphid populations; but the aphids' mobility and the availability of suitable plant growth obviate the predators' impact, in particular when aphid populations were numerous from the beginning.

Other natural enemies contribute to the control of aphid populations, but they often act later than the syrphids and only accelerate the aphid's population crash during the first days of summer. This is notably the case with regard to *Entomophthora* spp. which do not cause any epizootics as long as temperatures remain below 20 °C.

Biological methods of aphid control

Parasites

Parasitism by species of the hymenopterous families Aphidiidae and Aphelinidae is perhaps the most commonly observed cause of aphid mortality in the field. Many, if not all, aphid colonies include some 'mummified' individuals, i.e. dead and eviscerated aphids containing a fully-grown parasite larva or pupa. Well over 300 species of Aphidiidae were listed by Mackauer & Starý (1967), with their distributions, known host records, and ecological data. Starý (1970) published an extensive account of the biology of the group. The taxonomy of the European species of Aphelinidae was reviewed by Ferrière (1965), who recognised some 50 species as parasitic on aphids. Hagen & van den Bosch (1968) prepared a critical analysis of the impact of parasites on their aphid hosts, including relevant aspects of their biology.

Several aphidiid and aphelinid parasites have been used with striking success in the biological control of introduced aphid pests. For example, *Aphelinus mali* Haldemann has contributed to the control of the woolly apple aphid, *Eriosoma lanigerum* Hausmann, in most apple-growing areas of the world (Howard, 1929). Equally effective were the introductions from the Middle East into California of *Praon exsoletum* Nees, *Trioxys complanatus* Quilis, and *Aphelinus asychis* Walk. against spotted alfalfa aphid, *Therioaphis trifolii* Monell (van den Bosch *et al.*, 1959a, b), from Iran into California of *Trioxys pallidus* Hal., against walnut aphid, *Chromaphis juglandicola* Kalt. (Frazer & van den Bosch, 1973), and from India into large areas of North America of

92

Aphidius smithi Sharma & Subba Rao (van den Bosch *et al.*, 1966; Campbell & Mackauer, 1973; van den Bosch, Lagace & Stern, 1967). Aphid parasites were also involved in recent biological control programmes against the greenbug, *S. graminum*, in the mid-western United States (Jackson, Rogers & Eikenbary, 1971; Archer *et al.*, 1974), against the sunflower aphid, *Aphis helianthi* Monell, in Oklahoma (Rogers *et al.*, 1972*b*), and against a complex of aphids on urban trees in California (Olkowski *et al.*, 1974).

Protelean parasites of aphids other than Hymenoptera are observed seldom in nature and have been little studied. Several species of gallmidges (Diptera: Cecidomyiidae) are associated with aphids, and Barnes (1954) summarised what is known about their biology. A recent finding of an *Endaphis* sp. in British Columbia, however, suggests that these endoparasites perhaps are not quite as rare as has been assumed. That midge was locally quite common as a parasite of *Euceraphis* of which it attacked both the fourth nymphal instar and the adult stage. In contrast to all other known aphid parasites, this parasite was gregarious, with up to 10 larvae developing in a single host individual (Foottit and Mackauer, personal communication).

The impact of parasites on their hosts can be assessed directly, by observations in the field, or indirectly, by various exclusion and feeding-trace techniques (DeBach & Huffaker, 1971; Kiritani & Dempster, 1973). A mathematical model of the complete host–parasite system would also permit parasite impact to be estimated from system responses to changes in parasite parameters. From investigations of these kinds several generalities have emerged that may contribute to an informed (and increasingly effective) use of aphid parasites in biological control programmes.

Fecundity, longevity, sex ratio

The normal fecundity of aphidiid parasites has been estimated to range from 100 to 400 eggs per female (Vevai, 1942; Hafez, 1961; Messenger & Force, 1963). Quilis Pérez (1930) reported that *Lysiphlebus fabarum* Quilis laid an average of 700 and a maximum of 1500 eggs. Mackauer (1971) compared the fecundity of several parasites of pea aphid under optimum laboratory conditions. He found that the most important, *Aphidius smithi*, laid an average of 774 eggs per female, but that fecundity was influenced by the number of potential hosts available (Mackauer & van den Bosch, 1973). Superparasitism increased and the total number of eggs laid declined when access to hosts was restricted. Most parasites do not appear to reach their potential fecundity in the field (Gilbert & Gutierrez, 1973; Campbell, 1974). Parasite 'loss' as a result of dispersal (van den Bosch *et al.*, 1967), time spent in searching for hosts, and an overall reduction in longevity are the main factors that contribute to a sub-optimal reproductive achievement.

Fox, Pass & Thurston (1967) and Mackauer (1973*b*) demonstrated that the rate of development of *A. smithi* in the pea aphid is influenced by host size and/or age. Development in a small or only marginally suitable host may require one to several days longer than in the 'optimal' host under otherwise identical conditions.

The sex ratio is another important factor that affects population growth and thus parasite impact. In biparental species the ratio of males and females varies within broad limits but usually is about 1 : 1. In *A. smithi* the sex ratio is influenced by the rate of oviposition (Wiackowski, 1962) but apparently not by host size or age (Mackauer, 1973*b*). Evidently, the sex ratio is a population phenomenon; however, there is little specific information available on the factors that regulate it in the field and, more importantly, whether or not all females can be presumed to mate and to contribute equally to producing the next generation. All, or nearly all, females of *Praon pequodorum* Vier. and of several *Aphidius* spp. parasitic on pea aphid collected by sweeping in the field were mated (Mackauer, unpublished). This implies that mate finding and fertilisation must be very efficient. A direct or indirect effect of low temperatures on the sex ratio of *A. smithi* was noted by van den Bosch *et al.* (1966), who found that females were more abundant than males in California alfalfa fields during early winter whereas males were more abundant than females during the period from December to March.

The aphid-attacking Aphelinidae do not differ fundamentally from aphidiids in fecundity, longevity, or sex ratio. Aphelinids usually are long-lived under laboratory conditions (Lundie, 1924) but this may not be the case in the field. There are, however, two aspects in the biology of Aphelinidae that are relevant to control and must be considered. One is the complete, or almost complete, ability shown by several species to discriminate between healthy and previously parasitised aphids (Lundie, 1924; Thompson, 1934). The other is host feeding which is common to many chalcids (DeBach, 1943; Flanders, 1953). Adult aphelinids destroy a considerable proportion of their aphid hosts by feeding on the haemolymph which exudes from oviposition wounds. Cate *et al.* (1973) estimated that *Aphelinus asychis* killed an average of 1.5 greenbugs per day per female in this manner. Host feeding on a similar scale was also observed in *A. mali* and *Mesidia nigra* (Hartley, 1922; Lundie, 1924; Lagace, 1969; Michel, 1967).

Host selection

Various aspects of host selection and specificity of Aphidiidae and Aphelinidae were reviewed by Mackauer (1961, 1967) and Starý (1964*a*). Many of these parasites exhibit a relatively narrow host range, often attacking only one or a few related aphid species or genera (Mackauer & Starý, 1967). Influences of the host habitat on parasite specificity are not always clearly indicated and, apparently, differ between species. The black bean aphid, *Aphis fabae*, is attacked by one parasite complex on its summer host plants and by a different one on its winter host plants (Hodek *et al.*, 1959). In this and in a number of similar cases (Starý, 1964a) the habitat plays an important role in the determination of the parasites' 'effective' host range. Several mechanisms have been suggested that may enable aphidiids to select a particular habitat. Read, Feeny & Root (1970) identified mustard oil, allyl isothiocyanate, as the odour that attracted *D. rapae* in the laboratory to cruciferous plants and to aphids feeding on them. Vater (1971) reported that *D. rapae* responded to optical stimuli, and in particular to green colours, in finding the host habitat, and suggested that shape and plant odours were of no importance. Actual host finding on the plant was by random searching (Hafez, 1961).

To date it has not been possible to identify and to measure the relative importance of the precise characteristics that identify an aphid to a parasite as a potential host. What has been done was to ascertain in the laboratory the influences of different environmental and host variables on the oviposition behaviour of the adult female and to compile, from field data, the consequences of *successful* parasitism, i.e. to evaluate host records. While these approaches yielded relevant information, they do not establish causal relationships. More informative are quantitative data on relative host preferences and differences in host suitability. Wilbert (1964) examined the host range and selection behaviour of *Aphelinus asychis*; among some 20 aphid species tested four species were acceptable for oviposition but unsuitable for parasite larval development. Griffiths (1960, 1961) demonstrated that the aphidiid *Monoctonus crepidis* Hal. is specific to species of *Nasonovia*; other aphids on lettuce, which is among the usual host plants of *N. ribisnigri*, were readily attacked but responded by encapsulation and by other immune reactions that caused the death of the developing parasites at an early embryological stage. Calvert & van den Bosch (1972) listed a series of dactynotine aphids as hosts of *Monoctonus paulensis* Ashm. in California; however, only *Macrosiphum* (*Sitobion*) *fragariae* Walk. played a major role in the population dynamics of the parasite, and was well synchronised with it, whereas other aphids were infrequently attacked and perhaps played a subsidiary role. Mackauer (1973b) compared the 'host acceptability range' of *Aphidius smithi* females with the 'host suitability range' of its larvae. In addition to its usual host, the pea aphid, this normally highly specific parasite readily attacked alien hosts when deprived of a more suitable choice. In fact the female's 'acceptability range' was considerably broader than what would have been predicted on the basis of host optimality.

Differences in performance on different hosts and the consequences of host change were examined by Michel (1970, 1971) and Monadjemi (1972). When *Aphelinius asychis* was transferred from its habitual 'laboratory' host, *Rhopalosiphum padi* L., to green peach aphid, its fertility declined from a mean of 214 offspring to only 24 in the first generation on *Myzus persicae*. In the second and third generation 143 and 245 offspring, respectively, were produced, indicating that adaptation to a new host was rapid. Interestingly, a transfer back to *R. padi* in the fourth generation caused another severe drop in mean fertility to 46 offspring.

Parasite impact

The percentages of parasitism in natural aphid populations may reach as high as 80–90 % (Frazer & van den Bosch, 1973). Usually it is much lower. For example, Shands *et al.* (1965) reported that a maximum of 0.63 % of green peach aphids on potato were parasitised in north-eastern Maine during a 12-year period. Estimates based solely on 'mummy' counts, however, do not adequately measure aphid loss due to parasitism, as parasitised aphids tend to migrate away from their host plants (Behrendt, 1968) and thus are missed by the usual sampling techniques.

Perhaps more important than aphid mortality are direct and indirect effects on aphid fecundity caused by parasitism and parasite activities. For example, pea aphid fecundity was reduced when aphids were harassed by searching and ovipositing parasites (Tamaki, Halfhill & Hathaway, 1970).

Parasitised aphids showed a significant reduction in fecundity and longevity as compared to normal individuals, the degree of reduction varying with host age at the beginning of parasitism (Arthur, 1944; Fox *et al.*, 1967; Hafez, 1961). Pea aphids parasitised by *Aphidius smithi* during an early nymphal instar produced few or no offspring, whereas parasitised young adults reached almost normal fecundity (Campbell & Mackauer, unpublished). This effect on the intrinsic rate of increase (r_m) and the doubling time (DT) of apterous and alate pea aphid is shown in Fig. 16. Lagace (1969) reported that embryos within adult viviparous *Iziphya punctata* can be parasitised successfully by *Mesidia nigra*. While such cases may be rare, as they probably result from chance hits by the ovipositor, a consequence of them is that the 'burden' of parasitism is deflected from a reproductive to a non-reproductive member of the population.

Many aphids respond to the presence of entomophagous insects by species-characteristic escape and defence reactions. Green peach aphids successfully escaped up to 65 % of attacks by the parasite *D. rapae*, the success increasing with aphid age and instar (Klingauf, 1967). Parasite presence appears to affect colony size and density in aphids that form aggregations on their host plants. For example, *Myzus persicae* and *Neomyzus circumflexus* Buckton (reared on an artificial diet) formed larger and denser colonies in the presence of parasites than when left undisturbed. Such behaviour, Klingauf & Sengonca (1970) suggested, protects aphids in the centre of the colony against parasite attack leaving only 'marginal' individuals exposed. Pea aphids, in contrast to green peach and cabbage aphids, became distributed at random over the

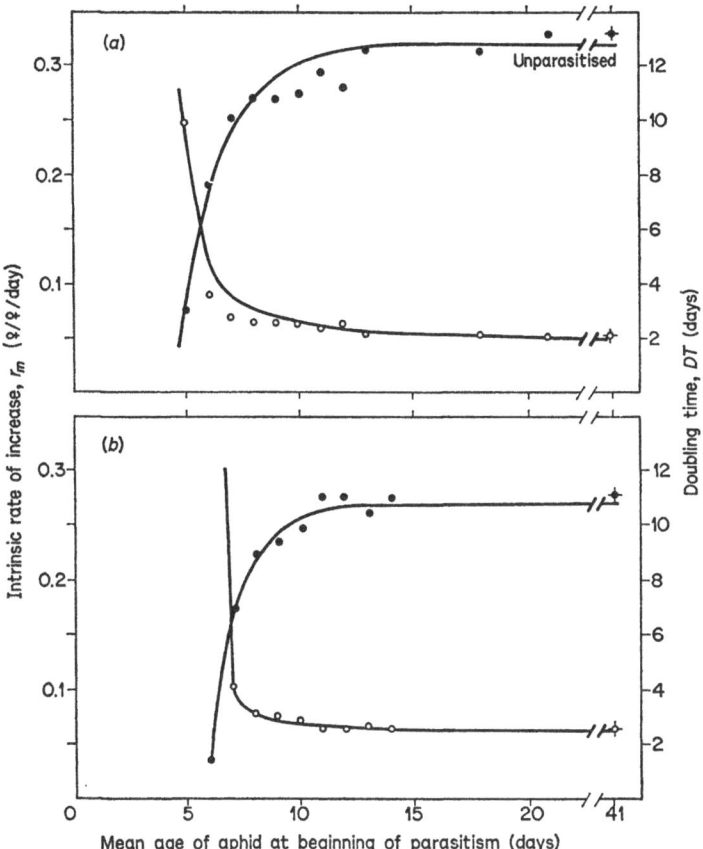

Fig. 16. The effect of parasitism by *Aphidius smithi* on the intrinsic rate of increase (r_m, ●) and the doubling time (DT, ○) of *Acyrthosiphon pisum*. (a) Apterous; and (b) alate pea aphids. The values of unparasitised aphids are included for comparison.

available host plants when disturbed (Tamaki *et al.*, 1970; Niku, 1972). Bowers *et al.* (1972) isolated and identified trans-β-farnesene as a pheromonal compound produced by several aphid species in the cornicle secretions when attacked by predators (Nault, Edwards & Styer, 1973). Obviously, any changes in aphid distribution will affect not only aphid/host plant interactions but also the relative success of parasites in locating hosts.

Aphid population models that included the effect of parasitism have been constructed for cabbage aphid, *B. brassicae*, on kale in Australia (Hughes & Gilbert, 1968; Gilbert & Hughes, 1971); for cowpea aphid, *Aphis craccivora* Koch, on subterranean clover in Australia (Gutierrez, Morgan & Havenstein, 1971; Gutierrez, unpublished); for thimbleberry aphid, *Masonaphis maxima* Mason, on thimbleberry in coastal British Columbia (Gilbert & Gutierrez,

M. Mackauer & M. J. Way

1973); and for pea aphid on alfalfa in the southern interior of British Columbia (Campbell, 1974). Although these models were capable of describing the parasites' performance, they did also show very clearly that present understanding of host/parasite interactions is incomplete. Perhaps not surprisingly, the models indicated that the timing of parasite attack was more important than parasite numbers. For example, the impact of *Aphidius rubifolii* Mack. on *M. maxima* was determined, largely, by the effectiveness of the first generation of *Aphidius* in the spring. The timing of attack, which depends on the parasite's threshold temperature for development in relation to that of its host, appears to be of fundamental importance in determining whether or not a parasite is an 'effective' natural enemy. Campbell *et al.* (1974) reported the temperature co-efficients of several aphids and of their parasites from different geographic areas. They also described a method for estimating developmental thresholds and times-to-adult under laboratory conditions. This method, although deficient because it does not consider the possible effects of rhythms on parasite emergence (Mackauer and Henkelman, unpublished), produces sufficiently accurate results to enable prediction of population growth on a physiological time-scale for use in the field.

The models showed that neither *D. rapae* in Australia nor *A. rubifolii* in coastal British Columbia seriously affected aphid population growth, demonstrating, once again, that parasites typically are 'croppers'. The relationship between *D. rapae*'s fecundity, number of emigrating offspring (which are transported within parasitised alate cabbage aphids to new habitats), and timing was such that the parasite's reproductive achievement was maximised *without* endangering the aphid (Gilbert & Hughes, 1971). According to the thimbleberry aphid model, which included the plant/aphid relationship, aphid population size would not change significantly even if the parasite disappeared completely or, alternatively, its initial numbers increased by a factor of 10 (Gilbert & Gutierrez, 1973). Evidently, other parasites have a far greater impact on their hosts than *A. rubifolii* appears to have. For example, *A. smithi* parasitises 80 % or more of pea aphid and can control its host on alfalfa in the southern interior of British Columbia, but the precise mechanisms by which this control is achieved are not as yet known (Campbell, 1974).

Conclusion

Two questions must be answered when considering the role and potential of parasites in the biological control of a particular pest aphid: one, what is the nature and scope of the damage; and, two, are there any parasites that are capable of suppressing the pest below its economic threshold? Different criteria will apply for 'direct' pests that damage crops only when present in large numbers, and 'indirect' pests, such as green peach aphid, which transmit plant diseases.

Myzus persicae *Sulz.*, *an aphid of world importance*

Huffaker, Messenger & DeBach (1971) considered some general characteristics of effective natural enemies. They noted that effective enemies usually possess a high reproductive capacity, have a short generation time, are dispersive, and are well synchronised with and relatively specific to their hosts. Many of these attributes as well as adaptability and genetic fitness denote also good colonising ability (Mackauer, 1972). Hymenopterous parasites of aphids, or at least those species that have been studied seriously, meet most, if not all, of these requirements. In addition, they can be mass-produced easily and at little cost in the insectary (Starý, 1964 *b*; Halfhill & Featherston, 1967).

Considering these advantages there is reason to be optimistic about a more frequent use of aphid parasites in biological control than in the past. The number of successful controls should increase against 'direct' crop pests, either through the introduction of parasites into areas where they are not or through mass-releasing laboratory-produced enemies at a critical phase in the aphid's life cycle. Much more use than hitherto could also be made of parasites in situations that are largely or exclusively under the control of man, as in glasshouses. McLeod (1937, 1940) was the first to report on the successful control of several aphid species, including green peach aphid, in greenhouses. Richardson & Westdal (1965), Wyatt (1970) and Hussey & Bravenboer (1971) noted similar examples, and Scopes (1970) described a technique for maintaining control of green peach aphid on year-round chrysanthemums.

However, there is at present no sound basis for expecting that a polyphagous, low-density pest on field crops, such as *M. persicae*, can be controlled by parasites. In fact published reports (Vevai, 1942; Shands *et al.*, 1965; Shiga, 1967, 1968; Leclant & Remaudière, 1970; Bonnemaison, 1971; Tao & Chiu, 1971; Tamaki, 1973) of parasite impact on green peach aphid are in broad agreement that natural enemies, with the possible exception of certain predators, do not significantly influence the aphid's population dynamics. The results of the international field project, discussed on p. 71–92 above, estimated aphid loss caused by parasitism to be generally less than 1 % (with a maximum of 6.2 %) and consequently negligible.

Although some 30 aphidiids and aphelinids are reported as parasites of green peach aphid, none appears to be specific to it or prefers *M. persicae* over other aphids (Mackauer, 1968); thus, none of these species meets one of the prime prerequisites for being an effective biological control agent. Mass-production and release of some of the oligophagous parasites could be attempted to augment existing natural control; however, these parasites would not necessarily concentrate their attacks on green peach aphid, which usually forms small and sparsely-distributed – and hence 'unattractive' (Pimentel, 1961) – colonies in the field. Finally, parasite strains that differ in their temperature requirements from indigenous, and presumably adapted, populations could be introduced (or bred in the laboratory) to manipulate the timing of attack. Again, the relevant methodology for assessing climatic variables and

4-2

99

their influence on biological control agents is available (Messenger, 1970) but their predictive power is as yet untested in field experiments.

In conclusion, the potential for biological control of green peach aphid on crops by parasites alone appears minimal. It may not be too difficult to increase parasite impact but this should be done in the context of an integrated control programme or of a pest management system.

Predators

The amount of work on different groups of aphid predators varies widely. A 1972 survey indicated that Coccinellidae were studied by 60 workers, Neuroptera by 43, Syrphidae by 31, Heteroptera by 21, Itonididae (= Cecidomyiidae) by 11, Arachnoidea by 9, Chamaemyiidae by 3, and Malachiidae by 1.

These predators were studied, to work out the ecology and behaviour of individual species, to determine the impact of predation on aphid populations, and to use predators for biological control of aphids either alone or in integrated control programmes.

Autoecological studies

Work on food and other environmental requirements of predators of different groups, especially the Coccinellidae, was reviewed by Hodek (1967, 1973), and Schneider (1969) reviewed the bionomics and physiology of aphidophagous Syrphidae. Brown (1972) studied four coccinellids preying on *S. graminum* in South Africa. Montes (1970) and Lim Sook Ming (1971) reported on some aphidophagous Coccinellidae in Chile and Malaysia, respectively. In addition, two previously neglected groups received some attention: the Itonididae (Uygun, 1971; Mayr, 1973) and the Lygaeidae–Geocorinae (Tamaki & Weeks, 1972 *a*, *b*).

The lack of a practical food on which large numbers may be reared in the laboratory has remained a barrier to the use of most coccinellids for control of aphids. Matsuka *et al.* (1972) were able to breed *Harmonia axyridis* continually on larvae and pupae of drone honey bees but the results with other predatory coccinellids were less encouraging.

Impact of predators

The effectiveness of predators in controlling aphids was reviewed recently by Hagen & van den Bosch (1968) and van Emden *et al.* (1969), and Hodek, Hagen & van Emden (1972) discussed some methods used to measure predator impact on prey populations.

Most of the studies on predator efficiency are based on a correlation between abundance of predators and aphid population changes.

Predators sometimes played an important role in the regulation of aphid numbers. Minoranskii (1967) concluded that a high abundance of natural enemies, especially of *Coccinella septempunctata* L. and *Adonia variegata* Goeze, was responsible for *Aphis fabae* not causing damage to sugar beet in the southern parts of the European USSR and in Japan. Stathopoulous (1967) found that *A. gossypii* Glover, on cotton in Greece, was significantly reduced by predators. Limited numbers of syrphid flies were effective in suppressing a potentially explosive population of *M. persicae* on peach trees in the State of Washington (Tamaki, Landis & Weeks, 1967). Inaizumi (1968) found that the numbers of *M. persicae* decreased immediately after blooming of potato plants because of increases in numbers of natural enemies. *M. persicae* is more intensely affected by natural enemies than is *A. gossypi*, which lives on the lower parts of the plant whereas *M. persicae* occurs in the upper parts where the density of natural enemies is higher.

Other authors are sceptical about the impact of predators. Evenhuis (1968) stated that predators, except perhaps coccinellids in early spring, are not very important in the natural control of *Rhopalosiphum insertum* Walk. on apple in Holland. *B. brassicae* and *M. persicae* are attacked mainly by the coccinellid *Hippodamia quinquesignata punctata* Kirby and the syrphid *Allograpta obliqua* Say in southern California but, according to Oatman & Platner (1969), these natural enemies cannot control the aphids adequately without the supplemental use of pesticides. Dunn & Kempton (1971) assumed that predators had only a minor effect on numbers of *B. brassicae* and *M. persicae* on Brussels sprouts in England.

Studies that demonstrate the impact of predators experimentally and that produce quantitative data about prey/predator relationships are extremely useful. Unfortunately they are still rather rare. The differential removal of predators of *B. brassicae* from Brussels sprouts in southern England by Pollard (1969) revealed that two syrphid flies, *Epistrophe balteata* and *Sphaerophoria scripta* L., were responsible for a rapid decline in aphid numbers in mid-August and demonstrated the importance of a cecidomyiid of the genus *Aphidoletes* as an aphid predator. The results of a comparative study by Pollard (1971) suggested that a very considerable diversity of habitats outside the crop as compared with a largely arable area made no difference to the extent of syrphid predation on *B. brassicae*. He concluded that only diversity within a habitat is an important influence on the amount of predation.

Tamaki & Weeks (1972*a*) tested the impact of three predators, the lygaeid *Geocoris bullatus* Say, the nabid *Nabis americoferus* Car., and coccinellid *Coccinella transversoguttata* Ws., on *M. persicae* on sugar beet in the greenhouse. Adults of *C. transversoguttata* destroyed all the aphids within five days when the predator/prey ratio was 1:34; *N. americoferus* or *G. bullatus* reduced the aphid population temporarily when the initial predator/prey ratio was 1:14. Another study (Tamaki & Weeks, 1972*b*) showed that *G. bullatus* was able to keep populations of *M. persicae* below the damaging level at a predator/prey ratio of 1:5. In this experiment the aphid population showed only a nine-fold increase within a month whereas the control population without the predator had over a 2000-fold increase.

Galecka & Kajak (1971) simulated predation by removing a certain fraction of *M. persicae* population on potatoes. The removal of less than 25 % of the population did not alter the normal dynamics of the colony and the removal of 10 % even had a stimulating effect. These results may suggest that minor activity of aphid predators is of no value in aphid control.

Attempts at biological or integrated control

Two Palaearctic coccinellids have been imported into the USA where they are subject of intensive study: *C. septempunctata* (Shands *et al.*, 1972*a*, *c*, *d*) and *Propylaea quatuordecimpunctata* Gangl. (Rogers *et al.*, 1972*a*). One species was imported into central Europe from the Far East: *Harmonia axyridis* (Voronin, 1968).

In north-eastern Maine Shands *et al.* (1972*d*) concluded from the analysis of 31 population curves of potato aphids that predators and entomogenous fungi were chiefly responsible for initiating 14 and 17 decreases, respectively, of the aphid population. Aphid control by predators was often much below a commercially-acceptable level because the population decreases were late in the season.

To increase the number of predators sufficiently early in the season to prevent the overpopulation of potato aphids, Shands *et al.* (1972*d*) released different stages of laboratory-bred chrysopid and coccinellid larvae in various combinations on small ¼-acre fields. The best all-season control, about 60 %, resulted from sequential introductions of both species in totals of 30 200 second and third instars of *C. septempunctata* and 85 100 second instars of *Chrysopa* spp. (mostly *C. carnea* Stephens) per acre from 23 June to 28 July (Shands *et al.*, 1972*b*).

In another set of experiments eggs of *Coccinella septempunctata* and *C. transversoguttata* were applied in a 0.25 % water solution of agar in varying numbers and at various time schedules to field plots of potatoes. The earliest treatment with only 13 500 coccinellid eggs per acre produced about 40 % greater reduction of the aphids than the last treatment with 23 100 eggs per acre. This demonstrated the value of early treatment. But neither treatment gave satisfactory control of aphids (Shands & Simpson, 1972). Shands *et al.* (1972*a*) considered *Coccinella* spp. to be more effective in controlling aphids in high-level treatments and *Chrysopa* spp. at low levels.

Syrphidae and Coccinellidae were also used directly for the biological control in orchards of the Rhône valley in southern France. Because of the great mobility of adults only syrphid larvae were released. The consequences of this are not clear. To increase the syrphid population, conservation measures are recommended; for example, to sow plant species which are the sources of pollen and nectar for adult syrphids or to protect plants with specific aphids which do not attack crops (e.g. *Arundo donax* with the aphid *Melanaphis donacis* Passerini) that can serve as alternative prey for predators.

Coccinellidae were used with success on apple. Although *Dysaphis plantaginea* Passerini was very abundant in spring, its population increase was successfully suppressed by the introduction of 20 *Adalia bipunctata* L. adults per five trees at 15-day intervals (Remaudière *et al.*, 1973).

A promising tool to increase the numbers and effectiveness of aphid predators is 'artificial honeydew', as developed by Hagen, Sawall & Tassan (1971). Adults of *Chrysopa* and *Hippodamia* spp. became concentrated in areas where various mixtures of yeast hydrolysates, sugar, and water were applied to the foliage of alfalfa. Syrphids were also attracted. *Chrysopa*, for which this food is sufficient for the maturation of eggs, consequently oviposited when the aphids were present in low numbers or even absent. The 'artificial honeydew' induced egg-laying in *Hippodamia* when not enough aphids were present in alfalfa to support oviposition. In a complete absence of aphids the coccinellids did not reproduce, but they did not leave the area. The method encouraged the chrysopids and coccinellids to lay eggs before or at the very beginning of the aphid build-up and thus prevented aphid populations from attaining damaging levels. Enough predator adults must, of course, be present naturally in the region for the food sprays to be effective (Hagen *et al.*, 1971).

Conclusions

There are certain areas that need emphasis in future studies.

(i) The predator groups that have been neglected, for example Chamaemyiidae.

(ii) Quantitative data on, for example, the predator/prey ratio that enables a predator to reproduce, or on the minimum prey density that prevents a predator from dispersing. Dixon & Russell (1972), for example, gave the minimum density of young *Drepanosiphum platanoidis* Schr. on sycamore leaves that enabled *Anthocoris nemorum* L. and *A. confusus* Reut. to survive. A similar example was described by Tamaki & Weeks (1972*a*).

(iii) Studies of food ecology that might contribute to the simplification of predator mass-rearing, as, for example the use mentioned by Hagen and his colleagues of yeast hydrolysates as artificial food for chrysopids and the use by Matsuka *et al.* (1972) of drone honey bee brood for rearing coccinellids.

To improve understanding of the effect of predators on aphid populations experimental studies must be increased. This may be stimulated by the manual of relevant methods (Hodek *et al.*, 1972). Despite the difficulties created by aphids having overlapping generations, mathematical modelling of aphid/predator systems should be attempted. They will probably show that 'the factor limiting the development of mathematical models with a biological meaning is the extent of our knowledge about the behaviour of parasites, predators and disease' (Varley & Gradwell, 1970), and thus stimulate the research on the predators themselves.

A biological control approach has to be largely empirical until the role of predators is completely understood.

Microbial pathogens

Fungal agents

Entomophthorales

Fungi of the genus *Entomophthora* are the most important pathogens that significantly effect aphid populations. Research during the past 15 years, in particular that of MacLeod, Hall, Yendol and Gustafsson (see references), has stressed the importance of these fungi and added to our understanding of the factors that influence their action.

Although the effect of *Entomophthora* spp. on aphid populations is closely dependent on climatic conditions, it has been shown (Gustafsson, 1969) that they possess many characteristics that facilitate the development of epizootics, i.e. a high reproductive capacity, short generation time, high dispersive capacity, and various adaptive ways of surviving under adverse conditions such as cold and dryness.

In addition to studies by Gustafsson (1965a) and MacLeod & Müller-Kögler (1973) on the systematics of *Entomophthora* spp., the ecology of those that parasitise aphids has recently been investigated by several workers. This research was mainly concerned with aphids on potatoes (Shands *et al.*, 1965, 1972b), *A. pisum* (Voronina, 1971; Wilding, 1968–1973), aphids on cereals (Dean & Wilding, 1971, 1973), *Capitophorus horni* (C.B.) on artichokes and *Aphis fabae* on beans (Robert *et al.*, 1972), aphids on non-cultivated plants and on peach (Remaudière & Michel, 1971; Thoizon, 1970), *B. brassicae* on cabbage (Hughes, 1963), and aphids on alfalfa (Hall & Dunn, 1957a, b).

Furthermore, several aspects have been investigated in the laboratory including: the physical requirements for the discharge and germination of the conidiospores and the development of resting spores (Voronina, 1964; Yendol, 1968; Wilding, 1969, 1971b); the experimental infection of aphids (Yendol, 1964; Thoizon, 1967; Gustafsson, 1969; Krejzová, 1972); and the physiology and culture of *Entomophthora* spp. (Gustafsson, 1965b).

Fifteen *Entomophthora* spp. are known to parasitise aphids. However, only three of them, *E. aphidis* Hoffm., *E. thaxteriana* Petch., and *E. planchoniana* Cornu, occur commonly. Two other species, *E. fresenii* Nowak. and *E. sphaerosperma* Fres., are appreciably less common, and the remaining 10 species occur only occasionally on aphids. The conidia of each species are morphologically distinct, but the resting spores of some species are indistinguishable. Usually, several species are found together attacking the same aphid population and occasionally even the same aphid specimen. Therefore, when one host contains both conidia and resting spores it must not necessarily be assumed that the two forms are of the same species.

The host range of *E. aphidis*, *E. thaxteriana*, *E. planchoniana* and *E. fresenii* is restricted almost exclusively to aphids; *E. fresenii* has been reported also as a pathogen of mites, but this record is considered doubtful by several authors. *E. sphaerosperma* infects insects of many orders.

The specificities of *Entomophthora* spp. differ for different aphid species. For example, *E. fresenii* is more frequently observed on species of the genus *Aphis* and of closely related genera than on other aphids. *E. sphaerosperma* has been recovered from various aphids, but certain species of *Myzus*, *Capitophorus* and *Sitobion* seem to be more susceptible than others to the fungus. *E. aphidis*, *E. thaxteriana* and *E. planchoniana* do not exhibit any obvious host preference. Shands *et al.* (1972b) reported that the rate of infection differs significantly in three aphid species on potatoes. In the field, the proportion of *M. euphorbiae* killed by *E. thaxteriana* is markedly higher than that of *M. persicae* or of *A. nasturtii*. Similar differences in the proportion of infection in different aphid species on the same crop were also noted by Byford & Reeve (1969) and by Dean & Wilding (1971, 1973). Further, Wilding (unpublished) found that the proportion of cereal aphids infected by *E. aphidis*, *E. planchoniana* and *E. thaxteriana* is much higher than that of *A. pisum* on nearby crops of alfalfa. In certain cases, however, the apparent difference in susceptibility between aphid species in the field is probably the consequence of differences in their behaviour, density or habitat (Remaudière, 1971).

The mortality caused by *Entomophthora* spp. in aphid populations increases during rainy periods, but little is known about the reactions of different species to weather conditions. Gustafsson (1969) reported that *E. fresenii* is most common in dry summers and *E. aphidis* and *E. thaxteriana* in wet summers. Thoizon (1970) indicated that *E. planchoniana* is less frequent above 1400 m than at lower altitudes, whereas *E. aphidis* appears not to be affected by altitude or by the factors correlated with it.

The spread of *Entomophthora* spp. in aphid populations is influenced by a complex of factors, the principal of which are the environmental conditions that prevail when the fungus is active outside its host, i.e. during sporulation, germination of the conidia, and infection of the host. These processes are only assured during periods when the relative humidity in the micro-environment of the aphid remains close to 100 % for at least 10 to 12 h each day, while the maximum temperature is near 20 °C, and probably for longer in cooler weather. Gustafsson (1969), however, showed that a period of 4 h at 100 % relative humidity was sufficient to infect aphids experimentally that had been sprayed with a conidial suspension of *E. virulenta*.

Wilding (1971*a*) suggested that epizootics caused by *E. aphidis* and *E. thaxteriana* develop more readily at relatively low temperatures because the life cycle of the aphid is then comparatively longer than that of the fungus. However, the development of epizootics does not depend solely on the relative length of the life cycle of the aphid and of the pathogen because their reproductive rates are very different and cannot be compared directly. The number of offspring of one aphid is greatly exceeded by the number of aphids that can be infected by the conidia produced on a single diseased aphid. At relatively high temperatures, the rapid growth of the host population can itself be an important factor in the spread of the disease provided that the ambient humidity remains suitable. These factors may explain the large differences noted in the mortality rates caused by *Entomophthora* in consecutive years. For example Wilding (1968–73) reported that the maximum percentages of infection by *E. aphidis*, *E. planchoniana* and *E. thaxteriana* of *A. pisum* on alfalfa in England were 48, 88, 10, 55, 35, and 22 for the years 1967 to 1972 respectively.

The development of the fungal attack during one year often shows one peak in the spring and a second one in the autumn (Thoizon, 1970; Wilding, unpublished) coinciding with the periods of greatest aphid abundance.

In the field, the rate of infection is correlated with the conidial concentration in the air (Wilding, unpublished). This concentration varies considerably during the day and is usually maximum between 0500 and 0700 h associated with humid air and sunrise (Wilding, 1970).

Generally, peaks of infection follow rainy periods. However, when rain follows a long period of dryness, epizootics tend not to develop because the concentration of the inoculum is too small. Remaudière & Michel (1971) showed that, in the Rhône valley, the development of epizootics depends on the number and frequency of days when the weather is suitable for the development of the vulnerable stages in the life cycle of *E. aphidis*. A period of 20 to 30 days with a relative humidity of less than 90 % for more than 12 h per day causes a decrease in the concentration of the inoculum which hinders the spread of the disease when conditions become favourable again. In southeast France, the relative humidity is usually above 90 % for 12 consecutive hours, an essential condition for the development of an epizootic, when there is heavy rain during the day. In western France, Missonier, Robert & Thoizon (1970) observed that the disease persists in the enzootic phase all the year round and suggest that this is because the aphids there are anholocyclic and the relative humidity nearly always remains above 90 % for at least 8 h per day.

Remaudière & Michel (1971) attempted to modify the microclimate in peach orchards in the Rhône valley by covering the peach leaves with a film of water for a period of 10 to 12 consecutive hours. However, it was not possible to raise the humidity sufficiently and the aphids were not infected. Attempts to raise the humidity in the herbaceous layer nearer the ground are more likely to succeed.

Myzus persicae *Sulz.*, *an aphid of world importance*

The effect of the aphid population density on the spread of the disease has received little attention. MacLeod, Cameron & Soper (1966) reported that the mortality of *Schizolachnus pini-radiatae* Davidson caused by *E. aphidis* does not exceed 20 % if the aphid density is below a threshold value of 60 to 90 individuals per 30 cm of pine branch, whereas it approaches 100 % at higher densities. Robert *et al.* (1973) showed that the epizootic spread of *Entomophthora* species in colonies of *Aphis fabae* on beans in Brittany, France, seems to depend on the presence of a minimum quantity of inoculum, which they estimated as 300 aphids killed by the fungi per stem; this value is not influenced by the size of the surviving aphid population.

The mode of sporulation of the different *Entomophthora* spp. also influences the transmission of the disease within an aphid colony. *E. thaxteriana*, *E. planchoniana* and *E. aphidis* propel their conidia for a distance of slightly more than 1 cm. Conidia that escape the electrostatic attraction of the plant may be carried a relatively long way in the wind to aphids far from the original colony. The conidia of *E. sphaerosperma* and *E. fresenii* are not propelled as far as those of the above species; some remain on the body of the dead aphids others fall on the plant. They produce secondary 'anadhesive' conidia which adhere to the first host that touches them. These two *Entomophthora* spp. seem, therefore, particularly adapted to multiply in dense aphid colonies. An increase in temperature increases the mobility of the aphids and, consequently, the probability of their becoming infected.

The effect of *Entomophthora* spp. can be spectacular during epizootics but it can also be important during the enzootic phase when the aphid population is starting to grow exponentially. Shands *et al.* (1972b) found that the rate of population growth of *M. euphorbiae* always slowed soon after the first aphids killed by *E. thaxteriana* had been found. In five of seven years, the fungi caused the premature limitation of aphid populations, before the effect of predators became measurable.

It is little understood how aphids first become infected in the spring, though it is known how several species of the fungi survive during the winter. The fungus can survive in dry air for several weeks as hyphae in the dehydrated aphid body; the fungus sporulates soon after the dead aphid is moistened by rain or dew (Wilding, 1973). A more effective way of preservation is found among those species, such as *E. fresenii*, *E. planchoniana* and *E. thaxteriana*, that produce resting spores. The latter form – under conditions not yet precisely determined – mainly in the autumn (Wilding, unpublished); but they also form in aphids infected in the spring, at the beginning of summer, and even during the winter (Remaudière, unpublished). Byford & Ward (1968) observed resting spores of *Entomophthora* sp. in dead individuals of *Phorodon humuli* Schr., on trunks of plum trees in June. Only conidia occurred on aphids killed on the leaves. The resting spores survive for many months if they are not submerged in water for long. Those of some species germinate following exposure to high humidity at 20 °C for 2–40 days, but others have never been observed to germinate. The resting spores of *E. aphidis* have been rarely found in Europe (Gustafsson, 1965*a*); the survival of this fungus during the cold season probably depends on the survival of the overwintering aphid population in areas or biotopes where severe frost does not occur. Currently there is no satisfactory explanation for the appearance of *E. aphidis* in the summer in continental areas.

E. aphidis, E. planchoniana and E. sphaerosperma infect many aphids in overwintering colonies along the northern French coasts. At temperatures near 0 °C, the development of the aphids and the incubation of the disease are considerably retarded. However, when aphids that were apparently healthy when collected were reared at 20 °C many were killed by the fungus within one or two days (Remaudière and co-workers, unpublished). In England, Byford *et al.* (1966) also found overwintering aphids infected by *E. aphidis*, *E. planchoniana*, and *E. thaxteriana* in mangold clamps. The fungi always produced conidia on the dead aphids.

Aphids damage crops directly and by spreading plant pathogenic viruses. The spectacular character of naturally occurring epizootics caused by *Entomophthora* spp. has been reported by many authors. In populations of pest aphids, such epizootics frequently occur only after the economic threshold for the crop has been exceeded; moreover, they may seriously disturb the eco-system through the indirect suppression of parasites and predators due to the sudden disappearance of their hosts and prey over large areas.

Observations suggest that these pathogens may have an important, though less obvious, effect in enzootic infections and may even check the rate of aphid multiplication at an early stage in the infestation of a crop. Research aimed at simulating the conditions under which this may occur and at selecting fungal strains that are optimally adapted to a particular host or environment, may lead to the development of a method that would prevent direct economic damage by aphids and reduce the number of winged aphids and therefore the potential for virus spread.

The possibility of using *Entomophthora* spp. for the biological control of aphids is encouraged by their adaptation to diverse environmental conditions and remarkable capacity for dispersal. One object of research into the bio-logical control potential of *Entomophthora* spp. is to provide the farmer with an active and stable inoculum compatible with plant protection procedures, but the development of farming methods that enhance the natural spread and survival of these pathogens would also be beneficial.

Other fungal pathogens

Fungi other than *Entomophthora* spp. rarely infect aphids in the field and have been little studied. Leatherdale (1970) recorded *Cephalosporium muscarium* Petch., *C. aphidicola* Petch., *Cladosporium aphidis* Thuem., *Hirsutella aphidis*

Petch. and *Paecilomyces farinosus* (Dickson ex Fries) infecting aphids in Britain. Recently, *Metarrhizium anisopliae* Metsch. was found infecting *Pemphigus* spp. on roots of *Aster tripolium* on salt marshes in England (Foster, personal communication).

Wilding (unpublished) showed that *Cephalosporium aphidicola*, which had been isolated from *Myzus persicae* from a glasshouse at the Glasshouse Crops Research Institute, Sussex, England, and maintained *in vitro*, was infective for *Acyrthosiphon pisum*, *Aphis gossypii*, and *M. persicae* when sprayed as a spore suspension on the aphids on plants at 20 °C. About 50 % of apterous adults of *A. pisum* and *M. persicae* were killed on 10–20 cm high broad bean and turnip plants, respectively, when each plant was sprayed with about 5×10^6 spores in 3 ml of water.

Development of *C. aphidicola* was inhibited *in vitro* by the systemic fungicides benomyl and triarimol but not by dimethirimol (Wilding, 1972). However, neither triarimol nor dimethirimol inhibited the fungus in aphids fed on plants treated with the fungicides. Plants treated with benomyl were aphicidal. These results suggested that *C. aphidicola* could be used to control aphids on plants when plant pathogenic fungi are being treated with triarimol or dimethirimol. Benomyl might appear best for control of both aphids and plant fungi, but it is not compatible with other biological control systems currently in use in the glasshouse.

Viral, bacterial, and protozoan agents

Aphids are rarely attacked in nature by non-fungal pathogens. Virus-like particles were described from aphid tissues by Moericke (1963) and Parrish & Briggs (1966); however, no evidence was presented that these inclusions were in fact pathogenic. A granulosis virus, which had been isolated from *Pieris rapae*, was highly toxic to *M. persicae* and *Rhopalosiphum padi* (Orlob, Seto & Sun, 1973). Both capsules, capsule protein, and free virions caused considerable aphid mortality when ingested or injected but not when applied as a spray. Infective particles were recovered from excreted honeydew, although the virus did not multiply in the aphid.

Bacterial epizootics were reported by Needham (1930) and Paillot (1930, 1931). Steinhaus & Marsh (1962) isolated small gram-negative rods from several aphid species that, in most cases, had caused a secondary bacteraemia. Recently, Mackauer & Albright (1973) assessed the susceptibility of pea aphid, *A. pisum*, to intrahaemocoelic infection by *Serratia marcescens* Bizio and other bacterial pathogens. Approximately 7×10^2 viable *Serratia* cells were sufficient to cause the death by septicaemia of fourth-instar nymphs within 10–21 h (Fig. 17). The value of the LD_{50} varied with aphid size and temperature. At room temperature, the LD_{50} for fourth-instars was estimated as 190 viable cells, with the 95 % confidence limits ranging from 123 to 288 cells.

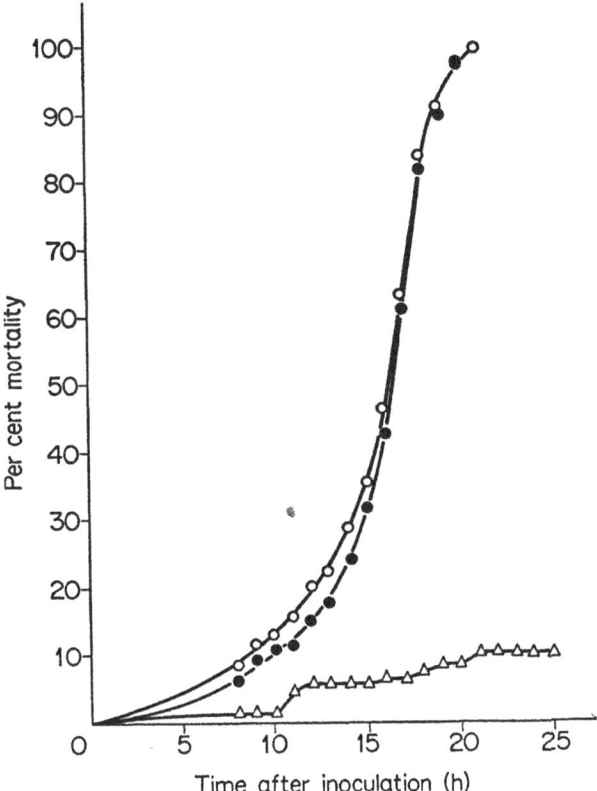

Fig. 17. Time–mortality curve of the survival time of fourth-instar *Acyrthosiphon pisum* nymphs following the intrahaemocoelic infection with 7×10^2 viable cells of *Serratia marcescens*. O–O–O, Observed mortality; ●–●–●, mortality adjusted for control mortality; △–△–△, control mortality. (Number of aphids = 130; 20 ± 1 °C; 50–60 % relative humidity.)

Among the Protozoa, only *Toxoglugea fanthami* Weiser has been recorded as a pathogen of aphids. It was described by Weiser (1961) from material found in the Malpighian tubules of *Aphis rumicis* L.

The utilisation of such 'potential' pathogens would seem to depend, among other things, on the development of methods of increasing their invasive powers. Aphids may become infected with bacteria (Srivastava & Rouatt, 1963) and virus particles (Orlob *et al.*, 1973) when probing or feeding. These micro-organisms, however, apparently are unable to invade the haemocoele, or at least not in numbers sufficient to cause a septicaemia. While it may be feasible to increase the rate of infection by combining, for example, bacterial sprays with chitinase to facilitate pathogen entry into the haemolymph (Smirnoff, 1971), such attempts may not be generally as practical as more reliable, more readily available, or more lasting control measures.

Genetic methods

Aphids are virtually unknown to the geneticist and there are many special features of their biology which hamper genetic studies and complicate the application of genetic control methods. They are therefore unlikely to have a place in the forefront of the development of genetic pest control.

From a purely practical viewpoint, aphids have both advantages and disadvantages as subjects for genetic control. They are extremely quick, cheap, and easy to rear and their parthenogenetic reproduction makes it possible to rear countless individuals all of similar or identical genotype, so that any derived mutant that one wishes to introduce into a population may be rapidly mass-produced in the laboratory. Natural aphid populations characteristically undergo great seasonal fluctuations in numbers and are often at a low ebb in late summer, so it may be possible to flood them with introduced aphids just prior to the sexual phase without immediate risk of increasing economic damage.

However, there are some formidable obstacles, especially when some of the important pest species are considered, that offset those advantages. Many of these aphids are heteroecious, migrating from secondary host plants for an undetermined distance in an undetermined direction back to a primary host plant for sexual reproduction. Could the timing and location of releases of laboratory-reared aphids, sterile males for example, ever be conducted with enough precision to enable the released aphids to compete successfully on the primary host with natural immigrants? The answer to this question may be 'no', or 'only with a limited number of species', but it is clear that we need to know far more than we know now about the seasonal movement of aphids, especially in relation to climatic conditions before we can answer this question with any degree of possible accuracy. The current state of knowledge was reviewed by Johnson (1969). More extensive use than hitherto of synoptic meteorology, as applied to the problems of studying and predicting locust swarm movements (Rainey, 1972), and the use of electrophoresis to 'fingerprint' populations and trace their movements (Whitten, 1970) are possible ways of speeding future progress in the study of aphid migration and dispersal. Obviously such work has a far wider significance than solely as a feasibility study for possible genetic control methods.

Conventional ideas for genetical pest control are based on sexual reproduction; however, in some aphid species sexual forms are unknown. Several important pest species, including *M. persicae* and *B. brassicae*, seem to have a facultative holocycle and do not rely on the sexual phase for their continuance from year to year. Steffan (1972) ruled out the possibility of genetical control for all anholocyclic or facultatively anholocyclic aphid species, but this conclusion may be a little premature. Very little is known of how selective pressures operate within aphid populations, but one might expect the competition between clones within a population to be intense due to the similarity of their requirements. Competitive displacement may be a powerful force altering the genotypic composition of aphid populations in the absence of sexual reproduction, and it could conceivably be turned to advantage when introducing laboratory-reared genotypes.

Aphids have, as a rule, only one sexual generation a year under natural conditions. This is a considerable disadvantage in inheritance studies and will impede progress towards any control measures that necessitate detailed genetic information. Breeding programmes are inevitably of a long-term nature, especially with heteroecious species that have their life cycles synchronised with the phenology of the primary host. Such programmes are off-seasonal and not necessarily time-consuming, however, and might be fitted into many research schedules. Cytogenetic studies are also difficult because aphid chromosomes lack localised centromeres or other distinguishing morphological features so that translocations and other chromosomal changes are difficult to detect. Modern techniques of labelling and banding chromosomes such as tritiated-thymidine autoradiography (Rogers, 1967), quinacrine fluorescence and differential Giemsa staining (Caspersson *et al.*, 1970) may overcome this problem.

While recognising the considerable difficulties of applying genetic control methods to aphids, it is worth noting some recent ideas that are as yet largely untried on any insect but that might be particularly applicable to aphids.

Aphids, in common with other Hemiptera and with Lepidoptera, have chromosomes with diffuse centromeres. Aberrations induced by X-rays in such holocentric chromosomes can survive cell division unchanged and can be transmitted from one generation to the next where they may interfere with meiosis. In other words, sterility can be inherited. The advantages of the use of inherited sterility over the conventional sterile-male technique for control of Lepidoptera were discussed by Knipling (1970), and similar benefits could apply to aphids. Release of partially sterilised male aphids at the correct time in autumn of one year would affect the sexual viability of the next year's aphids. Further releases in subsequent autumns could gradually build up sterility in the natural population even allowing for considerable dilution by cytologically normal immigrants.

Aphids survive adverse conditions and maintain flexibility by virtue of their elaborate system of phenotypic polymorphism. Potentially lethal situations, such as that of an apterous, parthenogenetic individual on a dying, annual secondary host plant, are avoided by means of genetic switch-mechanisms triggered by environmental signals. Inability to respond to these signals would normally be fatal. A mutant gene causing such inability, that is, a conditional lethal gene, would probably have no other adverse effect on the individual. Possible ways of using dominant conditional lethal genes in insect control were discussed by Klassen, Knipling & McGuire (1970). Conditional lethal mutations are probably common in aphids; facultative anholocycly is a conditional lethal trait when it occurs in regions that have severe winter climates. The chief problems are to investigate the nature and dominance of the genetic determinants of such traits and to find the best ways of introducing the genes into field populations. One proposal is to link a dominant conditional lethal gene with a gene for insecticidal resistance so that spraying after release will strongly favour the introduced genotype (Wehrhahn & Klassen, 1971).

Breeding plants for aphid-resistance

The advantages of insect-resistant crop varieties are well known (e.g. Painter, 1968; Pathak, 1970). Resistance has involved mostly oligophagous aphids but resistance to *M. persicae*, probably the most polyphagous aphid, has been found in several crops, as summarised below.

Potatoes

Radcliffe & Lauer (1966, 1967, 1970, 1971 *a*, *b*) screened large collections of wild *Solanum* species in the field and identified many sources of resistance to *Myzus persicae* and also to *Macrosiphum euphorbiae*. Resistance to these two aphids was usually, but not always, correlated in the *Solanum* stocks tested (Radcliffe & Lauer, 1970). Gibson (1971*a*) found that the resistance of three *Solanum* species differed with aphid species and with the part of the plant. Cultivars of *Solanum tuberosum* differ in susceptibility to *M. persicae* and *M. euphorbiae* according to their growth habit and earliness (Taylor, 1962).

High levels of resistance to *M. persicae* in *Solanum* species of the series 'Bulbocastana' and 'Pinnatisecta' are unlikely to be useful due to difficulties in crossing these species with *S. tuberosum*, but other species, e.g. *S. sanctae-rosae* and *S. stoloniferum*, are promising sources of resistance (Radcliffe & Lauer, 1971*b*). The F_1 and subsequent generations of crosses between certain wild *Solanum* species and *S. tuberosum* contained plants as resistant to *M. persicae* as the wild parent (Quisumbing, Lauer & Radcliffe, 1970), suggesting that this resistance can be incorporated into *S. tuberosum* cultivars. Resistance due to sticky hairs which trap aphids of all species occurs in *Solanum berthaultii*, *S. tarijense* and *S. polyadenium* (Gibson, 1971*b*). F_1 hybrids between *S. tuberosum* and *S. berthaultii* developed hairs whose secretion gummed down aphids (Gibson, 1974), and the possibilities of exploiting this resistance are being studied.

Sugar beet

Inherited resistance to both *M. persicae* and *Aphis fabae* in sugar beet, *Beta vulgaris*, was reported by Russell (1966) and is currently under investigation at Cambridge. The nature of the resistance has not been elucidated, although it is clear that a number of mechanisms are involved (H. J. B. Lowe, unpublished). The mode of inheritance of resistance to either aphid has not been determined, but appears to be polygenic (Lowe & Russell, 1969). As with *M. persicae* and *M. euphorbiae* on potato (Radcliffe & Lauer, 1970), resistance to *M. persicae* on sugar beet is often associated with resistance to *A. fabae* (Lowe, 1972*a*), but particular beet stocks may be markedly more resistant to one species than to the other (Russell, 1966). Selected stocks with resistance to both aphids have been developed (H. J. B. Lowe, 1973*b*). Resistance to aphids occur as quantitative differences between sugar beet stocks, varying in all degrees between the most resistant and the most susceptible stocks. Screening of wild *Beta* species revealed no resistance greater than that observed in sugar beet or sea beet, *B. vulgaris* ssp. *maritima* (H. J. B. Lowe, unpublished). Resistance, especially to *M. persicae*, is affected by environmental factors. Under poor lighting or with pot-bound plants, induced resistance prevented the proper assessment of inherited resistance (Lowe, 1974*a*). Inherited resistance was not expressed in young seedlings of some, but not all, sugar beet stocks, so that plants must be tested after at least four true leaves have developed (Lowe, 1972*b*).

On sugar beet, the performance of *M. persicae* from different sources, mostly sugar beet fields in east England, varied considerably. Several clonal aphid cultures were used to test a range of sugar beet stocks differing in resistance to aphids. Resistance assessments of the sugar beet stocks were similar in all cases, but the *M. persicae* clones differed in their vigour on sugar beet (Lowe, 1974*b*). Similar differences in the readiness of *M. persicae* to colonise sugar beet can be induced in some clones by rearing on different host plants (H. J. B. Lowe, 1973*a*). No specific relationships between resistant sugar beet stocks and clones of *M. persicae* were detected.

Sugar beet crops are often sprayed with aphicides to reduce virus yellows by controlling vector populations of *M. persicae*, and to control *A. fabae* which is sometimes sufficiently abundant to damage the crop. Combination of resistance to *M. persicae* with resistance to inoculation of the beet yellowing viruses and virus tolerance, as in the experimental variety VT 95 (Russell, 1972), should give good control of yellows in many seasons. In field experiments (H. J. B. Lowe, 1972*b*, 1973*b*), moderate levels of resistance to *M. persicae* in the presence of natural enemies provided a useful degree of control of introduced aphid populations and virus yellows, especially in large plots. Thus if the various kinds of resistance contributing to control of virus yellows are combined with resistance to multiplication of *A. fabae* (Lowe & Russell,

1969) and tolerance of direct feeding damage (Russell, 1966) the need for aphicides should be progressively reduced.

Tobacco

Thurston (1961) recorded good resistance to *M. persicae* in some wild *Nicotiana* spp. and less extreme resistance in *Nicotiana tabacum* stocks. Crosses and backcrosses of selected resistant *N. tabacum* with a susceptible variety were intermediate in resistance. Resistance in some wild *Nicotiana* spp., particularly *N. gossei*, is caused by the toxic secretions of leaf hairs which contain nicotine and other alkaloids and act as a contact insecticide (Thurston & Webster, 1962; Thurston, Smith & Cooper, 1966). Within the tobacco plant nicotine is mainly in the xylem and is avoided by *M. persicae* which feeds from phloem (Guthrie, Campbell & Baron, 1962). Burk & Stewart (1969) tested a larger number of *Nicotiana* species, substantially confirming Thurston's (1961) results and identifying five other species resistant to *M. persicae*, of which three are related to *N. gossei*. Abernathy & Thurston (1969) found that young plants of *Nicotiana* were susceptible to *M. persicae*, and that resistance developed as the plant aged, coincidentally with increases in the number and secretory activity of the leaf trichomes. They reported good resistance to *M. persicae* in crosses and backcrosses of resistant tobacco lines TI 698 and No. 40 with the susceptible Ky 12, which is a high-quality variety with resistance to a number of fungal pathogens (Stokes & Valleau, 1968).

Other crops

Resistance to aphids has been recorded in several brassica crops, usually against the oligophagous aphid species associated with them, as these tend to develop heavier infestations than does *M. persicae*. Resistance to *B. brassicae* in swedes var. NZ Resistant also affected *M. persicae* and caused similar reductions in populations of both aphids (Lamb, 1953). However brassica stocks resistant to the oligophagous aphids are often susceptible to *M. persicae* (Lammerink, 1968 *a*, *b*; Dunn & Kempton, 1969; Jarvis, 1969).

The development of aphid-resistant lettuce for use under glass is currently under investigation in Holland; differences between different varieties in resistance to *M. persicae*, one of the aphids concerned, have been detected (de Ponti, personal communication). At Cambridge, numbers of *M. persicae* in the glasshouse were usually larger on the lettuce variety Borough Wonder than on Avondefiance (H. J. B. Lowe, unpublished), these varieties being susceptible and resistant respectively to *Pemphigus bursarius* L. (Hardman & Wheatley, 1970). However, the resistance of Avondefiance was much less effective against *M. persicae* than against *P. bursarius*.

Chrysanthemum cultivars, used for the commercial production of blooms under glass throughout the year, differ considerably in suitability for *M. persicae* (Wyatt, 1965; Markkula, Roukka & Tiittanen, 1968). The resistant varieties allowed better control of *M. persicae* by the parasite *Aphidius matricariae* because the aphid populations increased more slowly and also attained smaller peak numbers than on susceptible varieties (Wyatte, 1970). However, the significance of resistant chrysanthemum varieties as a control measure has been lessened by new chemicals which have controlled *M. persicae* resistant to other insecticides and by improved methods of using *A. matricariae* (Scopes, 1970).

Conclusions

Resistant varieties have not played a major part in the control of *M. persicae* but the possibility of breeding for resistance has been demonstrated in several crop species. Resistance to *M. persicae* appears to be of two kinds: first, resistance that depends on external defence mechanisms (Gibson, 1971*b*; Thurston & Webster, 1962); and, second, 'internal' resistance that depends on mechanisms which mainly affect the feeding behaviour and nutrition of the aphids and are thus usually closely linked to the physiology of the host plant. Inheritance of resistance to *M. persicae* is often controlled by many genes and involves several different mechanisms, as in sugar beet (Lowe & Russell, 1969). Experience with sugar beet demonstrates that useful progress in breeding aphid-resistant varieties is possible even where resistance is complex.

Resistance to *M. persicae* also tends to affect other aphid species (Lamb, 1953; Radcliffe & Lauer, 1970, 1971*a*; Lowe, 1972*a*). In contrast, resistance to oligophagous aphids is often effective against only one species (Hormchong & Wood, 1963; Painter, 1968) but is less influenced by the environment than is resistance to *M. persicae* (van Emden, 1969*a*). Biotypes capable of overcoming resistance have been found in several oligophagous aphids (Lammerink, 1968*a*, *b*; Pathak, 1970) but have not been reported in *M. persicae*.

Resistance to *M. persicae* has been found, often readily, in all crops where a thorough search has been made, although levels of resistance are sometimes lower than those used against oligophagous aphids. In breeding resistant varieties the aim should be to combine different mechanisms of resistance to *M. persicae* in the same variety, together with resistance to viruses transmitted by *M. persicae* and to other pests requiring simultaneous control measures. Such varieties could form the basis of improved systems of integrated control.

Integrated control of *Myzus persicae*

Despite the labyrinthine complexities of the dynamics of almost all pest species, there are usually one, or a few, 'mortality' factors that are of overriding importance to any one species. It is by the manipulation of such factors that integrated control becomes practicable. A model aimed at managing a pest species may therefore have entirely different parameters from those required of a model that aims to explain the population dynamics of a species (Way, 1973*b*).

Integrated control of the green peach aphid, *M. persicae*, is feasible under certain circumstances despite the difficulties in understanding the aphid's dynamics in a comprehensive way, as discussed on pages 54–7. In fact successful integrated control measures against *M. persicae* were developed in very different circumstances essentially by the application of relatively simple and straightforward ecological information. The following two examples may serve to illustrate the point and, at the same time, suggest similar approaches.

Control in glasshouses

M. persicae can exist in a glasshouse as a virtually self-contained population, especially where there are overlapping crops, as on year-round chrysanthemums. In these circumstances, the effects of immigration, emigration, host plant conditions, and weather are no longer unpredictable; in fact temperature and humidity are strictly controlled, and aphid dispersal is limited to plants within the artificial environment. Wyatt (1965) described how a stable equilibrium population is established. As was noted by McLeod (1937, 1940), *M. persicae* can be controlled, and even be exterminated, by the aphidiid parasite *A. matricariae* under favourable temperature conditions in a greenhouse Such control, however, usually is incomplete and of a short-term nature. This applies in particular when discrete parasite generations enable aphid resurgences during periods of parasite scarcity. A continuous and well-distributed population of *A. matricariae*, with overlapping generations, can be established readily by the introduction at regular intervals of chrysanthemum cuttings pre-infested with heavily-parasitised aphids (Scopes, 1970). The parasite population can be maintained in that fashion at a level sufficient to prevent aphid increases from initially small numbers to damaging levels. From such observations Scopes & Biggerstaff (1973) developed an integrated control programme against *M. persicae* that gave economically satisfactory results. The programme combined parasite introductions with the use of resistant chrysanthemum cultivars and the occasional application of a selective insecticide, pirimicarb, which did not harm the parasite. The programme can be broadened, and control enhanced under certain circumstances, by including fungal pathogens of aphids (Hussey & Bravenboer, 1971; pp. 104–10).

Control of viruliferous M. persicae *on sugar beet*

The green peach aphid is the only important vector of sugar beet yellows viruses in much of Western Europe. Most of the damage is done early in the season by aphids that migrate to beet crops from overwintering host plants that are virus sources. A control programme which succeeded in limiting the transmission of beet yellows viruses integrated the following procedures (Hull, 1967): as many viruliferous aphids as possible were destroyed at their source; virus-free zones were created by keeping other sources of virus sufficiently far from crops at risk to prevent aphids from colonising them; and viruliferous colonists that did succeed in reaching beet crops were killed with insecticides when required. These measures have been very successful, yet control depended on methods that were almost entirely different from those which succeeded against green peach aphid in glasshouses. For example, at present, no conscious attempt is made to preserve (or increase the impact of) natural enemies to control sugar beet aphids, and no use is made of selective aphicides.

Conclusions

The IBP work on *M. persicae* has provided insight into two aspects of the aphid's life history and population dynamics that are relevant to control and that should now be examined in more detail. The first is host plant resistance in the widest possible sense. Earlier, pages 54–7 and 63–7, the unique sensitivity of this species to the physiological condition of its host plant was highlighted. Breeding for relatively limited increases in plant resistance could dramatically alter the status of *M. persicae* as a pest. Despite the fact that new biotypes adapted to the 'resistant' crop varieties are likely to appear with time, varietal control should be explored more fully than hitherto, especially as this may enhance the impact of other mortality factors. There is also much scope for methods which render existing plant varieties less suitable for aphid growth, notably through appropriate fertiliser treatments (van Emden, 1966) and the use of plant growth regulators (van Emden, 1969 *b*). Manipulation of the host plant, though insufficient alone to achieve control, could be invaluable when combined with other measures.

The other potentially important consideration is that of life cycle and biotype variation (pp. 57–67). Further elucidation of the host plant preferences, if any, of certain biotypes could suggest fundamental changes in control procedures based on cropping practices. It might be possible to control the aphids at the source rather than after they have spread and colonised the crop at risk. At present, such controls are not being attempted because they are unspecific in that they involve the broad-scale application of insecticides. In the same context, there is only inadequate information on the role of weeds as a source of aphids attacking different crops and on the facts that influence the distances covered by migrating alatae.

No immediate improvement is apparent that would drastically increase the impact of natural enemies on the green peach aphid. As mentioned above (pp. 92–104), aphid density is usually too low to attract and retain sufficient numbers of parasites and predators on field crops. Some improvement could be achieved by the use of food sprays to induce oviposition by syrphids and coccinellids. Inundative releases of laboratory-produced parasites conceivably could help to blunt aphid population growth at a critical period, e.g. on non-crop or alternative crop plants prior to the production of winged migrants. The introduction and inoculative release of parasite strains, which have been selected for their temperature requirements, is entirely feasible now and may be useful in specific situations to alter the timing of parasite attack. Entomogenous fungi (pp. 104–10) eventually could replace insecticides in many control programmes but additional research is required into efficient methods for storage and application of fungal spores.

In conclusion, it can be said that natural enemies, host plant resistance, environmental manipulation, and insecticides can all be useful components of an integrated control programme against *M. persicae*. Whether they are utilised alone or in combination with each other will depend on the composition of the pest complex and on the kind and extent of damage to the crop in a specific situation; no generally applicable solution can be suggested here. The aphid's naturally low density and other seemingly impossible difficulties are not sufficient cause for belittling (or not attempting) ecologically-sound control measures. These programmes can be developed through a continuing experimental analysis of component factors which are then integrated, step by step, into an existing control schedule. The fact that this approach was successful on some field crops, e.g. sugar beet where a mean of less than 0.5 aphid per plant may economically damage the crop, augurs well for the integrated control approach in other situations. Although comprehensive models of a pest situation will be very helpful in planning appropriate control strategies, economically acceptable control is not, or only rarely, dependent upon achieving a complete understanding of the dynamics of the interacting system. In fact economically acceptable integrated control of the green peach aphid, more likely than not, can be achieved by the imaginative manipulation of particular life cycle characteristics and requirements of the species rather than by a profound understanding of its population dynamics.

5. Rice stem-borers

Coordinator: K. YASUMATSU

Rice, *Oryza sativa* L., an important graminaceous food crop of Asiatic origin, is grown extensively in Asia. It is a highly adaptable plant, that is grown under diverse ecological conditions in temperate, sub-tropical, and tropical regions, at altitudes ranging from sea level to 4000 ft (1200 m). Although the plant is usually aquatic or semiaquatic, it is planted in dry-land conditions in some areas.

Insects associated with the rice plant also appear to be adaptable to diverse ecological situations. With few exceptions, the major species are found in nearly all rice-growing areas. The stem-borers, *Chilo suppressalis* (Walk.), *C. polychrysa* (Meyrick), *Tryporyza incertulas* (Walk.), *T. innotata* (Walk.) and *Sesamia inferens* (Walk.) are the principal lepidopterous pests of rice.

The rice paddy is the oldest and most important agricultural ecosystem of Asia, and the crop has been cultivated at least since 5000 B.C. Ninety per cent of the total rice area of the world is found in Asia and much of this has never received pesticide applications. One of the major current problems is how to control rice pests in such a way as to prevent disruption of the ecosystem and creation of environmental problems.

Since the selection of the biological control of rice stem-borers as one of the projects for inclusion in the International Biological Programme (IBP) in 1965, extensive field research has been conducted on the biology and ecology of rice stem-borers and their natural enemies. The object of this work has been to establish a rational basis for integrated control of rice stem-borers, especially in South and East Asia.

Three meetings were held during the project under the sponsorship of IBP: the first in Fukuoka (1968) to develop a methodology handbook for the study of rice stem-borers and their natural enemies (Nishida & Torii, 1970); the second in Bangkok (1969) to discuss the importance of conservation of natural enemies of rice stem-borers, the feasibility of integrated control of these pests, and the development of sampling methods for qualitative and quantitative population estimation. The results of the field work were presented at the final meeting in Canberra (1971).

The work on rice stem-borers has resulted in a number of publications during the past seven years, and has also been responsible for the establish-

Contributors: C. S. Li, T. Nishida, M. D. Pathak, G. H. L. Rothschild, T. Torii, T. Wongsiri, K. Yano, K. Yasumatsu.

K. *Yasumatsu*

ment of the International Association for Biological Control of Rice stem-borers (IABCR) with the object of exchanging information among those interested in this field.

Because of the complexity and diversity of the problems and a shortage of research workers, not all the information necessary to develop comprehensive integrated control programmes for rice stem-borers could be accumulated. Nevertheless, this synthesis report should contribute to the development of rational control methods for rice stem-borers that involve minimum usage of insecticides.

No attempt is made here to undertake a complete review of the literature. Instead the emphasis is on unpublished information that might be useful in the development of control measures.

The study area

The study area in this investigation included the major rice-growing countries of Asia, namely the Philippines, Malaysia, Thailand, India, Sri Lanka, and Japan (Fig. 18). Information was also available from Northern Australia. The chief researchers in these areas were K. Yasumatsu, T. Nishida, G. S. Lim, M. D. Pathak, V. P. Rao, H. Fernando, K. Aizawa, H. Hirashima, S. Momoi and C. Watanabe.

This rice area extends from the equatorial belt northward into the southern fringe of the temperate region, thus embracing tropical, sub-tropical and temperate climatic zones. Although rice is grown during the warmest time of the year, there are considerable seasonal differences in temperature within these zones. Rainfall and photoperiod also influence both rice and its stem-borers. The rice varieties grown differ within the zones and the choice of variety depends upon a number of factors, such as adaptability to the environment and consumer preference. The rice pest fauna is, however, remarkably uniform throughout the entire region.

Taxonomy and distribution of rice stem-borers and their natural enemies

Fifteen species of rice stem-borers are known from south and east Asia and Australia. *Chilo auricilia* (Dudgeon), *C. partellus* (Swinhoe), *C. polychrysa*, *C. suppressalis*, *T. innotata*, *T. incertulas*, and *S. inferens* are of greatest economic importance. Four other species, *Ancylolomia chrysographella* (Kollar), *Bathytricha truncata* (Walk.), *Maliarpha separatella* Ragonot, and *Niphadoses palleucus* Common are next in importance. *C. hyrax* Błeszyński was frequently captured in light traps together with *C. suppressalis* in Japan but it was later shown that this pyralid does not attack the rice plant, although it may occasionally oviposit on it. *Scirpophaga chrysorrhoa* Zeller exhibited

122

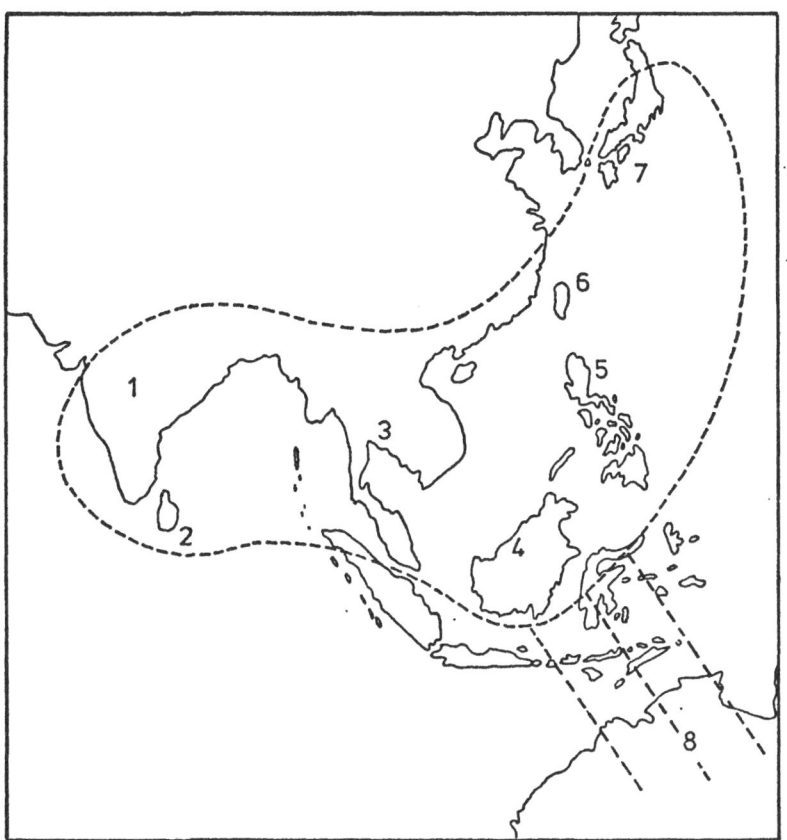

Fig. 18. Study areas of rice stem-borers in Asia. 1 to 8: India, Sri Lanka, Thailand, Sarawak, Philippines, Taiwan, Japan, Australia.

similar behaviour. Keys for identification of the larvae, pupae and adults of eight species were published in *IBP Handbook* no. 14 (Nishida & Torii, 1970).

More than 300 species of parasites and predators of rice stem-borers were recorded during this investigation. Not all are specific to rice stem-borers and no attempt was made to undertake comprehensive taxonomic studies of them. Preference has been given to the superfamilies Chalcidoidea and Ichneumonoidea, because of their apparent importance in regulating numbers of the stem-borers. Keys are published periodically in the *IABCR News* in addition to those included in the *IBP Handbook*.

The eggs of marsh flies, Sciomyzidae, as well as rice stem-borers were parasitised by *Trichogramma* spp. The presence of sciomyzid egg-masses is thus an important factor in the maintenance of populations of *Trichogramma*

123

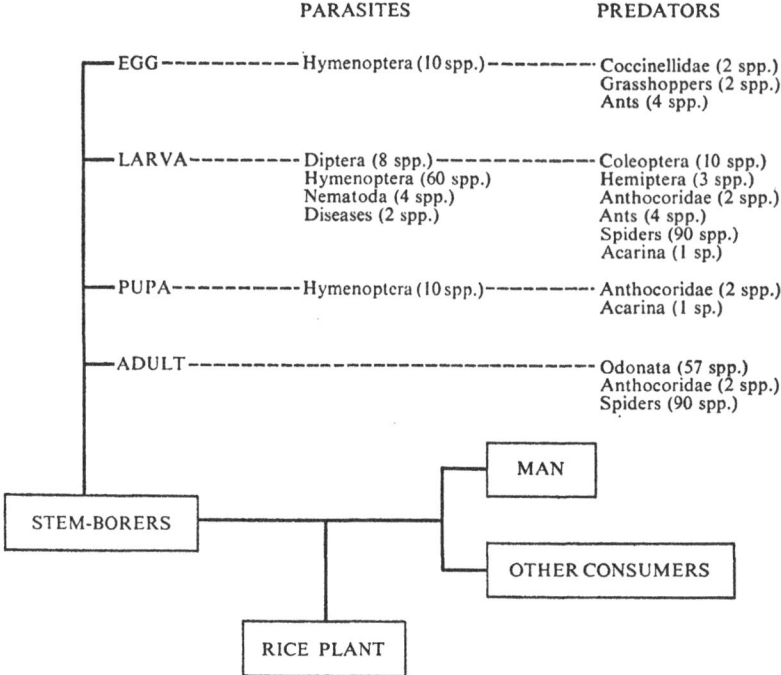

PARASITES   PREDATORS

EGG ------------- Hymenoptera (10 spp.) -------- Coccinellidae (2 spp.)
                                                Grasshoppers (2 spp.)
                                                Ants (4 spp.)

LARVA -------- Diptera (8 spp.) ----------- Coleoptera (10 spp.)
               Hymenoptera (60 spp.)   Hemiptera (3 spp.)
               Nematoda (4 spp.)   Anthocoridae (2 spp.)
               Diseases (2 spp.)   Ants (4 spp.)
                                            Spiders (90 spp.)
                                            Acarina (1 sp.)

PUPA --------- Hymenoptera (10 spp.) -------- Anthocoridae (2 spp.)
                                             Acarina (1 sp.)

ADULT ------------------------------------- Odonata (57 spp.)
                                            Anthocoridae (2 spp.)
                                            Spiders (90 spp.)

MAN

OTHER CONSUMERS

STEM-BORERS

RICE PLANT

Fig. 19. Simplified food chain of the rice ecosystem with reference to rice stem-borers.

in rice paddies, especially during the fallow period when stem-borer eggs are not present. A key was also published for identification of the Sciomyzidae occurring in rice paddies (*IBP Handbook* no. 14).

The rice ecosystem

Description of the rice ecosystem

It is necessary to have a good understanding of the rice ecosystem to develop rational control measures against rice stem-borers. In its simplest form the ecosystem may be looked upon as a food chain of varying complexity as shown in Fig. 19.

The origin of the rice ecosystem is unknown but rice is a crop of great antiquity, having been planted for some 7000 years at least (Grist, 1959; Solheim, 1971). It is generally accepted that rice was first cultivated somewhere within an area extending from Indonesia to South China and from Bangladesh to Vietnam (Fig. 20).

When man first began growing rice, the insect component of the ecosystem may have been limited, consisting of a few phytophagous, predaceous and parasitic species. However, as he began to expand rice production into more

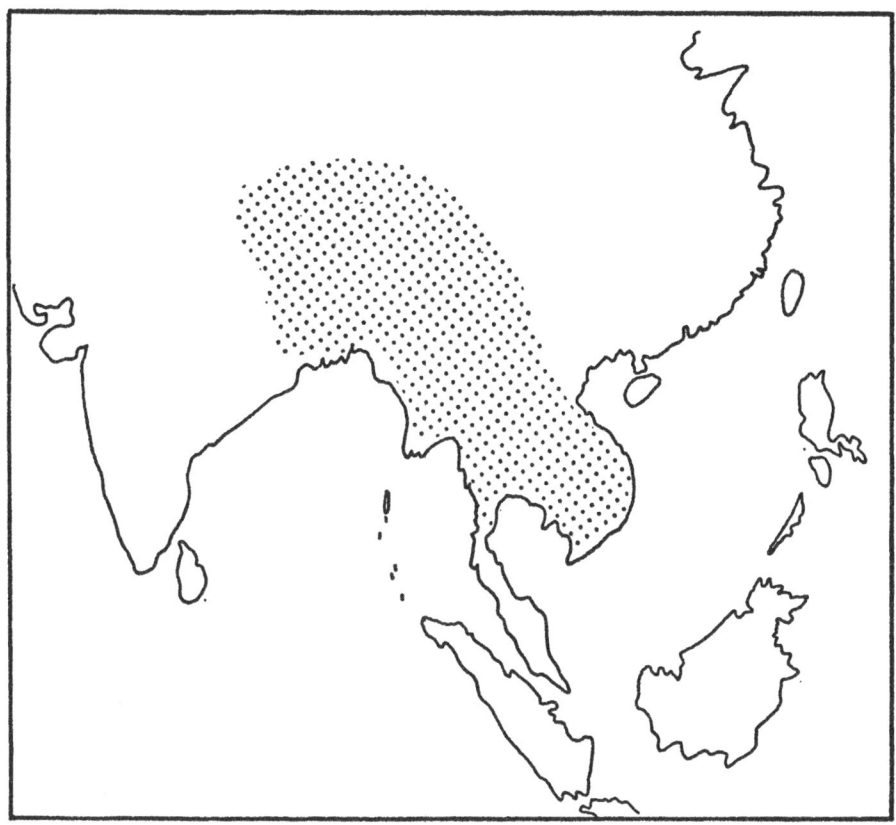

Fig. 20. The area encompassing the centre of origin of rice stem-borers and their natural enemies.

varied situations and using different rice varieties, the rice ecosystem probably became more complex. Today it is evident that the areas where rice has been grown the longest have the most complex faunal composition. For example, rice is a recent crop in the USA and the system is much simpler than that of northern Thailand where rice has been grown for many centuries.

Cultivated rice is not a perennial plant and its life cycle varies with variety from three to six months. The plant grows and matures rapidly in much the same manner as pioneer plants in a vegetational succession. After reaching maturity, the grain is harvested and the plant residues are destroyed. The rice ecosystem is therefore a very unstable one.

K. Yasumatsu

Species diversity

Field sampling of rice with a sweep-net was undertaken to determine variations in the diversity of insect species (Nishida & Torii, 1970). In general, the diversity varies with the age of rice cultures in the area, the cultural practices employed, and the stage of growth of the crop. In areas where rice has been grown only for short periods the diversity is low. The most important cultural factor influencing diversity is the use of insecticides. In areas where insecticides have been used continuously the diversity is low and the species present are mainly leafhoppers and rice stem-borers. The age of the plant also has a marked effect on diversity. Since rice is planted on cleared land devoid of vegetation, the diversity is low initially and increases as the plant matures to the booting stage.

Although the rice ecosystem does not attain stability with respect to all interacting components, an increasingly more complex food web develops rapidly, especially in areas undisturbed by pesticides. Leafhoppers, rice stem-borers, and their egg parasites, grasshoppers, spiders, and dragon flies appear in the early stages of crop growth; larval and pupal parasites of stem-borers appear later.

Phenology of the rice crop and associated insects

The seasonal appearance and disappearance of the rice ecosystem depends on man: his activities in turn are influenced by climatic factors. In temperate areas, time of planting is determined by temperature and rainfall, whereas in the tropics rainfall alone is of prime importance.

The varied arthropod fauna is able to exploit these different planting patterns. For example, when rice is harvested and the land is ploughed, the arthropods move into adjacent areas which are often marshlands. They reinvade rice when a fresh crop is planted. Where double-cropping is practiced there are constant faunal interchanges between the rice fields and surrounding areas throughout the year.

The rate at which rice stem-borers colonise newly planted paddies varies, but egg-masses of borers are often found one to two weeks after the rice plants have been transplanted.

Natural enemies of stem-borers appear early in the crop cycle. The first to appear are egg-parasites including *Trichogramma* spp. and *Tetrastichus* spp. Larval and pupal parasites of the stem-borers appear when the rice matures. Spiders may move into the paddies as early as two weeks after transplanting.

Rice stem-borers are well adapted for survival in the off-season when the crop is absent. In the temperate zone they enter diapause before winter, and in the tropics they aestivate during the dry season.

126

Ecology of rice stem-borers

As there is much published information on the ecology of rice stem-borers, especially from the temperate zone, the object of this section is to summarise information relevant to tropical areas, particularly on aspects that were not previously discussed in the literature (Torii, 1971c).

Centres of origin of rice stem-borers

From evidence obtained on the biology of rice stem-borers we can speculate on the original home of these insects. They exhibit varying degrees of host plant specificity; some species are nearly monophagous, whereas others are oligophagous or even polyphagous. A species that is almost completely specific to rice is *Tryporyza incertulas. C. suppressalis* is found in only two or three host plants in addition to rice. Other species, including *S. inferens* and *C. polychrysa*, have wider host ranges. The differences in host specificity among the species may indicate different centres of origin. This is further indicated by other biological differences. For example, *C. suppressalis* is distributed from the tropics far northwards into the temperate zone where it enters diapause during the winter. It seems to prefer well-drained paddies, and its larva remains high in the stalk at harvest time. On the other hand, *T. incertulas* is distributed from the tropics to southern areas of the temperate zone. It appears to prefer poorly drained situations and its larva bores towards the roots even in deep water. These differences suggest that *C. suppressalis* originated in a dry, temperate area and *T. incertulas* in a damp tropical situation.

Parasite specificity also supports the postulated differences in centres of origin. The egg-larval parasite, *Chelonus munakatae* (Munak.) is specific to *Chilo suppressalis* but occurs only in temperate areas where its host is present, whereas the egg-parasite *Tetrastichus schoenobii* Ferr. is specific to *Tryporyza incertulas* and its distribution coincides with that of its host.

Borer numbers in long-established rice areas

In areas near the centre of origin of the rice crop the populations of rice stem-borers are low. Natural enemies may contribute to this situation (Rothschild, 1970). Another important factor may be varietal resistance. Many rice varieties are resistant to attack by rice stem-borers and have been selected by farmers over many centuries. Varietal differences in the rate of decomposition of the stubbles may also influence population size. The stubbles of certain varieties do not ratoon and decompose rapidly. Survival of stem-borers such as *T. incertulas* is poor in such varieties.

Borer-free areas

Although rice stem-borers are widely distributed in Asia, there are areas where they are absent or are present in exceedingly low numbers. Field surveys made on the upper slopes of Mt Hikosan, Kyushu, Japan, revealed no borer infestation although *C. suppressalis* was present at lower altitudes.

Borer-free areas have been located in the northern foot-hills of Katmandu valley, Nepal, and in the mountainous area of Palung. No rice stem-borers were found during surveys in Shillong, Dehra Dun, and Kashmir, in India. Large rice-growing areas devoid of borers apparently exist in Iran along the southern coast of the Caspian Sea. Conditions seemed favourable for at least one or more species of rice stem-borers in all of these areas (Nishida, personal communication) and reasons for their absence remain obscure.

Co-existence of species

In some areas several species can be found within the same rice field, whereas in others only single species occur. In northern parts of Japan the only species present is *C. suppressalis*, but both *C. suppressalis* and *T. incertulas* are found in the southern parts. In tropical areas several species often occur in the same rice paddy.

The interrelationships among the species of rice stem-borers are not clearly understood. Two species may occupy a single stem but they do not compete unless the larvae of both species enter the same internode (Rothschild, 1971). Although there appears to be not direct competition between the species of stem-borers, a particular borer species may predominate in a given area. In Sarawak the commonest species are *T. incertulas* and *C. suppressalis* (Rothschild, 1971), whereas in Thailand, *T. incertulas* is the most common (Nishida & Wongsiri, 1972). Although *T. incertulas* is the predominant species in Thailand, there are localities in the north where other species, such as *C. suppressalis* and *C. polychrysa* are most abundant. There may be a shift in abundance from one species to another within a given area. It has been suggested that such changes are caused by the use of insecticides, but this phenomenon has also been noted in localities where insecticides have not been used. Perhaps changes in the efficiency of their respective natural enemies or other cultural factors, such as changes in rice varieties, time of planting, fertilisation, and cropping systems, are involved.

Stem-borer infestations in wild rice

Wild rice species occur in tropical and subtropical areas of Asia. Some are very small, inconspicuous plants growing in irrigation ditches and swamps. Others are large, approximately the same size as cultivated rice. It has been observed that wild rice is attacked to a lesser extent than the cultivated forms. In the swamp lands of southern Nepal rice grown in the immediate vicinity of wild rice is infested with borers but the latter remains free of attack. It seems that selection has made cultivated rice susceptible.

Natural control of rice stem-borers

There are still large areas where the use of insecticides has not interfered with the activities of natural enemies of rice stem-borers. Control by such natural enemies is not always adequate, and it may be necessary to augment the existing parasite and predator fauna of rice paddies. Such assistance is particularly needed in areas which have received regular applications of insecticides. It may also prove desirable to utilise the combined action of several species of natural enemies to obtain satisfactory control (Yasumatsu, 1967 *a*, *b*). Because of the rapid growth of the plant, the micro-environmental conditions within the rice paddies undergo rapid changes in light, humidity, and temperature, and as well, alterations occur in the age structure and density of the stemborers. In addition, the composition and density of the total fauna also changes with the growth of the rice plant.

Records were kept of disease incidence in the borers. Epidemics are not as readily observed as those which occur in foliage-feeding Lepidoptera because the larvae are concealed within the stalks. Data obtained from various localities indicate that mortality caused by diseases is usually very low, e.g. Rothschild (1971). It may, however, prove possible to utilise microbial agents that can be cultured on a large scale. Several micro-organisms are known to be infectious to rice stem-borers, e.g. the well-known nematode–bacterium complex, DD-136 of *Neoaplectana carpocapsae* Weiser and its associated bacterium *Achromobacter nematophilus* Poinar & Thomas. Torri (personal communication) has sprayed dilute suspensions of this nematode on rice stubbles infested with *C. suppressalis* and obtained encouraging results.

Native home of natural enemies

Workers in biological control are often interested in the native home of a pest because the probability of finding effective natural enemies is believed to be greater in such an area. This belief stems from the concept that the most effective natural enemies have evolved through long association with the pest.

Unfortunately the centre of origin of a pest is often difficult to locate and this is true of the various rice stem-borers. Since certain stem-borers are almost specific to the rice plant we can assume that the native home of these species is the same as that of the host plant. As mentioned previously, botanists and geneticists consider the rice plant to be endemic in an area bounded by Cambodia, Thailand, Burma, North India, Nepal and South China. Our studies have shown that the arthropod fauna of rice, including the stem-borers and their natural enemies, is richest in this general area and is poorer in bordering or more distant areas.

Evaluation of natural enemies

One of the questions that is often raised is how much control is exerted by natural enemies? The percentage of parasitism at a given time does not necessarily give a good picture of the beneficial effects of natural enemies. While again not infallible, a key factor analysis based on life-table data is far more useful (Rothschild, 1970). Only limited information has been obtained in the tropics. In Sarawak the mortality during the egg to pupal stage is as high as 99 % in *T. incertulas, C. suppressalis* and *S. inferens*, and is largely caused by natural enemies (Rothschild, 1971). The rice paddies of Sarawak are not treated with pesticides and larval borer populations rarely exceed 30000 per acre (75000 per ha). In other countries such as Japan and the Philippines, where insecticides are regularly used, the larval numbers range from 80000 to 200000 per acre (200000 to 500000 per ha). While the difference may not be due entirely to the effects of the insecticides on the natural enemies, this is a suspected contribution. Observations in other South-East Asian countries indicate that the situation with respect to natural enemies is similar to that in Sarawak. Thus, great caution must be exercised in the use of insecticides until the mechanisms regulating borer numbers are more fully understood.

Most investigations have shown that egg parasitism of stem-borers is higher than that of other stages. Studies in Thailand revealed the reason for high egg parasitism in contrast to low larval and pupal parasitism. Moths emerge over a prolonged period after the rice harvest and continue to lay eggs on various non-host substrates on which the larvae are unable to develop. Egg parasites attack these eggs and those of alternative hosts, thus building up and maintaining high numbers throughout the period when the rice is not being grown. When the next crop of rice is planted, there is already a large population of egg parasites present. On the other hand, larval and pupal parasites are able to find very few hosts after harvest so that their numbers then decline with the consequence that their populations are low at the beginning of the next crop.

Conservation of natural enemies

In Asia the rice stem-borers and their natural enemies are generally in their native environment. As shown in Tables 13–15, there are many natural enemies of these borers. The role of individual species is not well understood, and it is assumed that natural enemies are partially responsible for the con-

Table 13. *The distribution of primary parasites of rice stem-borers in various ecological zones*

Ecological zone	Species
Tropics only (16 species)	*Temelucha stangli* (Ashm.), *Amauromorpha accepta metathoracica* Ashm., *Isotima javensis* (Rohw.), *I. dammermani* (Rohw.), *Goryphus apicalis* Holm., *G. mesoxanthus maculipennis* (Cam.) (Ichneumonidae)
	Spathius helle Nix., *Rhaconotus oryzae* Wilk., *R. schoenobivorus* (Rohw.), *Chelonus* sp., *Apanteles schoenobii* Wilk. (Braconidae)
	Syntomosphyrum israeli Kur., *Tetrastichus ayyari* Rohw., *T. israeli* (Mani & Kur.), *Trichospilus diatraeae* Cher. et Marg., *Elasmus albopictus* Craw. (Eulophidae)
Sub-tropics only (1 species)	*Eriborus sinicus* (Holm.) (Ichneumonidae)
Temperate only (11 species)	*Temelucha biguttula* (Munak.), *Eriborus terebrans* (Grav.), *Lampronota mandschurica* (Uch.), *Scambus annulitarsis* (Ashm.), *Agrothereutes lanceolatus* (Walk.), *Gambrus ruficoxatus* (Son.), *G. wadai* (Uch.) (Ichneumonidae)
	Chelonus munakatae (Munak.), *Microgaster russata* Hal., *Apanteles chilonis* (Munak.) (Braconidae)
	Sympiesomorpha chilonis Ishii (Eulophidae)
Tropics–sub-tropics (12 species)	*Temelucha philippinensis* (Ashm.), *Amauromorpha accepta schoenobii* (Vier.), *Xanthopimpla flavolineata* Cam., *X. modesta* (Smith), *X. pedator* (Fab.), *X. stemmator* (Thunb.), *Centeterus alternecoloratus* Cush. (Ichneumonidae)
	Stenobracon nicevillei (Bingh.), *Tropobracon schoenobii* (Vier.), *Apanteles flavipes* (Cam.) (Braconidae)
	Tetrastichus schoenobii Ferr. (Eulophidae)
	Telenomus dignoides Nix. (Scelionidae)
Sub-tropics–temperate (5 species)	*Enicospilus sakaguchii* (Mats. et Uch.), *Diadegma akoensis* (Shir.), *Itoplectis narangae* (Ashm.), *Gregopimpla kuwanae* (Vier.) (Ichneumonidae)
	Bracon onukii Watanabe (Braconidae)
All zones (9 species)	*Trathala flaviorbitalis* (Cam.), *Xanthopimpla punctata* (Fab.), *Ischnojoppa luteator* (Fab.) (Ichneumonidae)
	Bracon chinensis Szépl. (Braconidae)
	Trichogramma japonicum Ashm., *T. australicum* Gir., *T. chilonis* Ishii (Trichogrammatidae)
	Telenomus rowani (Gah.), *T. dignus* (Gah.) (Scelionidae)

sistently low populations of borers found in certain areas. Care should be taken to conserve these natural enemies wherever possible through judicious use of pesticides and other means.

The presence of other vegetation among and around the paddies may be important for the maintenance of natural enemies. The practice of burning rice stubble after harvest over large areas merits investigation since it may be undesirable.

Larval or pupal parasites of rice stem-borers do not appear to be as effective as egg-parasites. This may be due to the greater concealment of the later

K. *Yasumatsu*

Table 14. *The distribution of primary parasites of rice stem-borers in various ecological zones with reference to the host stage attacked*

Ecological zone	Species attacking	
	Egg	Larva + pupa
Tropics (37 species)	48, 51, 52, 53, 54, 55, 56	2, 3, 4, 11, 12, 15, 16, 17, 22, 23, 24, 25, 26, 27, 28, 29, 31, 32, 33, 34, 36, 38, 39, 40, 41, 42, 43, 45, 46, 47, 50
Sub-tropics (29 species)	48, 51, 52, 53, 54, 55, 56	2, 4, 5, 6, 7, 12, 14, 15, 18, 22, 24, 25, 26, 27, 28, 29, 31, 33, 34, 41, 42, 49
Temperate (25 species)	51, 52, 53, 54, 55	1, 5, 6, 8, 9, 10, 14, 15, 18, 19, 20, 21, 24, 28, 30, 33, 34, 35, 37, 44

Ichneumonidae
  1 *Agrothereutes lanceolatus* (Walk.)
  2 *Amauromorpha accepta schoenobii* (Vier.)
  3 *A. a. metathoracica* Ashm.
  4 *Centeterus alternecoloratus* Cush.
  5 *Diadegma akoensis* (Shir.)
  6 *Enicospilus sakaguchii* (Mats. et Uch.)
  7 *Eriborus sinicus* (Holm.)
  8 *E. terebrans* (Grav.)
  9 *Gambrus ruficoxatus* (Son.)
 10 *G. wadi* (Uch.)
 11 *Goryphus apicalis* Holm.
 12 *G. basilaris* Holm.
 13 *G. mesoxanthus maculipennis* (Cam.)
 14 *Gregopimpla kuwanae* (Vier.)
 15 *Ischnojoppa luteator* (Fab.)
 16 *Isotima dammermani* (Rohw.)
 17 *I. javensis* (Rohw.)
 18 *Itoplectis narangae* (Ashm.)
 19 *Lampronota mandschurica* (Uch.)
 20 *Scambus annulitarsis* (Ashm.)
 21 *Temelucha biguttula* (Munak.)
 22 *T. philippinensis* (Ashm.)
 23 *T. stangli* (Ashm.)
 24 *Trathala flaviorbitalis* (Cam.)
 25 *Xanthopimpla flavolineata* Cam.
 26 *X. modesta* (Smith)
 27 *X. pedator* (Fab.)
 28 *X. punctata* (Fab.)
 29 *X. stemmator* (Thunb.)

Braconidae
 30 *Apanteles chilonis* (Munak.)
 31 *A. flavipes* (Cam.)
 32 *A. schoenobii* Wilk.
 33 *Bracon chinensis* Szépl.
 34 *B. onukii* Wat.
 35 *Chelonus munakatae* (Munak.)
 36 *Chelonus* sp.
 37 *Microgaster russata* Hal.
 38 *Rhaconotus oryzae* Wilk.
 39 *R. schoenobivorus* (Rohw.)
 40 *Spathius helle* Nix.
 41 *Stenobracon nicevillei* (Bingh.)
 42 *Tropobracon schoenobii* (Vier.)
Eulophidae
 43 *Elasmus albopictus* Craw.
 44 *Sympiesomorpha chilonis* Ishii
 45 *Syntomosphyrum israeli* Kur.
 46 *Tetrastichus ayyari* Rohw.
 47 *T. israeli* (Mani & Kur.)
 48 *T. schoenobii* Ferr.
 49 *T. sesamiae* Yosh.
 50 *Trichospilus diatraeae* Cher. et Marg.
Trichogrammatidae
 51 *Trichogramma australicum* Gir.
 52 *T. chilonis* Ishii
 53 *T. japonicum* Ashm.
Scelionidae
 54 *Telenomus dignoides* Nix.
 55 *T. dignus* (Gah.)
 56 *T. rowani* (Gah.)

stages and to lack of food in the form of pollen and nectar, which is perhaps more required or less available at time of need by the larval and pupal parasites, combined with the unfavourable consequences of harvesting indicated above. Parasites such as *Apanteles chilonis* (Munak.), *A. flavipes* (Cam.), *Bracon* spp. including *chinensis* Szépligeti, *Tropobracon schoenobii* (Vier.), *Temelucha biguttula* (Munak.), and *Xanthopimpla* spp. visit nectar- and pollen-

132

Table 15. *Predators of rice stem-borers in various countries*

Species	Stage attacked	Countries	References
Agrionidae			
Ischnura terresiana Tillyard	Adult	Australia	Li (1970)
Tettigoniidae			
Conocephalus longipennis (Haan)	Egg	Malaysia (Sarawak)	Rothschild (1971)
Anaxipha spp.	Egg	Malaysia (Sarawak)	Rothschild (1971)
Anthocoridae			
Xylocoris galactinus (Fieber)	Overwintered larva, pupa, newly emerged adult	Japan	Chu (1969)
Lyctocoris beneficus (Hiura)	Overwintered larva, pupa, newly emerged adult	Japan	Chu (1969)
Coccinellidae			
Micrapis discolor (Fab.)	Egg	Thailand	Yasumatsu (personal communication)
Micrapis vincta (Gorham)	Egg	Thailand	Yasumatsu (personal communication)

bearing plants to obtain food necessary for ovarian development. The conservation or planting of such plants in the vicinity of the paddies may prove worthwhile.

Transfer of natural enemies within Asia

Biological control in the classical sense often involves the importation of natural enemies into areas where the pest has been introduced and its important natural enemies are absent. It has been noted that natural enemies are unevenly distributed from locality to locality often within short distances and with apparently uniform conditions. Perhaps such features as the time of burning of stubble or the continuity of the surrounding vegetation may be involved. Until more information is available about the circumstances accounting for such differences, the value that might be gained by transfers of natural enemies is only speculation. Transfers in some situations might succeed and in others fail without any apparent reason. However, if close studies were made in association with such transfers, useful programmes might be developed.

K. *Yasumatsu*

Feasibility of integrated control

Increase in the human population will lead to increased demand for rice which can only be achieved by more intensive production. Considerable emphasis may be placed on pesticide usage in efforts to raise yields and much care will be needed if the adverse effects of chemicals upon non-target organisms are to be kept at a minimum.

Rational rice stem-borer control is based on the integration of natural control with artificial means, some aspects of which are summarised below.

Table 16. *Lindane and egg parasitism of* Tryporyza innotata, *December 1965–April 1966, Tortilla Flats, Australia (Li, 1972)*

Borer generation		Average number of egg-masses completely parasitised	Average number of egg-masses partially parasitised
1		10.0	20.0
2		23.3	46.7
*3	(Lindane applied)	21.4	64.3
†4	(Lindane applied)	20.8	70.8

* Application of lindane (active ingredient 2.12 kg/m²).
† Application of lindane (active ingredient 2.47 kg/m²).

Use of selective insecticides

In the present study systemic insecticides used in granular form did not appear to harm the natural enemies of rice stem-borers. The following experiment for the control of *T. innotata* was carried out by Li (1972) in Australia. Emulsifiable lindane concentrate was allowed to drip into irrigation water that entered the rice fields, with the result that the borer larvae were controlled in the leaf sheaths and stems, and there was no decrease in egg parasitism in successive borer generations following two applications of lindane (Table 16). Such treatments conserve natural enemies to a greater extent than do foliar applications.

Monitoring pest infestations

One method of reducing the use of insecticides and of conserving the natural enemies employs regular monitoring of the numbers of stem-borers. This may be done by periodic assessment using sequential sampling or other methods (Nishida & Torii, 1970). Insecticides are then applied only when borer damage exceeds the established threshold level which may vary with different situations. Monitoring of populations of the natural enemies of the borers is also important and can also be done by sequential sampling.

Crop resistance

The combined use of resistant rice varieties and of natural enemies is one of the most desirable ways of controlling rice stem-borers (Yasumatsu, 1972). It is necessary to utilise varieties immune to insect attack. Incomplete or partial resistance, in combination with the activity of natural enemies, can reduce and maintain borer populations below economic levels. Some degree of resistance to stem-borers exists in most long-established rice varieties.

Multiple cropping of rice

The scarcity of certain natural enemies in areas where single cropping is practised is partly caused by the absence of rice plants for long periods. The multiple cropping of rice may therefore provide a more favourable environment for natural enemies, especially egg-parasites (Lim, 1970). The question has, however, been raised as to whether or not rice stem-borers would be more serious pests with multiple cropping than with single cropping. Perhaps other factors more favourable to the borers could offset the increased favourability to the parasites. There is little information concerning this. Work by Kamran & Raros (1969 *a, b*) in the Philippines indicated that there was little difference in borer and natural enemy numbers between a multiple cropping area (IRRI, Laguna) and single cropping sites. However, only small rice plots were planted at Laguna and the results may not have been indicative of large-scale situations.

Rice production and rice stem-borer damage
Economic thresholds

One of the important aspects of rice stem-borer control is the concept of economic threshold level of borer abundance. Determination of the threshold levels for different situations is extremely difficult because of the many variables that influence such levels. These range from the variety of rice grown to the sociological and economic problems of those societies that depend on rice. In developed areas, such as the USA, the damage threshold is set at a lower level than in most Asian countries. It is impossible to provide accurate threshold levels for all situations.

Ideally, the best method of determining the threshold level is to grow different varieties in areas of varying borer incidence and then to obtain a regression between borer density (or infestation indices) and yield loss. From the value of the crop it is possible to relate economic loss to borer density. It is then possible to define borer density or threshold level in economic terms.

A crude but practical approach to determine a general economic threshold level of rice stem-borers was to ask farmers their opinion on whether or not the damage done by rice stem-borers to their crop was of economic significance. The paddies were then sampled to determine the percentage of hills with damaged stalks. When the results were combined with information from agronomists and entomologists it was found that the economic threshold level was reached when 10 % of the hills had at least one infested stem. This value was used in the sequential sampling programme discussed earlier (Nishida & Torii, 1970; Torii, 1971 *a*, *b*). However, the losses incurred in a 10 % infested field of Indica and of Japonica rice are not equal. Thus, 10 % infestation of Indica rice may not lead to serious damage, whereas this level may be too high for Japonica rice, especially for low-tillering varieties. The tentative threshold level must therefore be adjusted according to varieties.

Surveys conducted in many South-East Asian rice areas indicated that damage to Indica rice varieties is generally below 10 %. In such circumstances the farmer would probably get more return per dollar input from fertiliser usage than from insecticide treatment.

Assessment and prediction of borer damage

There are a number of ways of assessing damage by rice stem-borers. Some workers express damage or loss as in terms of percentage of damaged stalks or percentage of damaged hills, others in terms of percentage of white heads. Damage has also been expressed as a percentage of decrease in yield. The lack of uniformity has made it difficult to compare losses between countries.

A simple method of assessment is needed so that a decision to apply insecticides can be made rapidly. A modification of the sequential sampling test (Wald, 1947) was found to be suitable for this purpose (Nishida & Torii, 1970). This is based on the prior established relationship between the degree of infestation at a given time in the growth of the crop and the associated decrease in yield at harvest. The percentage of infested rice hills is adopted as a standard to measure the degree of infestation. The figures obtained are then ranked into categories: (1) low infestation where the percentage of infested hills (P) is less than 5 %; (2) moderate infestation where the range of P lies between 10 and 20 %; and (3) high infestation where P is greater than 40 %. An 'acceptance' or a 'rejection' limit ('requiring' or 'not requiring' control) is given to each percentage level in the sequential sampling test. The numerical relationship between the percentage of infested hills (P) and the average number of infested stems per hill (μ), the value of which is directly connected with a decrease in the rice crop yield, is given by the formula

$$P = 1 - e^{-a\mu^b}, \tag{1}$$

where a and b are constants and e the base of a natural logarithm. This formula was originally proposed by Kono & Sugino (1958) for infestation by the second generation of stem-borers, and a table of values was constructed for estimates of P ranging from 1 to 99 %. The values of μ have now also been tabulated for the first generation (Torii, 1970, 1971 a, b). Predictions of infestation can be made through the combined use of these tables and the sequential sampling test, after making the following assumptions: The number of infested stems in a paddy field depends upon the number of healthy larvae present and their ability to disperse. The mean number of healthy larvae is given by the number $(n-d)/t$, of larvae surviving per infested stem, where n is the number of live larvae found in the number (t) of infested stems sampled, and d is the number of dead individuals found. As regards dispersion, it is known that a single larva attacks at least one stem by the middle or latter half of each generation. The average number, N, of healthy larvae that survive per hill during each generation can then be given by the product of $(n-d)/t$ and μ. The latter refers to the mean number of infested stems per hill, and can be found in the table computed from formula (1). From these figures it is possible to derive the estimates of limiting values (N_1) of N corresponding to the three classes of infestation noted earlier. The values of N_1 will indicate the number of living larvae per hill, which represent the number of larvae available to infest plants in the next generation. The influence of environmental constraints on N_1 values can be considered by extending the calculations as follows:

$$\frac{N_1 ef}{m+f} \prod_{k=1}^{\infty} (1 - W_k/100) \tag{2}$$

for the three categories of infestation, where e is the expected number of eggs laid, $m{:}f$ is the sex ratio of male to female, and W_k the k-th component of the particular environmental constraints studied. These limiting values of expression (2) correspond precisely to the limiting values of the mean number of infested stems per hill (μ), and because of the relationship between P and μ, are also equivalent to the values of P in a table computed from formula (1). (See Torii, 1971 b, p. 196.) Thus, let P_{01} be the percentage infested hill to be predicted in the middle and/or the late period of the next generation, then, in accordance with the economic degrees of infestation, the limiting values of P_{01} can be given by the values of P in the numerical table according to the infestation category.

The expected percentage of infested hills by the middle or latter period of any generation can also be predicted from information obtained by sequential sampling towards the end of the preceding generation, using the appropriate tables for the first or second generation.

Reliable techniques for the quantitative prediction of rice stem-borer incidence have been lacking in the past. The methods just outlined should facilitate the process of forecasting borer incidence and this is a basic prerequisite for rational rice stem-borer control. It is, however, not yet known if the techniques are applicable in tropical situations where borer generations overlap widely.

6. Armoured scale insects

Coordinators: P. DEBACH & D. ROSEN

Background

The project

In spite of the proportionately high degree of successful biological control projects against scale insects, as compared with other pest groups, numerous serious economic problems still exist in many countries. However, present information indicates that some of those problems may be solved by a concerted, cooperative attack utilising current knowledge and technology, while at the same time accumulating new bio-ecological information which could well lead to solutions of additional ones.

There is hardly a country in the world today where scale insects do not constitute problems. They are notorious as invaders of new territories and it is thus to be expected that additional problems will continue to arise and that the already large insecticidal load, both economic and contaminative, will continue to increase unless alternative solutions are found.

The IBP project on biological control of armoured scale insects has provided much of the essential basic information and methodology, and has provided coordination for many major projects throughout the world from 1967 to 1973. Some 70 scientists in over 30 countries have collaborated in this project. An appreciable portion of this work is ongoing. The practical results of some of the recent armoured scale projects appear highly promising. For example, studies stimulated by IBP led directly to the discovery of a promising new parasite of the snow scale, *Unaspis citri* (Comst.), in Hong Kong, where the scale was found to be kept under effective biological control. The parasite was imported into Florida, where it is now established and has already controlled snow scale in the original release grove. Losses each year in Florida from this scale range from (US) $20000000 to 40000000. Numerous other countries also suffer from the same pest so total savings could be several times the potential Florida figure.

Although armoured scale insects were the principal targets of this extensive project, certain coccid and aleyrodid pests were included.

The first conference of the IBP Working Group on biological control of armoured scale insects was held at Riverside, California, in March 1968 during

Contributors: D. P. Annecke, L. C. Argyriou, P. DeBach, R. H. Gonzalez, D. Rosen, G. J. Snowball, M. Tanaka.

the First International Citrus Symposium. The outlines of the IBP project were discussed, and several papers were presented at the Citrus Symposium by members of the Working Group.

The second conference was held in Rabat, Morocco, in October 1970 as a joint working conference with the International Organisation of Biological Control (IOBC) Working Group on biological control of citrus scale insects and brought together nearly all specialists in biological control of armoured scale insects in the Mediterranean and Middle East areas. The discussions were mostly devoted to the development of standardised research procedures and to progress reports on various projects.

A half-day symposium covering the IBP project on biological control of armoured scale insects was included in the programme of the Twelfth Pacific Science Congress, Canberra, Australia, August 1971. Eight papers were presented by collaborators from Argentina, Australia, California, El Salvador, Israel, Japan and New Caledonia, and important informal discussions took place. Abstracts were published in the *Twelfth Pacific Science Congress, Record of Proceedings*, vol. I, pages 191–200.

In connection with the IBP Scale Insect Project, the IBP Soviet National Committee has issued a volume, *Host–parasite Relations in Insects*, as their contribution to the USSR IBP Project. This 130 page volume is composed of a series of contributions by 11 scientists. It was issued by the Academy of Sciences of the USSR, 1972.

A paper reviewing the biological control of coccids by introduced natural enemies (DeBach, Rosen & Kennett, 1971) was included in a symposium on biological control held in Boston in December 1969 during the annual meeting of the American Association for the Advancement of Science, and was published as Chapter 7 in the book *Biological Control*, edited by C. B. Huffaker.

An extensive chapter on biological control of Diaspididae (Rosen & DeBach, 1975*b*) is included in a book reviewing all case histories of biological control projects in the world, edited by C. P. Clausen. The chapter reviews biological control efforts made in various countries against some 20 species of armoured scale insect pests, and summarises the life histories of natural enemies.

An IBP sponsored article for the *Annual Review of Entomology* on 'The biology and ecology of armoured scales' by J. W. Beardsley (University of Hawaii) and R. H. Gonzalez (University of Chile) has been completed (Beardsley & Gonzalez, 1975). A companion review on 'The biology and ecology of natural enemies of armored scale insects' is being prepared by P. DeBach (University of California) and D. Rosen (Hebrew University of Jerusalem).

A methodology handbook for biological control of armoured scale insects is also being prepared by D. Rosen and P. DeBach. Since that handbook will not be completed for quite a while, a short discussion of the main points of methodology was presented at the Pacific Science Congress and has been published recently by Rosen (1973), to be circulated among all collaborators.

140

A monographic biosystematic revision of *Aphytis* is also nearing completion and will hopefully be submitted for publication in 1976. A progress report has been published by Rosen & DeBach (1975*a*).

The armoured scale insects and their natural enemies

The armoured scale insects, Diaspididae, are the largest and most specialised family of scale insects (Coccoidea), and constitute one of the most economically important and destructive groups of pests to agriculture.

One to six generations may develop annually, and the winter may be passed in the egg, larval or adult stages, or a combination of all stages. In multivoltine species, considerable overlapping of broods is common during warm seasons. Dispersal is mainly passive: crawlers may be carried by wind, water, insects, birds or other animals; many species have been widely distributed by commerce on infested plants. Active dispersal by crawling larvae is rather limited.

Some species are monophagous or oligophagous, but many are extremely polyphagous. Trees and shrubs are most frequently infested. Due to their great reproductive capacity, survival ability, and the difficulty in chemical control, many species are highly destructive pests of fruit trees and ornamentals. Any above-ground parts of the host plant may be infested.

Injury to infested plants may result from both direct effects of feeding and toxic effects of injected saliva. Severe attack on the leaves, stem, branches, and shoots can cause discolouration from loss of chlorophyll, deformation and splitting, retardation of growth, and general weakening of the plant. In severe cases branches may be killed and there may be crop loss, extensive defoliation, and even eventual death of the plant. The market value of infested fruits is often much reduced. Fruits attacked early may be small and deformed, or may drop prematurely (Ferris, 1942; Bodenheimer, 1951; Baranyovits, 1953; McKenzie, 1956; Borkhsenius, 1963; Rosen & DeBach, 1975*b*).

Chemical control of armoured scale insects is often difficult and problematic and is frequently followed by recurring infestations of the target pests and by outbreaks of non-target organisms.

Fortunately, armoured scale insects, as well as related coccids, aphids and aleyrodids, seem to be rather more amenable to control by natural enemies than are many other groups of organisms, because of their sedentary habit, colonial distribution, and relative population stability on perennial host plants. No wonder, therefore, that such insects have attracted more research emphasis than other groups. In fact, introduction projects against Homoptera comprise about two-thirds of all successes (partial, substantial or complete) in biological control, and more than four-fifths of all introduction projects that resulted in effective control. Armoured scale insects alone have accounted for about one-fifth of all complete successes in biological control (DeBach *et al.*, 1971; Rosen, 1973).

Predators of armoured scale insects belong mainly to the coleopterous families Coccinellidae (*Chilocorus, Cryptognatha, Lindorus, Orcus, Pentilia, Pharoscymnus* etc.) and Nitidulidae (*Cybocephalus* spp.). Coccinellids often lay their eggs singly beneath empty scales. The adult beetles consume their prey completely, whereas the larvae usually bite out a hole in the integument of the scale insect and suck out the fluid contents. Many coccinellids have some attributes of efficient predators. They are voracious feeders in both the larval and adult stages, have well-developed searching capacities, and their longevity, fecundity and rate of development are favourable in comparison with those of armoured scale insects. However, they often are dependent on high host densities, and are not highly host-specific. Specialised, highly effective predators of armoured scale insects such as *Cryptognatha nodiceps* Marsh., a predator of coconut scale, *Aspidiotus destructor* Sign., are rare among the Coccinellidae. Overemphasis on predators may lead to failure of otherwise promising biological control projects (Clausen, 1958; Hagen, 1962; Hodek, 1967; Rosen, 1973; Rosen & DeBach, 1975*b*).

Parasites have accounted for most of the success in the biological control of armoured scale insects. They include mainly species of the hymenopterous families Aphelinidae (*Aphytis, Aspidiotiphagus, Coccophagoides, Physcus, Prospaltella, Pteroptrix* (= *Casca*)), Encyrtidae (*Comperiella, Habrolepis* etc.) and Signiphoridae (*Signiphora*). They are usually much more restricted in their host preferences than the predators, many species being strictly monophagous or narrowly oligophagous.

Endoparasites have rarely proved capable of bringing about complete biological control of armoured scale insects by themselves. Species of *Prospaltella* have been very useful in the control of white peach scale, *Pseudaulacaspis pentagona* (T.T.), and San Jose scale, *Quadraspidiotus perniciosus* (Comst.) in Europe and South America, and *Comperiella bifasciata* How. has been credited with controlling yellow scale, *Aonidiella citrina* (Coq.), in certain California habitats. In other projects, endoparasites have usually proved to be inferior to ectoparasites but may be important complementary mortality factors. In certain species, the males develop as hyperparasites, sometimes exclusively on female larvae of their own kind (Flanders, 1959). Special procedures, such as successive releases of mated and unmated females, have to be adopted for the colonisation of these species. The biology of the male of certain species, such as *Pteroptrix chinensis* (How.), is still unknown and this has prevented their use in biological control (Flanders, Gressitt & Fisher, 1958). In certain cases, endoparasites may be necessary to complement the work of ectoparasites in bringing about successful biological control, as in the control of olive scale *Parlatoria oleae* (Colvée) and California red scale, *Aonidiella aurantii* (Mask.), in California (DeBach *et al.*, 1971; Rosen, 1973; Rosen & DeBach, 1975*b*).

Ectoparasites are by far the most effective, most promising natural enemies of armoured scale insects. They are confined to the aphelinid genus *Aphytis*, which now comprises more than 80 species. All the known species of *Aphytis* develop ectoparasitically on the body of armoured scale insects, beneath the covering scale. Pupation takes place beneath the empty scale, and the adult parasite emerges through a hole that it chews in the covering scale. The host is attacked only if its body is free beneath the scale. In addition to the hosts killed by parasitism, numerous scale insects may be killed by host feeding by the adult *Aphytis* female following drilling into the scale body with the ovipositor.

Several *Aphytis* spp. have been successfully employed in biological control programmes against important pests, such as the Florida red scale, *Chrysomphalus aonidum* (L.), the purple scale, *Lepidosaphes beckii* (Newm.), the California red scale, *Aonidiella aurantii*, the dictyospermum scale, *Chrysomphalus dictyospermi* (Morg.), and the olive scale, *P. oleae*. Moreover, naturally occurring species of *Aphytis* keep many species of armoured scale insects at low population densities. Such species should be regarded as prime candidates for introduction into regions where their hosts cause economic damage (Rosen & DeBach, 1975 a).

Although armoured scale insects are often amenable to biological control this is not always practical because other pest organisms in the same agro-ecosystem may require regular treatments with non-selective pesticides. Whenever possible, selective pesticides should be used rather than the more potent, but non-selective, organophosphates and carbamates. The effects of all commercially available pesticides on the main natural enemies of armoured scale insects need much more study and methods for evaluation of such effects have been developed by Bartlett (1953) and Rosen (1967 a).

In order to reduce the general amount of pesticides used, reliable economic thresholds need to be established for the various scale pests. This, in turn, is closely related to the problem of developing realistic marketing standards. Biological control is often impractical because certain marketing organisations impose unnecessarily stringent standards. Sometimes one scale on a fruit will cause it to be culled although the quality of the fruit is not impaired. Effective integrated control has been achieved on citrus in Israel (Harpaz & Rosen, 1971) and South Africa (Bedford, 1973), where scale insects are major pests in the citrus ecosystems.

Previous biological control accomplishments

Prior to the initiation of the present IBP project, 22 species of armoured scale insects have served as subjects for biological control in various parts of the world. These projects, recently reviewed in detail by Rosen & DeBach (1975 b), are listed in Table 17. Practical results worldwide have been outstanding. 18 of the 20 target species listed in Table 17 and 96 % of all projects (54)

143

Table 17. *Projects on biological control of armoured scale insects by introduced natural enemies before the initiation of IBP (data from Rosen & DeBach, 1975b)*

Scientific name	Common name	Crop attacked	Place of infestation	Principal natural enemies imported			Control results*
				Name	Type	Place of origin and date	
Aonidiella aurantii (Mask.)	California red scale	Citrus	California	*Aphytis lingnanensis* Comp.	Ectoparasite	China 1947	S
				A. melinus DeBach	Ectoparasite	India, Pakistan 1956	
				Prospaltella perniciosi Tower	Endoparasite	China 1947	
				Comperiella bifasciata How.	Ectoparasite	China 1941	
			Australia	*A. chrysomphali* (Mercet)	Endoparasite	China 1905	P
				C. bifasciata How.	Endoparasite	California 1942	
A. citrina (Coq.)	Yellow scale	Citrus	Greece	*A. melinus* DeBach	Ectoparasite	California 1962	s
			California	*C. bifasciata* How.	Endoparasite	Japan 1924–5	s
Aspidiotus destructor Sign.	Coconut scale	Coconut, other palms	Fiji	*Cryptognatha nodiceps* Marsh.	Predator	Trinidad 1928	C
			Mauritius	*Chilocorus politus* Muls.	Predator	Java 1937	C
			Principe	*C. nigritus* Muls.	Predator	Ceylon 1939	C
			New Hebrides	*Cryptognatha nodiceps* Marsh.	Predator	Trinidad 1955	C
			Bali	*Rhizobius pulchellus* Montr.	Predator	Unknown 1964	
A. nerii Bouché	Oleander scale	Lemon	Greece	*Aspidiotiphagus citrinus* (Craw)	Endoparasite	Java 1934	P
				Aphytis melinus DeBach	Ectoparasite	California 1962	s
Carulaspis minima (T.T.) and *Lepidosaphes newsteadi* (Sulc)	Bermuda cedar scales	Bermuda cedar	Bermuda	A complex	Predators	Various, 1946–51	F
Chrysomphalus aonidum (L.)	Florida red scale	Citrus	Israel	*A. holoxanthus* DeBach	Ectoparasite	Hong Kong 1956	C
			Mexico	*A. holoxanthus* DeBach	Ectoparasite	California 1960–2	C
			Florida	*A. holoxanthus* DeBach	Ectoparasite	California 1960	C
			South Africa	*A. holoxanthus* DeBach	Ectoparasite	California 1962	C
			Seychelles	*Chilocorus nigritus* (Fab.)	Predator	India 1938	C
C. dictyospermi (Morg.)	Dictyospermum scale	Coconut	Greece	*A. melinus* DeBach	Ectoparasite	California 1962	s
Comstockiella sabalis (Comst.)	Palmetto scale	Palmetto	Bermuda	A complex	Parasites	Florida 1925–9	S
Ischnaspis longirostris (Sign.)	A coconut scale	Coconut	Seychelles	*C. distigma* Klug	Predator	East Africa 1936	C
				C. nigritus Fab.	Predator	India 1938	
Lepidosaphes beckii (Newm.)	Purple scale	Citrus	California	*A. lepidosaphes* Comp.	Ectoparasite	China 1948–9	P
			Texas	*A. lepidosaphes* Comp.	Ectoparasite	California 1952	P
			Florida	*A. lepidosaphes* Comp.	Ectoparasite	California 1958	C
			Mexico	*A. lepidosaphes* Comp.	Ectoparasite	California 1954–6	s
			Peru	*A. lepidosaphes* Comp.	Ectoparasite	California 1958	s
			Chile	*A. lepidosaphes* Comp.	Ectoparasite	California 1951–2	P
			Cyprus	*A. lepidosaphes* Comp.	Ectoparasite	California 1961	P
			Greece	*A. lepidosaphes* Comp.	Ectoparasite	California 1962	
			Israel	*A. lepidosaphes* Comp.	Ectoparasite	Hong Kong 1956?	C

Table 17 (cont.)

Scientific name	Common name	Crop attacked	Place of infestation	Principal natural enemies imported			Control results*
				Name	Type	Place of origin and date	
L. ficus (Sign.)	Fig scale	Fig	California	*A. mytilaspidis* (LeBaron)	Ectoparasite	France 1949	P
L. ulmi (L.)	Oystershell scale	Deciduous fruits	Canada (British Columbia)	*Hemisarcoptes malus* (Shimer)	Predator	Canada (New Brunswick) 1917	P
Parlatoria blanchardi (T.T.)	Parlatoria date scale	Date palms	Algeria (S. Oran)	*Cybocephalus palmarum* (Peyer.)	Predator	Algeria (oases) 1925	S
			Morocco (oases)	*Pharoscymnus numidicus* (Pic)	Predator	Algeria (oases) 1925	P
				A complex	Predators	Algeria, Morocco 1951-4	
P. oleae (Colvée)	Olive scale	Olive, deciduous fruits	California	*A. maculicornis* (Masi)	Ectoparasite	Iran 1951	C
				Coccophagoides utilis Doutt	Endoparasite	Pakistan 1957	
P. pergandii Comst.	Chaff scale	Citrus	California	*A. ?maculicornis* (Masi)	Ectoparasite	Unknown	C-S
Pinnaspis buxi (Bouché)	A coconut scale	Coconut, other palms	Seychelles	*Chilocorus nigritus* Fab.	Predator	India 1938	C
			Hawaii	*Telsimia nitida* Chapin	Predator	Guam 1936	C
P. minor (Mask.)	Cotton scale	Cotton	Peru	A complex	Parasites and Predators	Various, 1904-12	S
Pseudaulacaspis pentagona (T.T.)	White peach scale	Mulberry	Italy	*Prospaltella berlesei* (How.)	Endoparasite	USA 1906, Japan 1908	C
			Austria	*P. berlesei* (How.)	Endoparasite	Italy 1910-11	C
			Switzerland	*P. berlesei* (How.)	Endoparasite	Italy 1910	C
			Spain	*P. berlesei* (How.)	Endoparasite	Italy 1914	C
			France	*P. berlesei* (How.)	Endoparasite	Italy 1918	P
			USSR	*P. berlesei* (How.)	Endoparasite	Italy 1947	C
		Mulberry, peach	Argentina	*P. berlesei* (How.)	Endoparasite	USA, Italy 1909	C
			Uruguay	*P. berlesei* (How.)	Endoparasite	USA, Italy 1912-14	C
			Brazil	*P. berlesei* (How.)	Endoparasite	Italy 1916	P
		Papaya	Puerto Rico	*Chilocorus cacti* (L.)	Predator	Cuba 1938	P
		Oleander	Bermuda	*Aphytis diaspidis* (How.)	Ectoparasite	Italy 1924	P
Quadraspidiotus perniciosus (Comst.)	San Jose scale	Deciduous fruits	California	*P. perniciosi* Tower	Endoparasite	Georgia 1943	P
			USSR	*P. perniciosi* Tower	Endoparasite	California 1947, Far East 1955-62	P?
			Germany	*P. perniciosi* Tower	Endoparasite	USA 1950	P
			France	*P. perniciosi* Tower	Endoparasite	Germany 1953	P
			Switzerland	*P. perniciosi* Tower	Endoparasite	USA, Canada, China, USSR 1958	P

* C – *complete success*, refers to biological control being obtained and maintained over a fairly extensive area so that insecticidal treatment becomes rarely, if ever, necessary. S – *substantial success*, includes cases where economic savings are somewhat less pronounced by reason of the crop area being restricted, or by the control being such that occasional insecticidal treatment is indicated. P – *partial success*, when chemical control measures remain commonly necessary but less frequent, or cases where complete biological control is obtained only in a minor portion of the pest-infested area. F – *failure*.

145

undertaken showed some degree of control following importation of natural enemies. Failure occurred with two species. Of the attempted projects, 41 % were completely successful, 20 % showed substantial success and 25 % partial success. Thus, over three-fifths of the projects attempted resulted in very significant reductions in the original pest populations. Some highlights of the more significant projects are briefly discussed below.

Aonidiella aurantii. Continuous efforts to bring this serious pest of citrus under biological control in California have spanned eight decades, but substantial success has been achieved only during the last 25 years or so. Early introductions into California comprised predators, which proved ineffective. Subsequent efforts to introduce endoparasites failed, due to insufficient knowledge of their biology, of the systematics of host scale insects, and the effect of host plants on parasite development. Ectoparasites were largely ignored due to insufficient knowledge of the systematics of *Aphytis*. Finally, a complex of four parasites was established: *Aphytis lingnanensis* Comp. and the so-called 'red scale strain' of *Prospaltella perniciosi* Tower along the coast, and *Aphytis melinus* DeBach and the so-called 'Chinese race' of *Comperiella bifasciata* in interior areas. *A. melinus* is now considered the most effective known enemy of *Aonidiella aurantii*. The introduced *Aphytis* spp. have entirely displaced *Aphytis chrysomphali* (Mercet), formerly the only parasite of *A. aurantii* in California. Satisfactory biological control has been attained in untreated test plots in all major citrus-growing areas of California, and the more promising parasites have been transferred to various other countries. Effective biological control has been attained in some districts in California but general commercial adoption remains to be achieved. Most of the methods for evaluation of the effectiveness of natural enemies were developed in association with this project (Compere, 1961; DeBach *et al.*, 1971).

Aonidiella citrina was for many years a serious pest of citrus in California. Its successful biological control was a by-product of the campaign against the California red scale. A race of *C. bifasciata*, introduced from Japan against red scale, failed to develop in that host but was found instead to attack yellow scale. It proved specially effective in residential areas, and was largely responsible for the elimination of yellow scale from southern California and for its commercial control in the large San Joaquin Valley citrus area.

Chrysomphalus dictyospermi has been brought under successful biological control on citrus in Greece. *Aphytis melinus*, introduced against *Aonidiella aurantii*, has effected complete control of *C. dictyospermi*, although it has proved only moderately effective against red scale.

A. destructor has provided the most outstanding example of biological control of a diaspine scale by predators. *Cryptognatha nodiceps* (in Fiji and Principe), *Chilocorus* spp. (in Mauritius) and *Rhizobius pulchellus* Montr. (in the New Hebrides) have all proved very highly effective against that pest. Other coconut scales, *Ischnaspis longirostris* (Sign.), *Pinnaspis buxi* (Bouché) and the

Florida red scale have also been effectively controlled by predators in the Seychelles. The successful utilisation of Coccinellidae against diaspine scales appears to be restricted to those species with a relatively thin and readily penetrated covering (Clausen, 1940). All successes in the control of diaspine scales with predaceous Coleoptera to date have been achieved on palms-attacking species.

The Bermuda cedar scales, *Carulaspis minima* (T.T.) and *Lepidosaphes newsteadi* (Sulc) were exposed to some 50 species of predators imported into Bermuda from various parts of the world, but the project ended in failure. Parasites were virtually ignored in this work, possibly because of the earlier spectacular success against the coconut scale using predators.

Chrysomphalus aonidum, formerly the most serious pest of citrus in Israel, has been brought under effective control by *Aphytis holoxanthus* DeBach in all citrus growing regions of that country except a few groves in the hot Jordan Valley. *Pteroptrix smithi* (Comp.), an endoparasite introduced simultaneously along with *A. holoxanthus* from Hong Kong, had no appreciable effect in the initial control but has managed to persist and disperse at very low host densities. *A. holoxanthus* has more recently been successfully transferred to several other countries (DeBach *et al.*, 1971).

L. beckii is another serious pest of citrus. The control effected by *A. lepidosaphes* Comp. in California has made possible the development of integrated control of purple scale by strip treatment (DeBach & Landi, 1961). This parasite has been transferred to various other countries and has given good results in most (DeBach, 1971*a*).

The oystershell scale, *Lepidosaphes ulmi* (L.), is effectively controlled as a pest of deciduous fruit trees in British Columbia by *Hemisarcoptes malus* (Shimer), a predatory mite introduced from eastern Canada. This is the only record of a mite being highly useful in the biological control of diaspine scales, although this same predator has some effect on San Jose scale and in earlier years on olive scale in California. *Aphytis mytilaspidis* (LeBaron) is considered to be a highly important parasite of *L. ulmi* in eastern Canada.

The Parlatoria date scale, *Parlatoria blanchardi* (T.T.), is a serious pest of date palms which has been effectively controlled in North Africa by various indigenous predators transferred within that region.

P. oleae in California has been completely controlled by the 'Persian strain' of *Aphytis maculicornis* (Masi), probably a sibling species, and by the endoparasite, *Coccophagoides utilis* Doutt, introduced from Pakistan (Huffaker & Kennett, 1966; DeBach *et al.*, 1971).

Pseudaulacaspis pentagona became a serious pest of mulberry and threatened the silk industry of Western Europe at the beginning of the present century. *Prospaltella berlesei* (How.) was extensively distributed in several European and South American countries and has apparently effected satisfactory control.

147

Q. perniciosus is a worldwide and serious pest of pome fruits. The continuing campaign against this pest in Europe has relied mainly on a biparental form of *P. perniciosi*, an endoparasite introduced from North America and the Far East. Some impressive results have been obtained with this form which is different from the uniparental one attacking California red scale. *Aphytis* spp. are known to be effective against this scale in some areas.

Objectives of the IBP Project

The specific objectives of the project were:

(i) to correlate, analyse and evaluate available information on armoured scale insects and their natural enemies;

(ii) to conduct and coordinate basic studies on the biosystematics, ecology, biology and genetics of armoured scale pests and their main natural enemies;

(iii) to develop and standardise experimental procedures for field and laboratory trials with armoured scales and their natural enemies.

Closely related to the main IBP aims of gathering information were practical programmes including extending successful projects into new areas by further translocations of proved, effective natural enemies; search for new natural enemies suitable for importation into new areas; evaluation through the use of experimental check methods and periodic census data of the effectiveness of certain indigenous enemies in host population regulation; evaluating possible beneficial effects of periodic colonisation of laboratory-cultured natural enemies on host population regulation; and evaluating the effects of pesticide treatments on the natural enemies of armoured scales.

Major pest problems investigated under IBP included *Aonidiella aurantii* and *L. beckii* on citrus in Australia, California, South Africa and many other countries; *Chrysomphalus aonidum* in various countries, the rufous scale, *Selenaspidus articulatus* (Morg.), on citrus and other plants in Peru and other countries, *U. citri* on citrus in Florida, Mexico and several Central and South American countries, Australia, and southern Europe; *Q. perniciosus* on deciduous fruit trees in Europe, Asia and elsewhere, and *Parlatoria oleae* on olive and deciduous fruit trees in the USSR and elsewhere. At the same time, some related pests such as soft scale insects, mealybugs and whiteflies were also considered, notably the woolly whitefly, *Aleurothrixus floccosus* Quaint., on citrus in California, Mexico (Baja California), the Canary Islands, and southern Europe. In total these pests cause plant damage and crop losses amounting to hundreds of millions of dollars annually.

A number of publications relevant to the control of scales appeared in recent years. These include De Lotto's (1971) taxonomic studies on the species of *Saissetia* which provide new insights into the possible reasons of certain failures in the biological control of black scales in various parts of the world.

The discovery of a female sex pheromone of *Aonidiella aurantii* was reported by Tashiro & Chambers (1967) and was the basis for a series of papers by Tashiro *et al.* (1969), Rice & Moreno (1969 *a, b*, 1970), Wharten *et al.* (1970) and Moreno, Rice & Carman (1972). Williams (1970) reported extensively on the biology, ecology and economic importance of the sugar-cane scale, *Miscanthaspis tegalensis* (Zehnt.), in Mauritius, and Greathead (1972) on the dispersal of crawlers of that scale insect by air currents. Biological and ecological studies on *A. aurantii* and *A. citrina* in Australia were published by McLaren (1971), on *A. aurantii* in Australia by Willard (1972 *a, b*), on *P. pergandii* and its interaction with the tropical grey chaff scale on citrus in Israel by Gerson (1967 *a, b*), on *Qadraspidiotus maleti* (Vayss.) on olive in Morocco by Bénassy(1967), and on *P. ziziphi* Lucas on citrus in Tunisia by Sigwalt (1971).

The natural enemies of armoured scale insects

Parasites of the genus Aphytis

Research in systematics is particularly important in the field of biological control. The successful solution to several biological control projects has depended upon taxonomic studies which showed that what was thought to be only one species actually included several sibling species with different host preferences and biological capabilities.

Parasitic Hymenoptera of the genus *Aphytis* (Aphelinidae) include the most effective natural agents of population regulation of armoured scales. Recent research has shown that this is a rapidly evolving, highly specialised and rather discrete group of insects. Morphologically very similar sibling species are common and it has become apparent that morphological characters alone do not provide adequate indications of actual specific relationships within the genus.

Systematics and bionomics

The first step in the revision of *Aphytis* was redefinition of the genus in relation to closely allied genera, especially *Marietta* and *Marlattiella*. In previous classifications the genera in the *Aphytis* group of Aphelininae were separated by unreliable characters. Thus, *Marietta* was separated from *Aphytis* mainly by the mottled wings, body and appendages, and by the distinctive sculpture of the metanotum. *Marlattiella* was considered distinct mainly because of the peculiar, four-segmented antennae. *Aphytis* was considered to have unmottled wings and six-segmented antennae. A detailed comparative study proved these characters to be unreliable for generic classification. Bona fide species of *Aphytis* may have mottled wings, a heavily maculated body, and four to six segments in the antennae. The main diagnostic character separating *Aphytis* from related genera is the well-developed, relatively long propodeum with a

149

distinct, sculptured median salient, usually bearing marginal crenulae. On the basis of this and other valid characters, about 12 species were transferred from *Marietta* to *Aphytis* and an additional one from *Marlattiella* (Rosen & DeBach, 1970). The distinction between *Aphytis* and *Marietta* is especially important as all the known species of the latter are secondary parasites of scale insects. Thus, the dozen or so species that have now been transferred from *Marietta* to *Aphytis* may be regarded as potential new species for biological control.

Some 80 valid species are now recognised in *Aphytis*, including those transferred from other genera and at least 26 new species. About 18 of the 80 may be regarded as sibling species. Detailed descriptions and photomicrographs have now been prepared for most species.

The genus *Aphytis* is now regarded as including six more-or-less distinct groups of species: the *vittatus, chilensis, proclia, mytilaspidis, lingnanensis* and *chrysomphali* groups. There are, however, several species that at present cannot be assigned to any of these groups, as some appear to be intermediate and others aberrant. Some may in fact represent additional groups.

Several new potentially important natural enemies of injurious armoured scale insects were revealed by the taxonomic studies. They include new parasites of *Aonidiella aurantii, Parlatoria pergandii* Comst., *P. blanchardi, L. beckii, Q. perniciosus* and such of lesser economic importance.

The indications are that the genus *Aphytis* originated from a *Marietta*-like ancestor in the South Pacific region. It gradually extended its distribution into the Neotropical, Ethiopian, Oriental, Palearctic and Nearctic regions, probably in that order, and then evolved into paler, and eventually yellow forms. The *vittatus* group appears to be the most primitive, whereas the *mytilaspidis* and *chrysomphali* groups are perhaps the most specialised.

Although no area should be neglected in the search for additional species of *Aphytis*, the South Pacific, Oriental, Neotropical, and Ethiopian regions are probably the most promising (Rosen & DeBach, 1975a).

The life cycle of all known species of *Aphytis* is simple and has been found to be remarkably similar. All larvae are ectoparasitic on diaspine scale insects. Host preferences range from monophagous to oligophagous. Even the latter are restricted in the number of host species they attack. Detailed biological differences have been ascertained under precisely controlled experimental conditions for a number of species regarding length of life cycle, longevity of adults, fecundity, host specificity, extent of host-feeding, and temperature tolerances. Certain of these attributes are useful taxonomic characteristics.

An examination of the host selection behaviour of several *Aphytis* groups involved the physiological stimulus to oviposit under choice and no-choice situations of host availability, the manner of host discrimination through antennal and ovipositional receptors, and the role of various host stages and responses in selection by the parasite. Premature adult mortality occurs in one parasite species-group as a result of host feeding on non-preferred host scale insects. Such information may be useful in separating morphologically indistinguishable species of *Aphytis* which may be unrecognised potential biological control agents.

A comparative morphological study of the developmental stages of *Aphytis*, with special emphasis on larval characters, was made recently by Rosen and Eliraz (personal communication). No differences were found between male and female eggs and larvae, but the sexes may be readily distinguished in the pupal stage by the presence of a pair of minute sclerites at the tip of the abdomen of the female pupa. Third-instar larvae of representatives of the major groups of species may be separated by the number and arrangement of minute tubercles in the cephalic region, which may be sense organs, but these differences are not practical aids in quick identification.

Experimental investigations

The gradual accumulation and study of cultures of various species from different parts of the world has led to standardised crossing trials whenever a new culture is obtained. Crossing tests were started about 1957 and soon showed that previously unrecognised sibling species were quite common, so that virtually no species can be identified for certain on morphological grounds alone. It also led to the recognition that the older, early described species actually involved several species in each case. This applies to *Aphytis chilensis* How., *A. vittatus* (Comp.), *A. mytilaspidis*, *A. lingnanensis*, *A. chrysomphali*, and *A. proclia* (Walk.).

Hybridisation and sexual isolation

Hybridisation experiments were carried out between various strains and species of *Aphytis* to elucidate the degrees of reproductive isolation between them (Rao & DeBach, 1969 *a*, *b*, *c*). The species used were: *A. africanus* Qued., *A. coheni* DeBach, *A. fisheri* DeBach, *A. holoxanthus*, *A. lepidosaphes*, *A. lingnanensis*, *A. melinus*, and undescribed species referred to by code names or numbers, as *A.* 'khunti', *A.* 'R-65-23', and *A.* '2202'.

Copulation takes place quite readily between some species in the laboratory, whereas in others it had to be encouraged by mating inducers, and in still others it did not occur. Fertile hybrids were obtained in some cases and, although the fertility and sex-ratio were initially abnormal, there were improvements and even returns to normal in subsequent generations. Sterile hybrids resulted from a few crosses, demonstrating the genetic incompatibility of the species involved. The results of hybridisation experiments are summarised in Fig. 21.

From these experiments it was concluded that the species of *Aphytis* used could be divided into two major groups, namely, (i) the *lingnanensis* group

P. DeBach & D. Rosen

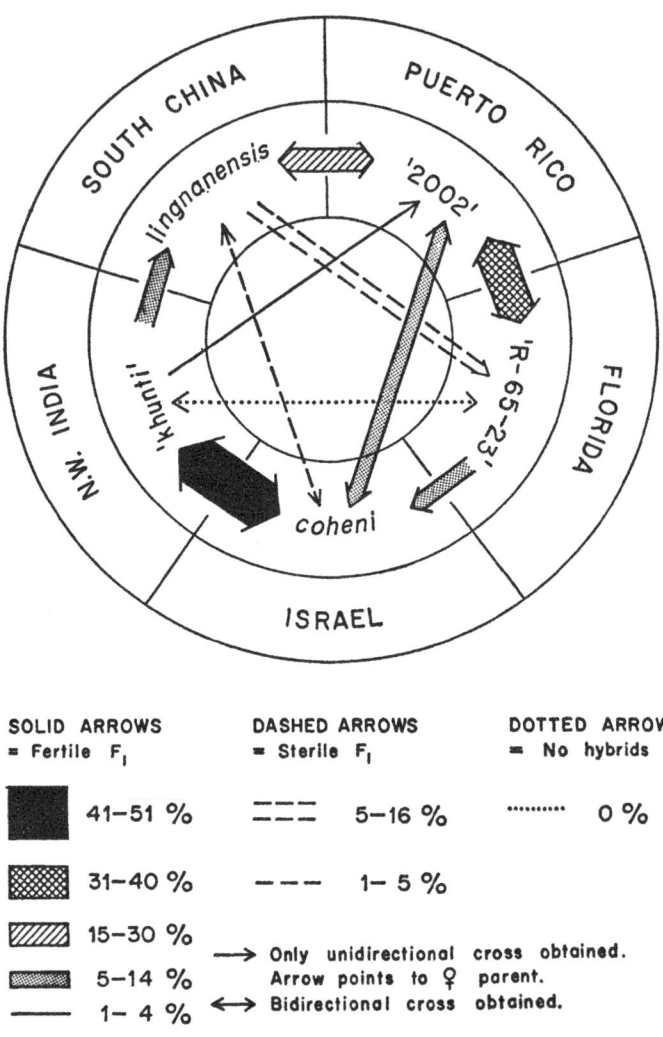

Fig. 21. Crossing relations within the *lingnanensis* group of *Aphytis* showing percent of F$_1$ progeny which are female and the nature of the F$_1$ hybrids. Normal intraspecies F$_1$ progeny production consists of 60–75 % females (from Rao & DeBach, 1969*a*).

comprising *A. lingnanensis*, *A.* '2002', *A. coheni*, *A.* 'khunti', and *A.* 'R-65-23', and (ii) the *melinus* group comprising *A. melinus*, *A. fisheri*, and *A. holoxanthus*. *A. africanus* and *A. lepidosaphes* could not be placed in either group, as they showed complete reproductive isolation from each other as well as from all the other species. Some hybridisation occurred within groups, but not between them. The spermathecae showed differences that also justified the separation of these species into the two groups: the spermathecae of members of the *lingnanensis* group are elliptical and larger than those of the *melinus* group which are nearly spherical.

As the unfertilised females of *Aphytis* species used in this study are arrhenotokous, the degree of success or failure of a heterogamic cross was based on the proportion of female progeny of the F_1 generation in addition to an examination of the fertility of the F_1 individuals. An index of degree of reproductive isolation was therefore devised, comparing the percent female progeny in the F_1 in a heterogamic cross with the percent female progeny in the relevant homogamic crosses.

The following conclusions were made on the basis of this study:

(i) *A. africanus* and *A. lepidosaphes* are valid species.

(ii) *A. melinus*, *A. fisheri*, and *A. holoxanthus* are sibling species.

(iii) Complex relationships exist in the *lingnanensis* group. *A. coheni* and *A.* 'khunti' hybridise readily in the laboratory. About half of the offspring is fertile. They show different crossing relationships with *A.* '2002', *A. lingnanensis* and *A.* 'R-65-23', indicating that they are separated from each other by considerable genetic differences. They are, therefore, considered to be 'semispecies' with respect to each other. The term 'semispecies' is used with respect to the genus *Aphytis* to indicate certain arbitrary upper and lower limits of reproductive isolation based on the coefficients of isolation (Rao & DeBach, 1969*b*). Using similar reasoning *A. lingnanensis* and *A.* '2002' are also considered to be 'semispecies' with respect to each other, as is *A.* '2002' with respect to *A.* 'R-65-23'. *A.* '2002' and *A. lingnanensis* are valid species with respect to *A. coheni* or *A.* 'khunti', as is *A.* 'R-65-23' with respect to *A. lingnanensis* and *A.* 'khunti' (Fig. 21) (DeBach, 1969; Rao & DeBach, 1969*a*).

Experiments with the *A. maculicornis* complex revealed that three cultures of diverse geographical origin are reproductively isolated and qualify for the status of valid species with respect to one another. Attempts to break their isolation through the use of female and male sex pheromones were only partially successful. Hybrids from interbreeding two cultures would back-cross and produce a limited number of viable progeny with the maternal culture but none with the paternal culture. The hybrids thus behave like 'semispecies' with respect to the maternal culture and good species with respect to the paternal culture. Experiments on the host preferences of these three cultures and their hybrids revealed the existence of significant differences in host preference.

Initial trials with two Greek cultures of *A. mytilaspisis*, one collected from *Diaspis echinocacti* (Bouché) and the other from *Carulaspis visci* (Schr.) indicate that partial reproductive isolation exists. Its degree was assessed quantitatively by performing multiple choice experiments that tested known numbers of females of both cultures with an equal number of males of one of the cultures. Examinations of the spermathecae for the presence of sperm indicated that these two cultures are probably 'semispecies'.

The newly discovered male sex pheromone is important in *Aphytis*, as the ultimate acceptance or rejection of a male by the female seems dependent on it. The male sex pheromone aids in assuring female acceptance and also quiets the female to permit copulation. A high specificity of the male sex pheromone is characteristic of *Aphytis* and may prevent hybridisation in nature.

These studies point out the possibility of producing new and valuable 'species' for biological control through interspecific hybridisation (Rao & DeBach, 1969 c).

Selection of strains for improved temperature tolerance

Experiments on artificial selection of strains for improved temperature tolerance were carried on for more than 100 generations. Various regimes of selective thermal pressure were imposed on *A. lingnanensis* to determine if tolerance to temperature extremes could be increased. One series of sub-cultures was periodically irradiated before selective pressure was applied. Striking and essentially permanent increases in tolerance to both high and low temperatures resulted, indicating another promising approach to the manipulation of parasite populations (White, DeBach & Garber, 1970).

Studies of uniparental species

Rössler & DeBach (1972 a, b) used a biparental arrhenotokous form and a thelytokous clone from a complex of sibling species described as *A. mytilaspidis*. Gene markers for eye colour introduced into the biparental population enabled the sexual relations of that form to be tested with the thelytokous form. It was found that the thelytokous females are in fact capable of mating, fertilising their eggs with sperm and producing fertile offspring. The various isolating mechanisms between biparental and thelytokous forms were tested in a series of experiments. The main barrier is poor insemination rate, which is only 20–30 % in heterogamic crosses compared to 80 % in homogamic crosses.

The biparental and thelytokous forms differ from each other in host preference. The latter parasitises both *D. echinocacti* and *Hemiberlesia lataniae* (Sign.) in the laboratory. The former form prefers *D. echinocacti* and will not thrive on *H. lataniae*. These traits are probably polygenic in origin and are redistributed in the hybrids.

The thelytokous form seems to be better adapted to high temperatures as the adults survive longer and the developmental period to adult emergence is shorter. The thelytokous population also seems to be more uniform in these traits due to a lower genetic variability caused by the abandonment of sexual processes.

Males are occasionally found in the thelytokous population which are

154

normal and functional, and their presence maintains a certain degree of sexual reproduction and gene flow.

Other natural enemies

Parasites

Research on the taxonomy and biology of Encyrtidae, Aphelinidae and Signiphoridae has contributed significantly to our knowledge of the natural enemies of armoured scale insects during the course of this IBP project. Notable contributions include: De Santis (1967), Krombein & Burks (1967), Gerson (1968), Nikolskaya & Yasnosh (1968), Avidov & Gerson (1968), Traboulsi (1968), Tachikawa & Valentine (1969a, b), Hoffer (1969), Tachikawa (1970), Ahmad (1970b), Annecke & Mynhardt (1970), Annecke & Insley (1970), Rosen & DeBach (1970), Benassy, Bianchi & Milaire (1971), Gerling & Bar (1971a, b), Trjapitzin (1971), Annecke & Insley (1971), Quezada, DeBach & Rosen (1973).

Quezada *et al.* (1973) reported extensively on the biology of the new species *Signiphora borinquensis* Quez., DeBach & Rosen, which is a primary parasite of armoured scale insects. They stressed that contrary to popular belief that most species of Signiphoridae are hyperparasites, many are primary parasites of armoured scales.

Two strains of *Comperiella bifasciata* were imported and colonised in California against diaspine scales. These two strains have long been known as the 'yellow scale strain' (from Japan) and the 'red scale strain' (from China). They interbreed freely in the laboratory. The former attacks and develops only in *Aonidiella citrina*. The latter develops well in *A. citrina* and *A. aurantii* and laboratory tests have indicated that the 'red scale strain' does better on *A. citrina* than does the 'yellow scale strain'. The perplexing aspect of this is that, whenever *Comperiella* was reared from the yellow scale in the field in southern California and tested in the laboratory, it has proved to be the 'yellow scale strain', even though the 'red scale strain' is present in the same area on *A. aurantii*. Arguing from laboratory fecundity studies, the 'red scale strain' should have displaced the 'yellow scale strain' in the field and competition cage tests show this to happen in the laboratory.

Experimental data indicate that the host preference and host suitability characteristics of these two strains are caused by single gene pair differences. A dominant gene appears to control the ability to develop in *A. citrina* and a recessive gene allows development to occur in both scales. Pure dominants and the hybrids develop only in *A. citrina*, whereas the pure recessive develops on both scales. The fact that two such closely related strains show different host preferences is of great interest to biological control research. This can also be viewed as a case of incipient speciation of considerable evolutionary interest and the mechanisms by which the 'yellow scale strain' keeps from

155

being displaced in the field by the 'red scale strain' are of significant ecological interest.

Predators

Extensive studies have been carried out particularly in Israel on the biology, ecology, and phenology of coccinellid predators of armoured scale insects. Kehat (1967, 1968 a) recorded the phenology of Coccinellidae associated with *Parlatoria blanchardi* on date palms, and Kehat & Greenberg (1970) surveyed the Coccinellidae in citrus groves. Kehat (1968 b) reported on the feeding behaviour of *Pharoscymnus numidicus* Pic, a predator of *P. blanchardi*. The food consumption of *Chilocorus bipustulatus* (L.), the most important predator of armoured scale insects in Israel, was studied by Yinon (1969), and the factors causing seasonal decline of that predator in citrus groves were investigated by Kehat, Greenberg & Gordon (1970) and by Applebaum *et al.* (1971).

Blumberg (personal communication) has studied the species of *Cybocephalus* in Israel. Ahmad (1970 a) has described a new species of *Pharoscymnus* from Pakistan.

Avidov, Blumberg & Gerson (1968) described the biology of *Cheletogenes ornatus* (C. & F.), an acarine predator of *P. pergandii*. Gerson (1971) recently reviewed the mites associated with armoured scales.

Ecology of natural enemies

Competitive displacement between ecological homologues

Studies on competitive displacement, as demonstrated by the interactions of the parasites of *A. aurantii* in California and as reviewed by DeBach & Sundby (1963) and DeBach (1966), were continued during the course of this IBP project in California, Greece, Israel and South Africa.

In southern California, *Aphytis lingnanensis* has been completely eliminated by *A. melinus* from the interior but not from coastal zones where *A. lingnanensis* remains strongly dominant and largely excludes *A. melinus*. In intermediate areas both species co-exist but this is really a distributional overlap due to dispersal from the zones in which each is dominant. The originally established species, *A. chrysomphali*, which once was common throughout southern California, has virtually become extinct because of competitive displacement by the other two. Only a few individuals of *A. chrysomphali* were found in 1972 in several isolated pockets immediately adjacent to the sea. *C. bifasciata* co-exists with *A. melinus* in interior areas, whereas *P. perniciosi* complements the action of *A. lingnanensis* along the coast (DeBach *et al.*, 1971). Homeostasis is being approached with these species in southern California.

In the Peloponnesus, *A. melinus* was introduced in 1962, and has since largely displaced *A. chrysomphali* from *A. aurantii* and *Chrysomphalus dictyospermi* populations on citrus (DeBach & Argyriou, 1967; Argyriou, 1975). In Israel, *A. melinus* appears to have displaced other parasites of the California red scale from certain important citrus-growing areas. However, *A. chrysomphali* has remained dominant in certain coastal areas, and *A. coheni* has maintained its stronghold in the hot interior valleys (Kamburov, personal communication).

Comperiella bifasciata and *A. holoxanthus* were introduced into the Western Transvaal to control *Chrysomphalus aonidum*. *C. bifasciata* was established, but was replaced in competition by *A. holoxanthus*. Both were subject to attack by *Marietta exitiosa* Comp., but it could not be shown that this secondary parasite had any harmful influence on the primary parasites. Predators were found to be far less effective than the parasites (Cilliers, 1971).

Similar studies were carried out by Quezada & DeBach (1971, 1973) on the natural enemies of the cottony cushion scale, *Icerya purchasi* Mask., in southern California. This pest is kept at low population levels in all areas investigated. The vedalia ladybeetle, *Rodolia cardinalis* (Muls.), is responsible for most of the control in the desert, whereas the parasitic fly, *Cryptochaetum iceryae* (Will.), exerts most control along the coast. In the interior, they share their prey in different proportions according to the season.

Effects of climate

Studies in the field and laboratory on the adverse effects of climate on *Aphytis lingnanensis* have been recently summarised by DeBach *et al.* (1971). These include one of the best examples of the interference with biological control by weather. Experimental field tests showed that average population densities of the host, *Aonidiella aurantii*, were not regulated by climate. The same type of tests showed that *A. lingnanensis* could successfully control *A. aurantii* in coastal and intermediate climatic areas of California, but not in the interior citrus areas where the efficiency of the parasite was reduced by the more severe climate.

Locally extreme, low winter temperatures, as well as high summer temperatures and low humidities caused high mortality of various stages of *A. lingnanensis*, greatly reduced the reproductive potential of adult females, adversely affected sex-ratios and at times even caused very localised extermination. However, weather did not limit the range of the parasite, but merely its efficiency.

These studies indicated that *A. lingnanensis* is not eradicated in extreme climatic areas, because shelter and genetically resistant individuals permit limited survival of a population under otherwise periodically intolerable conditions.

L. beckii was satisfactorily maintained under biological control by *Aphytis*

157

sp. in 'shadow habitats' in New Caledonia, that is on citrus growing under larger trees, but not in the sunny situations (Fabres, 1971).

Check methods for evaluation of natural enemies

Various check methods have been developed for evaluating the efficacy of armoured scale enemies (DeBach & Huffaker, 1971). Enemies can be excluded from experimental plots by mechanical, insecticidal or biological methods. Their role in the populations of armoured scale insects may then be demonstrated by following the development of pest densities in such plots in comparison with densities in plots where natural enemies are left undisturbed. Such 'plots' may consist of leaves, branches, entire trees or groups of trees. The simplest method is the use of a pesticide that is considerably more detrimental to the natural enemies than to the scale host. Comparisons can also be made between the host populations in closed (enemy-free) and open cages. A combination of the two methods, using paired, open sleeve cages, one of each pair being impregnated with a selective insecticide, is probably the best approach.

Under the auspices of IBP such check methods have been successfully used in several countries to demonstrate the potentialities of biological control for armoured scales.

Although highly satisfactory biological control of *A. aurantii* has been achieved in certain citrus districts of southern California, insecticidal treatment is still practiced in districts which studies indicate should be suitable for efficient regulation by established *A. aurantii* parasites. Means of phasing out these chemical treatments are urgently needed.

In Israel, the interrelations between *A. aurantii* and its natural enemies were studied on citrus using the insecticidal check method. Although natural enemies alone were unable to hold the pest populations below the thresholds of economic injury in the test plot (this was before the recent spread of *Aphytis melinus*), the results showed that natural enemies play a decisive role in retarding the increase of *Aonidiella aurantii*, especially at the onset of summer. A reduction of the natural enemy populations by chemical sprays at that period may be followed by a massive outbreak of the scales (Ben Dov & Rosen, 1969).

In Greece, experiments demonstrated the effectiveness of *Aphytis melinus* against *A. aurantii* (DeBach & Argyriou, 1967).

In South Australia, Brewer (1971) demonstrated, with the insecticidal check method, that *C. bifasciata* can stabilise the population of *Aonidiella citrina* on citrus at low levels and that it has little effect on *A. aurantii*. The difference in effectiveness is apparently because there is about 60 % mortality from encapsulation of immature stages of the parasite in *A. aurantii* compared with little or none in *A. citrina*. In coastal New South Wales, experiments in which low infestations of *L. beckii* and *U. citri* on citrus were shielded with mesh cages have indicated that natural enemies were not responsible for the infestations being low.

Biological control attempts and their consequences

The following are some results of practical biological control attempts made in various countries under the auspices of the IBP project. They are summarised in Table 18.

North and Central America

USA

Field studies on natural enemies of *A. aurantii, A. citrina, L. beckii* and *P. pergandii* in California were continued and further parasites were introduced and mass reared for liberation. *Aneristus ceroplastae* How. from Pakistan and Japan and *Microterys okitsuensis* Comp. from Japan were released against *Coccus pseudomagnoliarum* Kuw. in California. *Aleurothrixus floccosus* was first found in California in November 1966 infesting backyard citrus in San Diego, and shortly thereafter in adjacent Tijuana, Mexico. It has since been found elsewhere in California and attempts are being made at eradication of infestations outside the San Diego area. Biological control efforts were initiated in 1967 at Chula Vista. Two species of parasites, *Amitus spiniferus* Brèth. and *Eretmocerus paulistus* Hemp. were imported from Mexico in May and June of 1967 and *Amitus* was successfully established in San Diego. By mid-1968 *Amitus* had reduced the woolly whitefly to low levels on the original trees. *Eretmocerus* was successfully established in July 1968 (DeBach & Warner, 1969).

In 1969 the initiation of a chemical eradication campaign against *Aleurothrixus floccosus* ended biological control research at San Diego. However, biological control efforts were started at Tijuana, to reduce the hazard of spread to California and to evaluate biological control possibilities. In July 1969, *Amitus spiniferus, E. paulistus* and *Encarsia* sp. were imported from Sinaloa, Mexico and established at two sites in Tijuana. By June 1970 both *Amitus* and *Eretmocerus* were well established. By early August 1970 these sites were producing large numbers of parasites. Some of them were transferred to Spain and France.

In 1970, *Amitus* sp. (probably *spiniferus*), *Eretmocerus* sp. and *Encarsia* sp. were imported from El Salvador, and *Cales noacki* How. from Chile. *Amitus* from El Salvador was recovered as was *Cales* from Chile.

In Tijuana the *Aleurothrixus floccosus* population was reduced to one live whitefly or less per leaf as compared to the original observed infestation of 100 or more whiteflies per leaf, except on those trees infested by ants.

159

Table 18. *New cases and/or evaluations of biological control of armoured scale insects and related coccids by imported natural enemies following the initiation of the IBP project (1967)*

Scientific name	Common name	Crop attacked	Place of infestation	Principal natural enemies imported			Control results†
				Name	Type	Place of origin and date*	
Aonidiella aurantii (Mask.)	California red scale	Citrus	Argentina	*Aphytis melinus* DeBach	Ectoparasite	California 1966	E
			Australia	*A. melinus* DeBach	Ectoparasite	California 1961	S
			Chile	*A. melinus* DeBach	Ectoparasite	California 1966	E
			Cyprus	*A. melinus* DeBach	Ectoparasite	California 1959–61	P
				A. lingnanensis Comp.	Ectoparasite	California 1959–61	P
			Israel	*A. melinus* DeBach	Ectoparasite	California 1961–on	P
				Comperiella bifasciata How.	Endoparasite	California 1967–on	P
			Italy (Sicily)	*A. melinus* DeBach	Ectoparasite	California 1964	E
			Morocco	*A. melinus* DeBach	Ectoparasite	California 1966	E
				A. lingnanensis Comp.	Ectoparasite	California 1966	E
			South Africa	*A. melinus* DeBach	Ectoparasite	California 1964	E
				A. lingnanensis Comp.	Ectoparasite	California 1964	E
				Comperiella bifasciata How.	Ectoparasite	Australia 1966	E
				Chilocorus cacti (L.)	Predator	Texas 1966	E
			Swaziland	*Comperiella bifasciata* How.	Endoparasite	South Africa 1969	E
				Chilocorus cacti (L.)	Predator	South Africa 1969	E
			Turkey	*A. melinus* DeBach	Ectoparasite	Greece 1968 (ecesis)	E
				A. lingnanensis Comp.	Ectoparasite	California 1971–2	E
A. citrina (Coq.)	Yellow scale	Citrus	Australia	*Comperiella bifasciata* How.	Endoparasite	California 1964	C
			Turkey	*A. melinus* DeBach	Ectoparasite	Greece 1968 (ecesis)	E
Aleurocanthus woglumi Ashby	Citrus black-fly	Citrus	El Salvador	*Prospaltella opulenta* Silv.	Endoparasite	Mexico 1970–1	C
A. floccosus Quaint.	Woolly whitefly	Citrus	California	*Amitus spiniferus* Brèth.	Endoparasite	Mexico 1967–70	S (in progress towards C)
				Eretmocerus paulistus Hemp.	Endoparasite	Mexico 1967–70	
				Cales noacki Howard	Endoparasite	Chile 1970	
			Mexico (Baja California)	*Amitus spiniferus* Brèth.	Endoparasite	Mexico 1967–70	C
				Eretmocerus paulistus Hemp.	Endoparasite	Mexico 1967–70	
				Cales noacki How.	Endoparasite	Chile 1970	
			Spain	*C. noacki* How.	Endoparasite	California 1970–1	E
			France	*C. noacki* How.	Endoparasite	California 1971–2	E
			Chile	*Amitus spiniferus* Brèth.	Endoparasite	Peru 1967	P

160

Table 18 (*cont.*)

Scientific name	Common name	Crop attacked	Place of infestation	Principal natural enemies imported			
				Name	Type	Place of origin and date*	Control results†
Aspidiotus nerii Bouché	Oleander scale	Citrus	Italy (Sicily)	*Aphytis melinus* DeBach	Ectoparasite	California 1964	E
Chrysomphalus aonidum (L.)	Florida red scale	Citrus	Brazil	*A. holoxanthus* DeBach	Ectoparasite	California 1962	C
			Peru	*A. holoxanthus* DeBach	Ectoparasite	California 1965, 1967	C
C. dictyospermi (Morg.)	Dictyospermum scale	Citrus	Italy (Sicily)	*A. melinus* DeBach	Ectoparasite	California 1964	C
			Turkey	*A. melinus* DeBach	Ectoparasite	Greece 1968 (ecesis)	S
			Spain	*A. melinus* DeBach	Ectoparasite	California 1970	E
Dialeurodes citri (Ashm.)	Citrus whitefly	Citrus	California	*Prospaltella lahorensis* How.	Endoparasite	India, Pakistan 1969	E
Gascardia destructor (Newst.)	White wax scale	Citrus and ornamentals	Australia	*Anicetus communis* Ann.	Endoparasite	South Africa	E
				Paraceraptrocerus nyasicus (Comp.)	Endoparasite		
Lepidosaphes beckii (Newm.)	Purple scale	Citrus	Brazil	*Aphytis lepidosaphes* Comp.	Ectoparasite	California 1962	C
			Cyprus	*A. lepidosaphes* Comp.	Ectoparasite	California 1959–61	C
			Spain	*A. lepidosaphes* Comp.	Ectoparasite	California 1970	E
			South Africa	*A. maculicornis* (Masi)	Ectoparasite	California 1964	C
			USSR (Georgia)		Ectoparasite	California 1970–1	E
Parlatoria oleae (Colvée)	Olive scale	Olive	France	*Metaphycus helvolus* (Comp.)	Endoparasite	California 1969	E
			France (Corsica)	*M. helvolus* (Comp.)	Endoparasite	France 1971	E
Saissetia oleae (Olivier)	Black scale	Olive	Italy (Sicily)	*M. helvolus* (Comp.)	Endoparasite	California 1970–1	E
Selenaspidus articulatus (Morg.)	Rufous scale	Citrus	Peru	*Aphytis roseni* DeBach & Gordh	Ectoparasite	Uganda 1970	C
Unaspis citri (Comst.)	Citrus snow scale	Citrus	Florida	*Aphytis ?lingnanensis* Comp.	Ectoparasite	Hong Kong 1971–2	E

* In cases where importations were made before the IBP project began in 1967, evaluation of results was made under IBP auspices between 1967 and 1973.

† C – *complete success*, refers to biological control being obtained and maintained over a fairly extensive area so that insecticidal treatment becomes rarely, if ever, necessary. S – *substantial success*, includes cases where economic savings are somewhat less pronounced by reason of the crop area being restricted, or by the control being such that occasional insecticidal treatment is indicated. P – *partial success*, when chemical control measures remain commonly necessary but less frequent, or cases where complete biological control is obtained only in a minor portion of the pest-infested area. E – *established* and too early to evaluate or results unknown.

P. DeBach & D. Rosen

Another serious pest, the citrus whitefly, *Dialeurodes citri* (Ashm.) was discovered on backyard trees in San Diego in October, 1966, and later at Sacramento, Fresno, Bakersfield, and Tustin. Unsuccessful attempts at chemical eradication delayed biological control work until 1968. Attempts to establish a *Prospaltella* sp. from Japan were unsuccessful, but in 1969 *Prospaltella lahorensis* How. from Pakistan and India was successfully introduced and established. By 1972, this parasite had extended its range beyond original colonisation sites in Orange and San Diego counties with noticeable reduction of citrus whitefly populations and had overwintered successfully at release sites in Sacramento. Two coccinellids from Florida and a new parasite from India were also established during 1972.

U. citri has developed into a major pest of citrus in Florida in recent years. It infests 40–50 % of citrus groves, often causing economic losses of up to about 25 %. It is a serious pest of citrus also in Mexico, El Salvador, Peru, Argentina, Guadaloupe, Fiji, Australia and New Caledonia, and is found also in Brazil, Jamaica and elsewhere, and has recently invaded southern France. The scale had very few natural enemies in Florida or in any of the countries listed. Total parasitism has been very low. Research was started under IBP auspices and the scale was found to be under excellent biological control in Hong Kong. Parasites were brought during 1971 to 1973 to Florida. Several species of parasites were obtained, including *A. ?lingnanensis*, *A.* nr. *funicularis* Comp., *Aspidiotiphagus ?citrinus* Craw, and *Prospaltella* sp. *A. ?lingnanensis* is by far the dominant species and was established in January, 1973. Preliminary results appear to be very promising in that the first release grove became commercially clean by late spring of 1973.

Mexico

Chrysomphalus aonidum has been brought under effective biological control by *A. holoxanthus*, introduced in the early 1960s from California. Similar results have been achieved against *L. beckii*, with *A. lepidosaphes* introduced from California in the 1950s (Maltby, Jiménez-Jiménez & DeBach, 1968).

Recent work on the biological control of woolly whitefly in Baja California, Mexico has been reported under USA above.

El Salvador

The citrus blackfly, *Aleurocanthus woglumi* Ashby, invaded El Salvador about 1965, and by the end of 1971 had spread practically all over the country and was causing extensive damage. Natural mortality factors were assessed in studies carried out from 1969 to 1972. High temperatures and low humidity during the dry season (November to April) cause about 70 % mortality of eggs and nymphs, but the rainy season (May to September) permits a tremen-

162

dous increase of populations. Native predators and pathogenic fungi cause considerable mortality of developmental stages during that period, but their combined action is insufficient for economic control. *Prospaltella opulenta* was introduced from Mexico and Barbados in 1970 but was not recovered. In July 1971, a further shipment from Mexico was released. Only four months later, the orchard was completely cleaned of the pest, and large numbers of parasites were found. Further parasites were imported from Mexico in April, 1972, and released in several orchards. In six weeks the pest was under control in all orchards and the parasites were recovered by the thousands for distribution to other citrus groves. The pest is presently under excellent biological control (Quezada, 1972).

U. citri is the most severe armoured scale pest of citrus in El Salvador. Attempts to introduce the ladybeetle *Telsimia* sp. from Fiji failed (Quezada, 1972).

Aphytis lepidosaphes was found to have established itself fortuitously in El Salvador, and has apparently brought about satisfactory biological control of *L. beckii* there (DeBach, 1971 *a*).

South America

Peru

In the early 1960s *C. aonidum* heavily infested citrus in Peru, especially in northern coastal areas. *A. holoxanthus* was introduced from California in 1965 and 1967. In 1969 the scale became scarce and there was evidence *A. holoxanthus* was controlling it. In 1972, Beingolea (personal communication) reported the scale to be very rare and parasitism by *A. holoxanthus* to be high.

S. articulatus has been the most important pest of citrus in Peru. Importations in 1964 of the parasite *Prospaltella fasciata* Malen. and the predators *Lindorus lophanthae* (Blaisd.) and *Pentillia* sp. from the West Indies were unsuccessful, as was *A. melinus* from California in 1970. Meanwhile Africa was considered the likely centre of origin of the genus *Selenaspidus* and the search for suitable agents has since been concentrated there. Shipments of *Selenaspidus* sp. from South Africa in 1970 yielded *Metaphycus ?aspidiotinorum* Comp. and *Aphytis* sp., but these did not reproduce on *S. articulatus*. However, in 1970, an undescribed species of *Aphytis* was obtained from *S. articulatus* in Uganda. The parasite, subsequently described as *A. roseni* DeBach & Gordh, has become well established in Peru and by March 1973 it had attained high levels of parasitisation and had reduced the infestation to non-economic levels in at least one release grove. By June 1975 this project was reported as a complete success. Efforts are also being made to introduce *Habrolepis rouxi* Comp. from Uganda (Beingolea, 1973 and personal communication).

163

Aphytis lepidosaphes was introduced into Peru from California in 1958, and has effected substantial control of *L. beckii* in certain citrus areas.

Aleurothrixus floccosus is usually heavily parasitised in Peru where parasites are not adversely affected by insecticides used for other pests. In areas where upsets have occurred, effective integrated control has been achieved using high-pressure water washing combined with parasite activity.

Chile

Aphytis lingnanensis and *L. lophanthae* were introduced from California and established in 1966 on *Aonidiella aurantii*. More recently, *Aphytis melinus* was imported from California, and in 1970 was found to be well established in certain citrus areas. *A. lepidosaphes*, introduced earlier, was also found to be rather common on *L. beckii* in certain localities.

Amitus spiniferus was introduced from Peru to Chile and became well enough established in 1970 to add appreciably to the degree of control of *Aleurothrixus floccosus* formerly effected by *C. noacki* alone.

Argentina

Aonidiella aurantii and *Chrysomphalus aonidum* are among the most important pests of citrus. *Aphytis lingnanensis*, *A. melinus* and *A. holoxanthus* have been introduced from California and Chile and established in various citrus areas. *A. melinus* has become well established and has reduced *A. aurantii* to a certain extent (Crouzel, 1971; DeBach, personal communication). By 1973, this parasite had spread into Paraguay where it has become established. In 1970, *A. lepidosaphes* was found to be established on *L. beckii* in Tucuman (DeBach, personal communication).

Brazil

A. holoxanthus and *A. lepidosaphes* were introduced from California in 1962 against *C. aonidum* and *L. beckii*, respectively. In 1970 both pests were under effective biological control by the introduced parasites at the release sites (DeBach, personal communication).

The Pacific Region
Australia

Aonidiella aurantii has been a serious pest of citrus for many years. In the Mildura citrus area of Victoria, the 'red scale strain' of *C. bifasciata*, introduced in 1947 from California, *Aphytis chrysomphali*, introduced in 1954 from California and New South Wales, and *A. melinus*, introduced in 1961 from

164

California, have all become established, whereas *A. lingnanensis*, introduced in 1962 from California and mass released, failed to establish. Attempts to establish *P. perniciosi* from California were started in 1971 and are continuing. *C. bifasciata* is the most widespread parasite and *A. melinus* has increased its range at the expense of *A. chrysomphali*. Since the establishment of *C. bifasciata*, *A. chrysomphali*, and *A. melinus* in Victoria, the importance of *Aonidiella aurantii* has declined in all areas. In one location the incidence of *A. aurantii* declined from 300 to 1200 living scales per 150 leaves to less than 10. In 1972, most of the citrus acreage of Victoria was not sprayed at all, or was sprayed with mineral oil only (McLaren, 1972).

Studies on the effect of malathion on *Aphytis melinus*, *C. bifasciata* and *L. lophanthae* in South Australia indicate a low probability of integrating this widely used insecticide with the biological control of *Aonidiella aurantii* (Abdelrahman, 1973).

In inland New South Wales, *Aphytis chrysomphali*, introduced repeatedly since 1925, and *C. bifasciata*, introduced in 1943–4, have recently increased in abundance (Hely, 1968). In coastal New South Wales, changes in cultural practices aimed at reducing dust deposits in citrus groves have enabled native predators and the introduced parasites to reduce *Aonidiella aurantii* to low levels.

A. citrina has been brought under effective biological control in Victoria by predators, by the 'yellow scale strain' of *C. bifasciata*, introduced in 1964, and by the more recent establishment of *Aphytis* spp. This was confirmed by the experimental use of DDT which interfered with the introduced natural enemies and permitted *Aonidiella citrina* populations to explode.

C. aonidum is a serious pest of citrus in coastal Queensland and *U. citri* is the most important armoured scale in coastal citrus orchards of New South Wales but no effective natural enemies of either are present. The white wax scale, *Gascardia destructor* (Newst.), is an important citrus pest in coastal New South Wales and Queensland. Native coccinellid predators attack it, but seldom keep it at low levels. Since 1968, natural enemies have been introduced from South Africa and New Zealand. Two South African parasites, *Anicetus communis* Ann. and *Paraceraptrocerus nyasicus* (Comp.), are now established. The latter has considerably reduced the incidence of *G. destructor* on non-citrus plants, mainly oleanders, in the coastal area near the Queensland–New South Wales border (Snowball, 1972).

Q. perniciosus is a problem in South Australia, Victoria, New South Wales and Tasmania which are infested with codling moth and against which broad-spectrum sprays are applied. In orchards not receiving broad-spectrum sprays, *Q. perniciosus* does not usually become serious. Application of DDT always results in conspicuous, often damaging, increases in originally sparse scale populations, confirming the view that, under undisturbed conditions, natural enemies assist in regulating *Q. perniciosus* populations in pome fruit orchards. In Western Australia where codling moth does not occur, studies in an un-

P. DeBach & D. Rosen

treated orchard showed that fruit infestation by *Q. perniciosus* exceeded 6 % in only one year in four. Two parasites, *Aphytis* sp. and *Prospaltella* sp., and a predator, *Rhizobius lindi* Blackb., are known to be present. These are absent from sprayed orchards. However, the insecticides used in sprayed orchards normally keep *Q. perniciosus* at low levels except that sporadic outbreaks occur when insecticide applications are mistimed or omitted.

New Caledonia

L. beckii is kept at low levels on citrus in shaded habitats, such as coffee plantations, by natural enemies that include an undescribed species of *Aphytis*, and by competitors for space (lichens) (Fabres, 1971).

New Hebrides

Aspidiotus destructor appeared on Vate Island about 1962 and by early 1964 coconut palms were heavily attacked. The coccinellids *Cryptognatha nodiceps*, *Azya trinitatis* Marsh. and *Pseudoscymnus* sp. were introduced but did not become established. In May 1964 the coccinellid beetle *R. pulchellus*, a native of New Caledonia, was detected on Vate Island attacking *A. destructor*. It multiplied rapidly and by early 1965 eventually almost eliminated the scale (Cochereau, 1969).

Europe and Mediterranean region
Israel

The integrated control programme of citrus pests in Israel has been discussed in detail by Rosen (1967*b*, 1975) and by Harpaz & Rosen (1971).

Aphytis melinus had been repeatedly released against *Aonidiella aurantii* in Israel since 1961, but failed to become firmly established until late in 1967 when it was found to be the dominant parasite of *A. aurantii* in certain citrus areas on the West Bank of the river Jordan. This supposedly adapted strain has since been established over large citrus areas, displacing other parasites. *Comperiella bifasciata* has also extended its range in citrus groves, both by natural spread and by repeated releases.

C. aonidum has been kept under effective biological control in the coastal plain of Israel since *Aphytis holoxanthus* was established in the late 1950s. Recent studies showed that *Pteroptrix smithi* Comp., an endoparasite introduced simultaneously with *A. holoxanthus*, has been spreading in the small residual populations of the scale (Porath, 1969). As *A. holoxanthus* is unable to exert full control in the hot Jordan Valley, periodic releases have become routine during autumn, winter, and spring and have brought about satisfactory biological control of the scale on citrus and banana in that region.

Parlatoria pergandii, the tropical grey chaff scale *P. cinerea* Hadd., and various soft scale insects and mealybugs have recently risen to alarming pest proportions on citrus in Israel. Efforts are currently being made to introduce various natural enemies. *Metaphycus helvolus* (Comp.) from California and several species of *Metaphycus*, *Coccophagus* and *Scutellista* from Africa have been introduced against the black scale, *Saissetia oleae* (Olivier). *Anarhopus sydneyensis* Timb. from Australia and California has been introduced against the long-tailed mealybug, *Pseudococcus longispinus* (T.T.), on citrus and avocado.

In 1973, integrated control groves were set aside in all citrus-growing regions of Israel for the colonisation of natural enemies. Only selective pesticides are used in such groves, and the populations of pests and natural enemies are carefully monitored in them. Although serious setbacks have occurred, it is hoped that integrated control will eventually be established in most of the citrus acreage.

Greece

Studies on citrus scales have been continued in Greece under the auspices of IBP. *Aonidiella aurantii* is still a serious pest, but a decrease of infestations has been noticed, coincident with an increase in parasitism by the introduced *Aphytis melinus*. Meanwhile, that parasite has effected complete biological control of *C. dictyospermi* and the oleander scale, *Aspidiotus nerii* Bouché, on citrus. *P. perniciosi* and *C. bifasciata* were introduced from California in 1969 but failed to become established. *C. bifasciata* was again introduced from France in 1972 (Argyriou, 1975).

S. oleae has been one of the most serious pests of olive and citrus. Attempts at biological control started in 1962 (Argyriou & DeBach, 1968) when *M. helvolus* was introduced from California into Crete and the Peloponnesus. Releases of the parasite continued in 1968–9, and establishment was confirmed during 1969–71 (Argyriou, 1975).

Turkey

Studies on citrus scale insects and their natural enemies indicated that a considerable reduction in the populations of *C. dictyospermi*, *Aonidiella aurantii* and *A. citrina* occurred because of the combined action of the indigenous *Aspidiotiphagus citrinus* and *Aphytis melinus*, recently established fortuitously from Greece (Tunçyürek & Oncuer, 1975). *A. melinus* and *A. lingnanensis* were imported from California in 1971 and 1972, and the latter species has now become established.

P. *DeBach & D. Rosen*

Cyprus

Aonidiella aurantii is considered to be a key pest of citrus in Cyprus. Several species of *Aphytis* were introduced between 1959 and 1961 (Wood, 1963). In 1968, *Chilocorus circumdatus* Schon and *C. houseri* Weise were imported from India and released in the Morphou citrus area, but have not been recovered. In the same year, DeBach (personal communication) found *A. lingnanensis*, *A. melinus* and *A. coheni* to be well established and abundant in populations of *Aonidiella aurantii*. Also by 1968 the previously imported *Aphytis lepidosaphes* had reduced *L. beckii* populations to negligible proportions (DeBach, personal communication).

France

C. bifasciata and *P. perniciosi* were imported from California in 1971, and a culture of the former species has been maintained for release against *Aonidiella aurantii*. A study on the population dynamics of *L. beckii* is under way in the Alpes Maritimes. Attempts to introduce *A. lepidosaphes* from California in 1971 have apparently failed. *Q. perniciosus* has been a serious pest of deciduous fruit trees in France as well as several other countries in western and central Europe. Partial biological control resulted from the establishment of *P. perniciosi* (Rosen & DeBach, 1975b).

M. helvolus was introduced from California into the French Riviera in 1969 and into Corsica in 1971 against *S. oleae* (Panis, 1975).

Aleurothrixus floccosus has recently invaded southern France. Attempts to introduce *E. paulistus* from California in 1969 failed. In 1970, parasitised *A. floccosus* were sent from Peru, and in 1971 and 1972 shipments of *E. paulistus*, *Amitus spiniferus* and *Cales noacki* were made from California. The last species was established in 1971 (Onillon & Onillon, 1975).

Spain

Aleurothrixus floccosus has recently invaded southern Spain and has become a serious threat to the entire citrus industry, which is one of the mainstays of the Spanish economy. *E. paulistus*, *Amitus spiniferus* and *C. noacki* have been introduced from California in 1970 and 1971, and the last two have reportedly become established. The parasites were also introduced from California into the Canary Islands in 1971 against the same pest.

Italy

Aphytis melinus and *A. lingnanensis* were introduced from California into Sicily and the former species has become established on *C. dictyospermi* as well as on *Aonidiella aurantii* and *Aspidiotus nerii* on citrus. *C. dictyospermi* has apparently been reduced to low levels by the introduced parasite. *M. helvolus* was introduced from California in 1970 and 1971 and has become established on *S. oleae* on olive in Sicily, but no control has resulted as yet.

USSR

The 'Persian' *Aphytis maculicornis* was introduced from California into Georgia, USSR, in 1970 and 1971 against *P. oleae* and has apparently successfully overwintered and established in Tbilisi.

Morocco

A. lingnanensis and *A. melinus* were introduced in 1966 and established on *Aonidiella aurantii* (Bénassy & Euverte, 1967; Bénassy, 1969; Euverte, 1975). *Aphytis melinus* appears to be the more effective.

Tunisia

Studies on the biology and natural control of *S. oleae* have been carried out by Jarraya (1973). Indigenous natural enemies do not exert satisfactory control.

Africa south of the Sahara
South Africa

There is now ample evidence that the indigenous *Aphytis africanus* Qued. was originally the key parasite for the biological control of *Aonidiella aurantii* on citrus in South Africa and that it is capable of keeping it under satisfactory control in many groves that are under an integrated control programme. Several parasites and predators of *A. aurantii* have been introduced into South Africa in recent years (Annecke, 1969). *Aphytis lingnanensis*, *A. melinus* and *A. coheni* were released in large numbers in various localities. Although there were some initial recoveries soon after liberation, the species apparently did not become established and, until recently, the native *A. africanus* was thought to be still the major parasite. Subsequently *A. melinus* and *A.* nr. *lingnanensis* or *coheni* were unexpectedly recovered during 1971 to 1973. Preliminary surveys in 1973 indicate that *A. melinus* is present in all Zululand orchards (northern Natal). This is a surprise as none were liberated there, but

P. DeBach & D. Rosen

it once again demonstrates the remarkable dispersal ability of this species. In the hotter Transvaal Lowveld the proportions of *A. melinus* range from 70 % to 20 % and of *A. africanus* from 30 % to 80 % in individual orchards, with an occasional *A. lingnanensis* or *coheni* being present. A recent large sample from milder Nelspruit yielded about 75 % *A. melinus* and 25 % *A. africanus*. It would appear that *A. melinus* may be displacing *A. africanus* in certain areas of South Africa as it has other *Aphytis* in several other countries.

The biparental red scale strain of *C. bifasciata*, which was introduced from Australia in 1966, seems to be established in the Rustenburg area, but it is very scarce when infestations of *A. aurantii* are very low as a result of the activity of *A. africanus* which remains a superior parasite. *C. bifasciata* is also established in the Transvaal, as well as in Cape Province. The coccinellid, *Chilocorus cacti* (L.), introduced from Texas in 1966, has been remarkably effective against very heavy infestations of *Aonidiella aurantii* at both Rustenburg and Citrusdal, but it remains to be seen whether it will become permanently established as it seems to disappear once the scale is under control.

Chrysomphalus aonidum has been brought under excellent biological control by the introduced *Aphytis holoxanthus* which was established by 1972 in all the major citrus-growing regions.

L. beckii is mainly a pest in the Eastern Cape Province. The introduced *A. lepidosaphes* has proved to be just as effective as *A. holoxanthus* was on *C. aonidum*. Once established, *A. lepidosaphes* brought *L. beckii* under effective biological control within two years. High parasitism and predatory host feeding by *A. lepidosaphes* have been the main factors in the drastic decline of living scales (de Villiers, 1970).

Most of the main citrus pests can now be kept under satisfactory biological control in South Africa. Effective integrated control programmes have been developed for various citrus-growing regions, based on selective sprays, combined with ant control and, when necessary, recolonisation of the various parasites (Bedford, 1973). The beautifully integrated control programme in the Rustenburg high veld was seriously disrupted as a result of large scale use of Abate instead of the more selective tartar emetic bait for an unusually heavy citrus thrips outbreak in 1972. The few growers who didn't change programmes had groves clean of scales in 1973. Adjacent groves receiving the 1972 Abate treatment show increased populations of *Aonidiella aurantii*, *L. beckii*, brown soft scale and sooty mould fungus. At Letaba Estates there has been a population explosion of *A. aurantii* with an abnormal increase of some minor pests as well, as a result of a parathion spray programme plus aerial spraying with dimethoate. Up to four sprays within a year have been needed to hold *A. aurantii*.

Swaziland

A. aurantii is the key pest of citrus in Swaziland. Native parasites are capable of controlling it in certain areas, and in 1969 were complemented by *C. bifasciata* and *Chilocorus cacti* introduced from western Transvaal. The parasite became established and appears very promising (Catling, 1971). In 1972, *Aphytis melinus* was introduced from Israel.

The Orient

Japan

The Arrowhead scale, *Unaspis yanonensis* (Kuw.), is one of the most serious pests of citrus in Japan. Observations on its indigenous predators and parasites were reported by Yasumatsu (1971), Murakami *et al.* (1972) and Murakami (1975). In 1971–2 *A. lingnanensis* was introduced from Hong Kong.

Exchange of natural enemies

Live cultures of numerous natural enemies of armoured scales and related homopterous pests involving many thousands of specimens have been exchanged between various cooperating countries during the course of the IBP project. Following is a partial list of major natural-enemy shipments, arranged in alphabetical order.

Shipments from Australia
Aphytis chilensis: to California (1967, 1971).
A. holoxanthus: to California (1971).
A. lepidosaphes: to California (1968).
A. nr. *lingnanensis*: to California (1971).
Comperiella bifasciata: to California (1971).
Prospaltella perniciosi (red scale strain): to California (1971).

Shipments from Brazil
Encarsia sp.: to California (1971, 1972).
Eretmocerus sp.: to California (1971).
Prospaltella braziliensis (Hemp.): to California (1971).
Prospaltella sp.: to California (1971).
Scymnus (*Nephus*) sp.: to California (1971).
Signiphora sp.: to California (1971).

Shipments from California
Amitus spiniferus: to Spain (1970, 1971), France (1971, 1972), Canary Islands (1971) and Peru (1971, 1972).
Anagyrus pseudococci (Gir.): to Greece (1969) and Iran (1969).

P. DeBach & D. Rosen

Aphytis holoxanthus: to Argentina (1970), Mexico (1972) and Peru (1965, 1967).
A. lepidosaphes: to Mexico (1969), Spain (1969, 1970, 1971), Iran (1970), France (1970, 1971), Israel (1972), Argentina (1973) and Morocco (1973).
A. lingnanensis 'CC': to Turkey (1968, 1969, 1971) and Argentina (1971.)
A. lingnanensis 'CI': to Spain (1969, 1970, 1971), Iran (1970) and Turkey (1971, 1972, 1973).
A. maculicornis: to Pakistan (1968), USSR (1969, 1970) and Israel (1970, 1971, 1972).
A. maculicornis, 'Escondido form': to Israel (1968).
A. melinus: to Iran (1969), Spain (1969, 1970, 1971), Morocco (1970), Peru (1970), Yugoslavia (1970), Turkey (1971, 1973) and Egypt (1972).
A. mytilaspidis: to Israel (1972).
A. nr. *chrysomphali*: to Israel (1972).
Aphytis sp. '2002': to Pakistan (1968).
Cales noacki: to Spain (1970, 1971), France (1971, 1972) and Canary Islands (1971).
Chilocorus sp.: to France (1967).
Comperiella bifasciata: to Mexico (1968), Greece (1969), Argentina (1971) and France (1971).
Cryptochaetum iceryae: to Yugoslavia (1970).
Cryptolaemus montrouzieri Muls.: to Peru (1967).
Eretmocerus paulistus: to France (1969, 1971, 1972), Baja California, Mexico (1969), Spain (1971) and Canary Islands (1971).
Leptomastix dactylopii How.: to Greece (1969), Iran (1969), France (1971).
Metaphycus helvolus: to Greece (1968, 1969), France (1969, 1971), Iran (1969), Israel (1970, 1971, 1972), Italy (1970, 1971) and Yugoslavia (1970).
Prospaltella perniciosi: to Mexico (1968), Chile (1969), Greece (1969), Australia (1970), Israel (1970), France (1971) and Argentina (1971).
Prospaltella sp.: to Chile (1970).
Signiphora boringuensis Quez., DeBach & Rosen: to Pakistan (1968).

Shipments from the Canary Islands
Delphastus sp.: to California (1970).

Shipments from Chile
Amitus spiniferus: to California (1971, 1972).
Cales noacki: to California (1970, 1971, 1972), France (1970) and Spain (1970).

Shipments from El Salvador
Amitus spiniferus: to California (1970, 1972).
Aphytis nr. *lingnanensis*: to California (1970, 1972).
Encarsia sp.: to California (1970).
Eretmocerus clauseni Comp.: to California (1970, 1972).
Prospaltella sp.: to California (1970, 1972).
Signiphora sp.: to California (1970).

172

Shipments from Florida
Encarsia sp.: to California (1968).
Eretmocerus sp.: to California (1968).

Shipments from Greece
A. sp. nr. *chrysomphali*: to California (1971).
Aphytis sp. (*mytilaspidis* group): to California (1970, 1971).
Aphytis 'Z': to California (1971, 1972).
Prospaltella sp.: to California (1970, 1971, 1972).

Shipments from Hong Kong
A. diaspidis: to California (1972).
A. lepidosaphes: to California (1972).
A. lingnanensis: to California (1971, 1972) and Florida (1971, 1972).
A. nr. *fisheri*: to California (1972).
A. nr. *funicularis*: to California (1971, 1972) and Florida (1971, 1972).
Aspidiotiphagus ?citrinus: to California (1971, 1972) and Florida (1971, 1972).
Encarsia sp.: to California (1971).
Eretmocerus sp.: to California (1972).
Prospaltella sp.: to California (1971, 1972) and Florida (1971, 1972).
Pteroptrix chinensis (How.): to California (1971).
Pteroptrix wanhsiensis (Comp.): to California (1971).

Shipments from India
Catana chapini Kapur: to California (1972).
Catana parcesetosum Sic.: to California (1972).
Catana perdistinctus: to California (1972).
Chrysopa sp.: to California (1972).
Eretmocerus sp.: to California (1969).
Prospaltella 'O': to California (1969, 1970).
Prospaltella 'RJ': to California (1969, 1970).

Shipments from Israel
A. coheni: to Cyprus (1969).
A. melinus: to Cyprus (1969), South Africa (1972) and Swaziland (1972).
A. nr. *aonidiae*: to California (1970).
A. nr. *mytilaspidis*: to California (1970).
Compariella bifasciata: to Italy (Sicily) (1972).
Hungariella peregrina (Comp.): to Australia (1972, 1973).

Shipments from Jamaica
L. dactylopii: to California (1968).

Shipments from Japan
A. nr. *melinus*: to California (1971).
Coccophagus yoshidae Nak.: to California (1969).
Comperiella bifasciata: to California (1968, 1969, 1971, 1972).
Encarsia sp.: to California (1972).
Eretmocerus sp.: to California (1969, 1971, 1972).
Prospaltella spp.: to California (1968, 1969, 1970, 1971, 1972).

Shipments from Mexico
A. spiniferus: to California (1967–72).
Aphytis spp.: to California (1967, 1969).
Encarsia nr. *basicincta* Gah.: to California (1967–72).
E. paulistus: to California (1967–72).
Prospaltella sp.: to California (1967, 1969).

Shipments from New Caledonia
Aphytis n. sp.: to California (1968).

Shipments from Pakistan
Aneristus ceroplastae: to California (1968).
Aphytis spp.: to California (1968, 1970).
Comperiella spp.: to California (1968).
Encarsia sp.: to California (1970).
Eretmocerus 'PAK': to California (1970).
Prospaltella lahorensis: to California (1968, 1969, 1970).

Shipments from Philippines
Prospaltella sp.: to California (1968).

Shipments from South Africa
Chilocorus cacti: to Rhodesia (1972).
Comperiella bifasciata: to Swaziland (1969, 1970).
M. ?aspidiotinorum: to Peru (1970).

Shipments from Tahiti
A. chrysomphali: to California (1972).

Shipments from Texas
A. nr. *hispanicus*: to California (1969).

World list

One of the major aims of the IBP Project was to develop a census list of the principal armoured scale insect pests of the world and a list of their natural enemies. This information has never been summarised before. It is given in Tables 19 and 20.

Table 19. *Principal armoured scale insect pests of the world*

Tribe	Scientific name	Common name
Aspioditini		
1	*Aonidiella aurantii* (Mask.)	California red scale
2	*A. citrina* (Coq.)	Yellow scale
3	*A. orientalis* (Newst.)	Oriental red scale
4	*Aspidiotus destructor* Sign.	Coconut scale
5	*A. nerii* (Bouché) [= *A. hederae* (Vallot)]	Oleander scale
6	*Chrysomphalus dictyospermi* (Morg.)	Dictyospermum scale
7	*C. aonidum* (L.) [= *C. ficus* (Ashm.)]	Florida red scale
8	*Hemiberlesia lataniae* (Sign.)	Latania scale
9	*H. rapax* (Comst.)	Greedy scale
10	*Pseudaonidia duplex* (Ckll.)	Camphor scale
11	*Quadraspidiotus forbesi* (Johns.)	Forbes scale
12	*Q. juglansregiae* (Comst.)	Walnut scale
13	*Q. ostreaeformis* (Curt.)	European fruit scale
14	*Q. perniciosus* (Comst.)	San Jose scale
15	*Q. pyri* (Licht.)	False San Jose scale
16	*Selenaspidus articulatus* (Morg.)	Rufous scale
Parlatorini		
17	*Parlatoria blanchardi* (T. T.)	Parlatoria date scale
18	*P. cinerea* Hadd.	Tropical grey chaff scale
19	*P. oleae* (Colvée)	Olive scale
20	*P. pergandii* Comst.	Chaff scale
21	*P. theae* Ckll.	Parlatoria tea scale
22	*P. ziziphus* (Lucas)	Black parlatoria scale
Diaspidini		
23	*Aulacaspis rosae* (Bouché)	Rose scale
24	*Carulaspis minima* (T. T.) [= *C. caruelii* (T. T.) of authors]	Minute juniper scale
25	*Chionaspis salicis* (L.)	Willow scale
26	*Diaspis boisduvalii* Sign.	Boisduval scale
27	*D. echinocacti* (Bouché)	Cactus scale
28	*Epidiaspis leperii* (Sign.)	Italian pear scale
29	*Fiorinia theae* Green	Tea scale
30	*Lepidosaphes beckii* (Newm.)	Purple scale
31	*L. ficus* (Sign.) [= *L. conchiformis* (Gmel.) of authors]	Fig scale
32	*L. gloverii* (Pack.)	Glover scale
33	*L. malicola* Borkh.	
34	*L. pistaciae* Arch.	
35	*L. pistacicola* Borkh.	
36	*L. ulmi* (L.)	Oystershell scale
37	*Miscanthaspis tegalensis* (Zehnt.)	Sugarcane scale
38	*Phenacaspis pinifoliae* (Fitch)	Pine needle scale
39	*Pinnaspis aspidistrae* (Sign.)	Aspidistra scale
40	*P. strachani* (Cooley) [= *P. minor* (Mask.) of authors]	Lesser snow scale
41	*Pseudaulacaspis pentagona* (T. T.)	White peach scale
42	*Tecaspis asiatica* (Arch.)	
43	*T. prunorum* (Borkh.)	
44	*Unaspis citri* (Comst.)	Citrus snow scale
45	*U. euonymi* (Comst.)	Euonymus scale
46	*U. yanonensis* (Kuw.)	Arrowhead scale
Odonaspidini		
47	*Rugaspidiotus tamaricicola* (Malen.)	

Table 20. *A preliminary list of the most effective known natural enemies of selected armoured scale insects*

Armoured scale species	Principal natural enemies		
	Name	Type	Source
Aonidiella aurantii	*Aphytis africanus* Qued.	Parasite	South Africa
(Mask.)	*A. chrysomphali* (Mercet)	Parasite	Mediterranean?
California red scale	*A. coheni* DeBach	Parasite	Israel?
	A. lingnanensis Comp.	Parasite	South China, Hong Kong
	A. melinus DeBach	Parasite	India, Pakistan
	Comperiella bifasciata How.	Parasite	South China, Hong Kong
	Habrolepis rouxi Comp.	Parasite	South Africa
	Prospaltella perniciosi Tower (red scale form)	Parasite	South China, Hong Kong
A. citrina (Coq.)	*A. chrysomphali* (Mercet)	Parasite	Cosmopolitan
Yellow scale	*A. melinus* DeBach	Parasite	Pakistan
	C. bifasciata How.	Parasite	Japan
A. orientalis (Newst.) Oriental red scale	*A. melinus* DeBach	Parasite	India, Pakistan
Aspidiotus destructor Signoret	*A. lingnanensis* Comp.	Parasite	Cosmopolitan, Orient
Coconut scale	*Cryptognatha nodiceps* Marsh.	Predator	Trinidad
	Lindorus lophanthae (Blaisd.)	Predator	Australia
A. nerii Bouché [= *A. hederae* (Vallot)]	*A. chilensis* How.	Parasite	Cosmopolitan, Mediterranean
Oleander scale	*A. melinus* DeBach	Parasite	India, Pakistan
Aulacaspis rosae (Bouché)	*A. mytilaspidis* (LeBaron)	Parasite	Mediterranean
Rose scale	*A. proclia* (Walk.)	Parasite	Mediterranean
Carulaspis minima (T.T.)	*Aphytis aonidiae* (Mercet)	Parasite	Mediterranean
Minute juniper scale	*Aphytis mytilaspidis* (LeBaron)	Parasite	Mediterranean
Chionaspis salicis (L.) Willow scale	*A. proclia* (Walk.)	Parasite	England
Chrysomphalus aonidum (L.) [= *C. ficus* (Ashm.)]	*Aphytis costalimai* (Gomes)	Parasite	Brazil
	Aphytis holoxanthus (DeBach)	Parasite	Hong Kong
	Aphytis n.sp.	Parasite	Philippines
Florida red scale	*Comperiella pia* (Gir.)	Parasite	Queensland
	Habrolepis aspidioti Comp. & Ann. (= *H. fanari* Del. & Trab.)	Parasite	South Africa, Mediterranean?
	Habrolepis rouxi Comp.	Parasite	South Africa
	Pteroptrix smithi (Comp.)	Parasite	Hong Kong, Israel
Chrysomphalus dictyospermi (Morg.)	*A. chrysomphali* (Mercet)	Parasite	Mediterranean
Dictyospermum scale	*A. melinus* DeBach	Parasite	India, Pakistan
Diaspis boisduvalii Sign.	*Aspidiotiphagus citrinus* (Craw)	Parasite	Java
Boisduval scale	*Coccidencyrtus ochraceipes* (Gahan)	Parasite	United States, Bermuda
	Signiphora coquilletti Ashm.	Parasite	United States

Table 20 (*cont.*)

| Armoured scale species | Principal natural enemies | | |
	Name	Type	Source
Hemiberlesia lataniae (Sign.) Latania scale	*Aphytis diaspidis* (How.)	Parasite	Cosmopolitan
	A. mytilaspidis (LeBaron)	Parasite	Mediterranean
	Signiphora spp.	Parasite	Cosmopolitan, California
H. rapax (Comst.) Greedy scale	*A. diaspidis* (How.)	Parasite	California, cosmopolitan
	Signiphora spp.	Parasite	California, cosmopolitan
Lepidosaphes beckii (Newm.) Purple scale	*A. lepidosaphes* Comp.	Parasite	Hong Kong
L. ficus (Sign.) Fig scale	*A. mytilaspidis* (LeBaron)	Parasite	Italy
	Physcus testaceus Masi	Parasite	Italy
L. gloverii (Pack.) Glover scale	*A. hispanicus* (Mercet)	Parasite	Mexico
	A. lingnanensis Comp.	Parasite	Mexico, Orient
	Prospaltella elongata Dozier	Parasite	Puerto Rico?
L. malicola Borkh.	*Physcus testaceus* Masi	Parasite	Iran
L. ulmi (L.) Oystershell scale	*A. mytilaspidis* (LeBaron)	Parasite	Mediterranean
	Physcus testaceus Masi	Parasite	Mediterranean
	Prospaltella spp.	Parasite	Cosmopolitan
Miscanthaspis tegalensis (Zehnt.) Sugarcane scale	*Adelencyrtus miyarai* Tach.	Parasite	South Africa
	Physcus sp. nr. *seminotus* Silv.	Parasite	South Africa
	Physcus subflavus Ann. & Ins.	Parasite	South Africa
Parlatoria blanchardi (T.T.) Parlatoria date scale	*Aphytis* n.sp.	Parasite	Middle East
	Aphytis spp.	Parasite	Middle East
	Chilocorus bipustulatus L.	Predator	Iran
	Pharoscymnus numidicus Pic	Predator	North Africa
Parlatoria oleae (Colvée) Olive scale	*A. maculicornis* (Masi) (Persian form)	Parasite	Iran
	Coccophagoides utilis Doutt	Parasite	Pakistan
P. pergandii Comst. Chaff scale	*A. hispanicus* (Mercet)	Parasite	Cosmopolitan
	Prospaltella inquirenda Silv.	Parasite	Cosmopolitan
Phenacaspis pinifoliae (Fitch) Pine needle scale	*A. mytilaspidis* (LeBaron)	Parasite	California
	Prospaltella bella Gah.	Parasite	California
	Achrysocharis phenacapsia Yoshimoto	Parasite	California
Pinnaspis aspidistrae (Sign.) Aspidistra scale	*Aphytis mytilaspidis* (LeBaron)	Parasite	Cosmopolitan
	Aspidiotiphagus spp.	Parasite	Cosmopolitan
Pseudaonidia duplex (Ckll.) Camphor scale	*Aphytis cylindratus* Comp.	Parasite	Japan, Brazil
	Prospaltella spp.	Parasite	Cosmopolitan
Pseudaulacaspis pentagona (T.T.) White peach scale	*A. proclia* (Walk.)	Parasite	Cosmopolitan
	Prospaltella berlesei (How.)	Parasite	Cosmopolitan
Quadraspidiotus forbesi (Johns.) Forbes scale	*Prospaltella* spp.	Parasite	Cosmopolitan

Table 20 (*cont.*)

Armoured scale species	Principal natural enemies		
	Name	Type	Source
Q. juglansregiae (Comst.)	*A. mytilospidis* (LeBaron)	Parasite	California
	A. diaspidis (How.)	Parasite	Cosmopolitan
English walnut scale	*Coccidencyrtus ensifer* (How.)	Parasite	California
Q. ostreaeformis (Curt.)	*A. mytilaspidis* (LeBaron)	Parasite	Cosmopolitan
	A. proclia (Walk.)	Parasite	Cosmopolitan
European fruit scale	*Anabrolepis zetterstedtii* (Westw.)	Parasite	United States
Q. perniciosus (Comst.)	*Aphytis diaspidis* (How.)	Parasite	Cosmopolitan
	A. mytilaspidis (LeBaron)	Parasite	Cosmopolitan
San Jose scale	*Adelencyrtus inglisiae* Comp. & Ann.	Parasite	South Africa
	Habrolepis obscura Comp. & Ann.	Parasite	South Africa
	Prospaltella perniciosi Tower (San Jose scale form)	Parasite	California (Orient)
Selenaspidis articulatus (Morg.) Rufous scale	*Aphytis* n.sp.	Parasite	Uganda
U. citri (Comst.) Citrus snow scale	*A. lingnanensis* Comp.	Parasite	Hong Kong
U. euonymi (Comst.) Euonymous scale	*A. proclia* (Walk.)	Parasite	?

ACKNOWLEDGEMENTS

We wish to acknowledge the support of the following National Science Foundation grants: GB–30843X, 'The Role of Entomophagous Insects in Population Regulation of Armored Scale Insects in Major World Ecosystems'; GB–17829, 'Biosystematics and Phylogeny of Species of *Aphytis* (Hymenoptera: Chalcidoidea: Aphelinidae) of the World'; and GB–34718, 'The Principles, Strategies, and Tactics of Pest Population Regulation and Control in the Citrus Ecosystem'.

7. Spider mites

Coordinators: N. W. HUSSEY & C. B. HUFFAKER

In July 1967 a dozen specialists concerned with the biological control of spider mites or tetranychids met at Sutton Bonnington, England, to discuss ways in which this subject could be fostered under the auspices of the International Biological Programme.

It was clear that most of the work would have to be achieved within existing institutional programmes without additional funds. Consequently three areas of work were chosen that lent themselves to pursuit by existing personnel and research programmes. The first area was the development and compilation of information on the basic ecology and behavioural characteristics of the more important tetranychids and their natural enemies. The second area was research on the interactions between spider mites and their enemies in the controlled environments of growth cabinets or glasshouses. Studies on the effects of pesticides and cultural conditions on spider mite/natural enemy interactions constituted the third area. Huffaker, Van de Vrie & McMurtry (1969, 1970), McMurtry, Huffaker & Van de Vrie (1970), and Van de Vrie, McMurtry & Huffaker (1972) reviewed the current state of knowledge of the ecology of spider mites and their natural enemies as a starting point to this concerted effort.

The past four years have seen considerable progress in this important area of research and, by keeping those concerned aware of current developments, by the reorientation of some efforts, and by the provision of IBP-sponsored funds for new research in some instances, the IBP deserves a partial credit for this progress. This is especially true of the rapid increase in our knowledge of the effects of pesticides on phytoseiid mites, and of the importance of predators as exemplified by the commercial control of *Tetranychus urticae* (Koch) in glasshouses by *Phytoseiulus persimilis* Athias-Henriot.

Contributors: J. Boczek, L. C. Caltagirone, D. J. Calvert, F. Chaboussou, E. Collyer, Z. T. Dabrowski, E. A. Elbadry, D. L. Flaherty, W. Helle, H. J. Herbert, J. Hobart, S. C. Hoyt, C. B. Huffaker, N. W. Hussey, C. E. Kennett, J. E. Laing, J. A. McMurtry, E. Niemczyk, E. R. Oatman, W. P. S. Overmeer, L. Readshaw, J. G. Rodriguez, K. H. Sanford.

Biology of spider mites

Genetics

A comprehensive summary of the genetic variation of tetranychids was published by Helle & Overmeer (1973).

Genetic studies were mostly on *T. urticae* and *T. pacificus* McGregor largely because these species are easily reared. Chromosomal studies have been made on a number of genera and more than 50 species of Tetranychidae. The number of chromosomes ranges from $n = 2$ to $n = 7$. Important economic species examined usually have $n = 3$, with the exception of *T. tumidus* Banks, where $n = 6$ (Helle & Bolland, 1967; Helle, Gutierrez & Bolland, 1970). In mitotic stages the chromosomes are very small, and they do not exhibit special features. In *T. urticae*, as apparently in other tetranychids, the behaviour of the chromosomes during meiosis suggests a holokinetic nature (Pijnacker & Ferwerda, 1972). Radiation experiments provide further evidence that centromeres are not localised in view of the autonomous behaviour of the chromosomal fragments during mitosis.

All bisexual species so far examined exhibit a haplo-diploid sex determination. It is believed that unfertilised females of all species produce eggs, although in some the mating act stimulates egg production considerably (Gutierrez & Van Zon, 1973). Visible mutations have been found in some *Tetranychus* spp. The pattern of inheritance of these mutations is in accordance with haplo-diploidy (Helle, 1969a; Helle & Van Zon, 1970; Van Zon & Helle, 1967).

The sex ratio is not relevant to the amount of concealed genetic variability in a population (Crozier, 1970) but it influences the effective population size. Genetic drift in isolated populations increases with increases in the proportion of males. Genetic variation with respect to factors controlling the sex ratio of *T. urticae* was reported by Overmeer & Harrison (1969) and Mitchell (1972). Matings between lines with low and high sex ratios showed that the female determines the sex ratio of her offspring irrespective of her mate. Mitchell (1972) reported that control of sex ratio was polygenic, whereas Overmeer & Harrison (1969) found that a maternally inherited factor was involved.

The mating system of spider mites is a relatively unexplored field and deserves special attention with respect to the potential of applying genetic controls. Within a population mating is usually not random; sib mating presumably occurs frequently. It is clear that at a high degree of inbreeding genetic drift of populations will be enhanced. Studies aimed at the genetic control of spider mites (Smith, Boswell & Webb, 1969; Dieleman & Overmeer, 1972) revealed that genotypic assortative mating probably occurs regularly in this group. Occasionally the opposite was found, in that the allotypic males were reproductively more effective than the autotypic ones.

In accordance with theory, reproductive barriers within populations of a single species are common, e.g. in *T. urticae* (Boudreaux, 1963; Helle & Pieterse, 1965) *T. neocaledonicus* André (Gutierrez & Van Zon, 1973), and *Panonychus citri* McGregor (Boudreaux, 1963; Tanaka, Inoue & Kita, 1969) suggested that chromosomal rearrangements may underlie such complete or partial genetic incompatibilities. Usually, hybrid sterility or breakdown occurs, although hybrid non-viability in intra-specific crosses is known in *T. lombardini* Baker & Pritchard.

Van Zon & Overmeer (1972) showed that chromosome rearrangements could easily be artificially induced and fixed in a strain. Fixation of chromosome mutations is relatively easily accomplished in spider mites, both because of the haplo-diploid sex determination and also because of the possibility of mother × son matings. This suggests that different chromosomal races will arise frequently and spontaneously in nature.

The fact that interpopulation crosses within a species so often result in hybrid breakdown has stimulated work to determine if such a phenomenon might be useful for developing a non-chemical method of spider mite control (Helle, 1969 b). Genetic control of glasshouse populations through release of large numbers of males of an appropriate type might be feasible. Such a method might be desirable if the glasshouse population were resistant to acaricides. A prerequisite for genetic control is that the released males be reproductively incompatible with the endemic glasshouse strain. The method resembles the sterile male technique, although in this case the released males need not be treated. As several mutations for resistance to a number of acaricides are available, the males to be released can be provided with a resistance distinct from that of the glasshouse population. By genetic techniques it is also possible to provide these males with the desired genetic incompatibility and mating preference. By combining acaricide treatment and release of these males, complete eradication of a glasshouse strain might be accomplished. Recent studies have suggested the feasibility of such control, but have also revealed that resulting hybrid populations, though sterile, can cause considerable damage to crops.

Nutrition

The nutrition of spider mites has been much explored in recent years as a means of better understanding the role of the host plant physiology in the dynamics of pest species and the resistance or susceptibility of crop varieties to them, and to develop methods of culturing spider mites in the laboratory.

Early clues to the importance of host plant physiology in spider mite physiology and population dynamics came from laboratory studies on fertilisers and pesticides. Nitrogen absorption, for example, was shown to correlate positively with increased fecundity of some tetranychids. This work

181

was reviewed by Rodriguez (1964). Stoltz *et al.* (1970) showed that the total non-essential amino acids of strawberry clones was positively correlated with nitrogen, phosphorus and potassium treatments and with the clones *per se*. Essential amino acids, on the other hand, were not significantly correlated with either treatment or clone.

The work of Storms (1965, 1969) and Storms, Harrewijn & Noordink (1967) on host plant phsyiology and spider mite relationships is noteworthy. Storms was aware of ion ratio interactions and used this knowledge to devise methods and special apparatus to improve the artificial culture of plants. In his study on the relationship between mineral nutrition of apple rootstocks and reproduction of *T. urticae* he confirmed other findings (Rodriguez, 1964) that an increase in the total nitrogen-content in the leaves can increase egg production of spider mites. A comparison test using beans showed that the nitrogen–mite relationship was similar to that for apples. Females reared on leaves with a high nitrogen content lived 25 % longer than those reared on foliage of low nitrogen content.

Development of feeding systems

All existing publications on the development of an artificial feeding system and diet for tetranychids concern *T. urticae*. The mechanics of mite feeding presents a difficult problem because a liquid diet must be supplied via a membrane. The physical requirements of the membrane are difficult to meet and a suitable one has not yet been developed. The membrane is required, of course, to support the mites as well as to substitute for the leaf surface in the mechanism of feeding.

A flexible collodion membrane was used in the early studies (Rodriguez, Singh Pritam Seay & Walling, 1967). Walling, White & Rodriguez (1968), in their studies with fatty acids, used resin membrane made from Butvar B-76. More recently, a parafilm membrane has been used (Ekka, Rodriguez & Davis, 1971; Storms & Noordink, 1970).

Nutritional requirements

The composition of the amino acid fraction of an early test diet reflected that of bean leaf and *T. urticae* tissues (Rodriguez & Hampton, 1966). When the mites were fed a complete diet, labelled with [U-^{14}C]glucose for 24 h, and then analysed for amino acids, it was found that relatively high labelling was encountered in alanine, aspartic acid, cysteic acid, cystine, glutamic acid, glycine, proline, serine, and threonine. This indicated that the mite was capable of synthesising these amino acids from glucose and hence that they are nutritionally non-essential. Arginine, histidine, isoleucine, leucine, lysine, methionine, phenylalanine, tyrosine, and valine showed sufficiently low activity to be classified as essential in the diet of *T. urticae*. Because acid hydrolysis in sample preparation destroys tryptophan the essentiality of this amino acid was not determined; it is most probably essential (Rodriguez & Hampton, 1966). Moreover, later work showed that threonine separation (from serine) is not complete and that this amino acid should be classified as essential. These findings agree with what is known of the qualitative amino acid requirements of insects and also, for example, for the rat and the dog.

Fatty acid analyses of nymphal and adult mites taken from bean leaf disks showed that linolenic acid was the most abundant acid in every stage (Walling *et al.*, 1968). Other major acids were palmitic, palmitoleic, stearic, oleic, and linoleic. There was a remarkable similarity in the fatty acid composition of all growth stages: a progressive increase in the percentage of linolenic acid and a progressive decrease in the percentage of myristic, palmitic and palmitoleic acids.

The amount of fatty acid ranges from 0.5 to 1.5 % of the net weight of mites. Starvation for half the growth stage induced changes in the proportions of fatty acids, especially linolenic. The C_{18} series of fatty acids comprised about 84 % of the total present in the mites. The more unsaturated the C_{18} fatty acid the more rapid was its utilisation during starvation.

Substantial nutritional improvement in the lipid fraction of the diet was achieved by Ekka *et al.* (1971) who sought to alter the sterol/fatty acid fraction and to make other improvements that would enhance oviposition and viability of eggs. Walling *et al.* (1968) showed the existence of a prevalence of palmitic, stearic, oleic, linoleic, and linolenic acids, which comprised 91.8 % of fed and 86.9 % of starved adult females. Neither mite bodies nor bean leaf tissue yielded measurable cholesterol; however, leaf tissue yielded the phytosterols β-sitosterol, stigmasterol, and campesterol at 53, 46, and 5 % respectively of the total sterol in the bean leaf. The same composition of these sterols was found in the mites' bodies. In diets that tested sterols with fatty acids remaining constant, oviposition rate was increased by increasing the diet sterols to a level four times the original. The highest rate of oviposition occurred when the total concentration of fatty acids was half that of the original diet. Higher levels of fatty acids shortened the oviposition period. However, egg viability still remained low, even though adults survived more than 28 days. By increasing vitamin E, the oviposition rate increased from 0.343 to 0.649 egg per female per day and the oviposition period from 25 to 36 days. Later experimentation showed that it was not necessary to use more than 5 mg of vitamin E per 100 ml of diet. The pre-oviposition period was shortened from an average of 3.5 to 1.5 days and the viability of eggs was improved from 3.6 to 19.2 %.

Microbiological studies were conducted with *T. urticae* to ascertain whether there were associated micro-organisms and if so what were their functions (Sologic & Rodriguez, 1971). Twenty-five isolations were made from a greenhouse bean culture. Four organisms were found consistently and a fifth only occasionally. More than 30 generations of microbe-free mites were reared on bean seedlings with no apparent ill effects. Six antibiotic materials were incorporated into the diet at 1000 ppm without any harmful effects. It was concluded that the micro-organisms associated with *T. urticae* are not important to its well-being.

Diets and feeding behaviour

Storms & Noordink (1970) reported nutritional studies on *T. urticae* using ^{32}P as a label to measure ingestion during a 48-h period. By reducing the diet of Rodriguez & Hampton (1966) to $\frac{1}{16}$ its original concentration (excluding lipids) they reported relatively good uptake. More improvement came when they increased the pH to 10 and decreased the lipid fraction by half. Hence, the major fractions of this diet total 144 mg amino acids, 1500 mg sucrose, and 18.25 mg minerals per 100 ml water. Egg production was 2.1 eggs per female per 24 h. No data were offered on egg viability. The relatively high egg production was probably due to the fact that the young females that were transferred from leaf disks (even though they were starved for 24 h) had already fed on natural food. This also would account for the wide variability in oviposition that they mention. Although the diet was adequate to permit 'nymphal mites' to develop to adults, larvae would not moult when fed on it.

Feeding was stimulated in teneral females of *T. urticae* (Rodriguez *et al.*, 1967) through use of a green filter at 535 nm although protonymphs and teneral females differed in their reaction to wavelength. Food uptake was increased 24 % when a green filter was used instead of cool-white fluorescent light. It was noteworthy that protonymphs feeding upside down ingested 33 % more than those feeding upright, whereas in teneral mites the increase was only 16 %. Although it is possible that phototaxis to transmitted light influenced feeding, it is more plausible that hydrostatic pressure of the liquid against the membrane facilitated feeding.

Recently, the ingestion rates of individual amino acids were tested by deleting each from the complete diet fed to teneral *T. urticae* females. The diet was labelled with [U-^{14}C]glucose. When arginine, an essential amino acid, was omitted the intake was only about 15 % of the level on the complete diet at 6 h and 16 % at 24 h. Deletion of cysteine, aspartic and glutamic acid, the next three most influential but non-essential amino acids, reduced the intake to 18, 26, and 33 %, respectively, of the level of the complete diet. Table 21 gives the chemical composition of an improved diet. This diet is adequate to permit an occasional individual of *T. urticae* to develop from egg to adult, beginning with quiescent deutonymphs transferred to the parafilm-covered diet. Diet formulations and presentation have been reported elsewhere (Walling *et al.*, 1968; Ekka *et al.*, 1971).

Table 21. *Composition of basic synthetic diet per 100 ml diet (Ekka* et al., *1971)*

L-Amino Acids	(mg)	Water-soluble vitamins	(mg)
Alanine	150	Folic acid	0.34
Arginine	200	Inositol	6.00
Aspartic acid	200	p-Aminobenzoic acid	0.50
Cysteine	50	Pyridoxine HCl	0.10
Glutamic acid	180	Riboflavin	0.20
Glycine	200	Thiamine HCl	0.20
Histidine	100	Nicotinic acid	0.30
Isoleucine	100	Biotin	0.0001
Leucine	160	Ca-pantothenate	0.50
Lysine	120	Choline chloride	10.00
Methionine	60	B_{12}	0.002
Phenylalanine	60	Ascorbic acid	100.00
Proline	60	Lipids	(mg)
Serine	80	β-Sitosterol	0.50
Threonine	120	Stigmasterol	0.44
Tryptophan	40	Stearic acid	1.00
Tyrosine	50	Palmitic acid	0.50
Valine	120	Oleic acid	0.50
RNA	100	Linoleic acid	0.50
Sugars	(g)	Linolenic acid	3.00
Sucrose	3.0	Mineral salts	(mg)
Levulose	0.5	K_2HPO_4	37.50
Dextrose	0.5	$Na_2HPO_4.12H_2O$	10.60
Fat-soluble vitamins	(mg)	$MgSO_4.7H_2O$	15.60
A	3.0	$CaCl_2.2H_2O$	3.62
E	5.0	$CoCl_2.6H_2O$	0.08
Tween 80	10	NaFe	1.00
		NaMn	0.50
		NaCu	0.08
		NaZn	0.12

Hibernation

The seriousness of the spider mite problem in glasshouse cropping is governed by the number of mites overwintering in the structure. These hibernating populations create difficulties both by their numbers and by the pattern and timing of their reactivation in the Spring. Successful use of the predator *Phytoseiulus persimilis* on many British nurseries has confirmed these generalisations for it has virtually eliminated the summer population of spider mites on the growing crop so that none enter hibernation. Hence, in a following season, infestation of the new crops has been delayed for several months until mites were introduced from outside sources. Invasion is therefore much less important than hibernation.

Parr & Hussey (1966) reviewed the factors known to induce and terminate diapause in *T. urticae* and Hussey (1972) outlined the special implications of this hibernation for glasshouse culture.

From the viewpoint of control, the pattern of diapause termination is most important for both protected and outdoor crops, and the recent contributions of Helle & Overmeer (1973) and Geyspits, Sapozhinikova & Toranets (1971) are of special interest. It is clear that the widely-held belief that the diapausing sensitivity of a population of *T. urticae* is controlled only by external stimuli, such as photoperiod and temperature or food, is invalid. Dubynina (1965) first showed that diapause induction in successive generations is variable. When hibernation was induced in several different populations, no further diapause response was induced by short day lengths in the second generation, although, during later generations, the diapause tendency gradually returned. This response is apparently an adaptation to prevent premature hibernation in early spring when day length is still short. Similar results for both *T. urticae* and *Panonychus ulmi* (Koch) were reported by Razumova (1967). Geyspits *et al.* (1971), in a series of experiments at Leningrad, USSR, established nine photoperiodic response curves for a single mite population. Of these only one, obtained in February, was of the normal single-peak, long-day type (i.e. diapause is an adaption to winter conditions). All the rest were different although each exhibited two peaks typical of the short-day type. These peaks shifted with the season but one was consistently between 6 and 11 h, the other between 11 and 17 h. This response was further analysed at different temperatures. It was apparent that in January and March two weak peaks were formed at 25 °C, but at 18 °C a normal long-day response was exhibited. In August, on the other hand, the response was of the two-peak type at both temperatures. Geyspits *et al.* (1971) also investigated the effect of different reactivation conditions on the photoperiodic reaction of *T. urticae*. Diapausing mites were either reactivated by heat (25 °C) or following exposure to cold at 4 to 7 °C. The reactivated females laid eggs which were then exposed to different photoperiods. Despite the uniformity of the original material, two different reactions resulted. After cold there was a single peak, long-day response, whereas reactivation by heat produced a two-peaked curve similar to that normally associated with August conditions. The photoperiodic response is therefore dependent on the reactivation conditions affecting the preceding generation.

Such complexity of response shown by a single population, together with the cyclic reactivation from diapause as previously recorded by Dubynina (1965) when mites had entered hibernation in response to short photoperiods and low temperatures in contrast to the linear response to diapause induced by long photoperiods and food shortage, helps to explain the varied and complicated patterns of photoperiodic response commonly encountered in different populations of glasshouse red spider mite.

Biology and ecology of natural enemies of spider mites
Efficiency of spider mite predators

The efficiency of a predator is difficult to appraise in the absence of good methods to ascertain prey consumption in the field by a given predatory species. There are a number of general relationships concerning the efficiency of natural enemies in general which apply to analyses of predation on spider mites. The comparative powers of increase of predator and prey species and the prey consumption capabilities of the former are, of course, important. Moreover, specific behavioural interference in predator efficiency caused by disturbing effects of the prey has been postulated as rendering phytoseiids inefficient under high prey density situations. Each of these relationships is discussed here.

The broad potential of natural enemies for exerting economic control of tetranychids is suggested because before World War II spider mites were not major pests of crops and because they remain to this day unimportant where pesticides are not used (Huffaker *et al.*, 1970). There are two main hypotheses for explaining the changed situation: (i) an increased fecundity of spider mites arising from enhanced nutrition following improvements in fertilisation, pruning, and use of pesticides, and (ii) the adverse effects of modern pesticides on mite natural enemies. The two hypotheses are not mutually exclusive but much evidence suggests that the latter is sufficiently common to explain the more significant changes in status of tetranychids which embrace such a wide variety of crops, spider mites, and environments.

Evidence of the efficiency of particular predators in different situations has accumulated over many years from field studies as well as from numerous laboratory, greenhouse, and growth chamber experiments. The evidence derives largely from examples of control by predaceous insects and mites, and to a very limited degree pathogens, on such crops as apples and other deciduous fruits, grapes, strawberries, avocado, and citrus, and a complex of glasshouse plants. The genera *Tetranychus* and *Panonychus* have been most involved, but examples in the genera *Bryobia*, *Oligonychus*, *Eutetranychus*, and *Eotetranychus* are documented.

The enemies of these tetranychids are many, and include Acarina (especially Phytoseiidae), Araneida, Coleoptera, Diptera, Hemiptera, Neuroptera, Thysanoptera, and certain fungi (e.g. *Hirsutella* spp., *Entomophthora* spp.) and one or more viruses. Predators dominate in most situations, but virus or fungal disease may do so under certain conditions (McMurtry *et al.*, 1970). No bacteria or parasitoid-types of natural enemies have been reported.

There are five main characteristics of a natural enemy that determine its effectiveness: adaptability to the environmental conditions, searching capacity, capacity of increase relative to that of its prey, capacity for prey consumption, and other inherent properties, some interrelated with the above, such as synchronisation with the prey, prey specificity, degree of discrimination, ability to survive prey-free periods and certain behavioural traits that alter its performance as related to its prey's or its own density and/or dispersion (Huffaker *et al.*, 1971).

No single natural enemy need be highly endowed with all of these characteristics, but a high searching capacity is essential. Perfection in one characteristic may be precluded by countermanding requirements. For example, true monophagy is seldom found; rather, a tetranychid predator maximises its efficiency as a specialist feeder consistent with the need to survive year-round, which may, at some season, require an alternative prey species.

Comparisons between the intrinsic rates of increase of the prey species and the predator give, in isolation, no measure of efficiency of control because prey consumption by the predator offsets much of the potential increase in the prey. Furthermore, the functional response of a predator (prey consumption), particularly for the more efficient, highly prey-specific ones, presents a relatively small component of its regulatory power. The latter rests more in the rapidity of its numerical response, even though this is substantially derived from the functional response. The intrinsic rate of increase of a natural enemy is relatively more important in fluctuating environments (e.g. in pesticide treated fields) which induce 'escapes' from time to time, and relatively unimportant in truly stable situations where the predator population can (on average) only replace its own numbers each generation.

A good regulating agent must respond to increases in prey density at non-economic densities; otherwise, it will not be able to maintain the potential pest below damaging levels.

There is often, additionally, a high degree of resilience in the total density-dependent response because of the presence of a complex of predators. When a phytoseiid, which regulates its host at a low density, is ineffective (after a new planting, or following an unusual debilitating condition by weather or pesticides) other opportunistic predators having higher prey density thresholds for reproduction often come in and exert the control temporarily.

Two significant hypotheses have been advanced that challenge current practices in the use of natural enemies of spider mites: (i) that the occurrence of, or introduction of, any natural enemies other than the single 'best' species is detrimental to the control desired; and (ii) that phytoseiids can only control their prey at low densities, but cannot suppress densities that are already high because of interference effects from their own numbers or the abundant prey. The first hypothesis as a generality has been disproved many times, although in usual circumstances such an effect from multiple species action could prove to be detrimental. That this is at least very rare is evidenced by the enormous number of sequences of introduction of complexes of natural enemies in various parts of the world, with the later ones either having a negligible effect on or improving the level of control (Huffaker & Kennett, 1969; Huffaker *et al.*, 1971). Results from laboratory studies related to the second hypothesis are contradictory, and there is much evidence from field studies to indicate that phytoseiids can be efficient agents in suppressing high spider mite densities as well as in preventing such outbreaks from occurring.

To appraise the potential control efficiency of two examples, detailed studies of the biologies of *Typhlodromus occidentalis* Nesbitt and *Phytoseiulus persimilis* and of their host, *Tetranychus urticae*, were made under cyclical temperatures from 15 to 28.3 °C in a growth chamber, and life tables were constructed. Under the conditions of this study *P. persimilis* has an intrinsic rate of increase (r_m) of 0.219, a net reproductive rate (R_0) of 44.4, and a generation time (T) of 17.3 days. *T. occidentalis* has an r_m of 0.183 and multiplies 24.3 times in a generation time of 17.4 days. The r_m of *T. urticae* is 0.143, and its population multiplies 30.9 times in a generation time of 24 days. Each stage of *P. persimilis* has a higher rate of predation on the eggs of *T. urticae* than does the corresponding stage of *T. occidentalis* (Laing, 1968, 1969).

Assuming that the predators had unlimited searching abilities, arithmetic models were generated from these data that showed that either predator could quickly reduce a population of *T. urticae* when the initial ratio of prey to predator was 10:1. *T. occidentalis* theoretically annihilates the pest within 12 days whereas *P. persimilis*, because of its greater r_m and rate of prey consumption, annihilates the population within nine days. These theoretical events were confirmed by actual population interactions.

The inherent abilities of *T. occidentalis* and *P. persimilis* to regulate populations of *T. urticae* were compared in growth chambers programmed to give a diurnal cycle of 15.5 h of light and a temperature range of 15 to 28.3 °C.

Both predators can quickly overtake and reduce an increasing population of *T. urticae* below economic levels. However, *P. persimilis* reacts more quickly to increases in prey density than does *T. occidentalis* but sometimes over-exploits its prey on strawberries and does not survive the resulting prey scarcity. *P. persimilis* is thus less reliable in maintaining low prey densities. *T. occidentalis* requires less food and has a lower r_m than *P. persimilis* and, therefore, reacts more slowly to increases in prey density so allowing *T. urticae* to reach higher numbers than does *P. persimilis*. However, *T. occidentalis* did not over-exploit its prey to the same degree and thus it prevented an imbalance in the interaction for more than two years while maintaining the prey at low levels (Laing & Huffaker, 1969).

Several authors deny that phytoseiids can suppress high densities of tetranychids. Some have restricted their evaluation of the efficiency of natural enemies in general to either their functional or numerical response when, in reality, the two are closely related and must be considered together.

An important aspect of this controversy involves the possible occurrence of an interference component between predators and their prey. It has been suggested that high prey densities cause a disturbance in the functional and/or

189

numerical response of phytoseiid predators, and that this results in fewer prey being killed and/or fewer predator eggs laid as prey density increases (Chant, 1961; Mori & Chant, 1966). The decrease in functional response of the predator gives rise to a dome-shaped curve as described by Holling (1961). Some authors have suggested that this type of response would prevent tetranychids being regulated by phytoseiids at high prey densities.

However, in recent experiments with the phytoseiids *T. occidentalis*, *Amblyseius chilenensis* Dosse, and *P. persimilis* attempts were made to determine the extent of this interference component (Laing & Osborn, 1974). Adult females of each of the three species were placed in cells (enclosing a total surface area of 1 in.^{2}, equivalent to 6.45 cm^2) with densities of male two-spotted mites ranging from 10 to 300 mites per in.^{2}. The highest densities used were above those which would commonly occur in outbreak situations in the field and above the densities at which the host plant would limit the reproduction of the prey.

For each predator the functional and numerical responses can be described by a curve which exhibits a curvilinear rise to a plateau (Holling type 2 response). The functional response of *T. occidentalis* increased from an average of 4.6 prey killed per predator per day at a prey density of 10 mites per in.^{2} to an average of 12 prey killed per predator per day at a prey density of 150 mites per in.^{2}. The curve for the numerical response of *T. occidentalis* at these host densities was rather flat and ranged from 1.5 to 2.4 eggs per female predator per day. Daily prey consumption of *A. chilenensis* steadily increased from 6.2 mites at 10 prey per in.^{2} to 14 mites at 210 prey per in.^{2}. Egg production per predator per day increased from 1.6 eggs per day to a plateau at approximately 2.7 eggs over the same range of prey densities.

The form of the curves for the functional and numerical responses of *P. persimilis* were essentially the same as those for the other two predators. The functional response increased from 7.5 prey killed per predator per day at a prey density of 10 per in.^{2} to approximately 25 prey killed at 300 prey per in.^{2}. The numerical response increased from 2.1 eggs per predator per day to approximately 3.0 over the same range of prey densities.

Although the results were somewhat variable there was no consistent decrease in either the functional or the numerical response at any prey density that would suggest a true dome-shaped response. It was concluded that these three species exhibit no prey interference component at high prey densities. This is in contrast to findings of some authors, but agrees with what appears to happen in the field.

Feeding habits of phytoseiids

Phytoseiids are usually regarded as predaceous, but many species can feed, develop, and reproduce on a variety of non-prey foods. Except for the larvae of a few species, all stages of phytoseiids apparently feed (McMurtry & Scriven, 1964; Prasad, 1967; Croft & Jorgensen, 1969; Knisley & Swift, 1971). Some species possess an active feeding larval stage (Elbadry, 1968; Elbadry & Elbenhawy, 1968; Lee & Davis, 1968; Zaher, Wafa & Shehata, 1969) and may consume more prey during their development than do other species whose larvae do not feed.

Phytoseiids possess a hypostome armed with a pair of strong pointed projections, the malae, which are thought to comprise piercing organs. There is a positive correlation between the presence of well-developed malae and larval feeding.

Culture of phytoseiids

Phytoseiids have been cultured using pollen grains or tetranychid prey (McMurtry & Scriven, 1965; Kamburov, 1966; Swirski, Amitai & Dorzia, 1967; Kennett & Caltagirone, 1968; Rasmy, 1970). McMurtry & Scriven (1962, 1966) found that various types of media added to shipping or rearing containers aided survival, and in some cases allowed reproduction. Both sucrose and molasses increased the survival of all adult females and yeast hydrolysate combined with either molasses or sucrose sustained oviposition by *Amblyseius limonicus* Garm. and *A. hibisci* (Chant). The latter food combination provided sufficient nutrients for these species to develop to maturity but development was slow and mortality high. Shehata & Weiseman (1972) developed an artificial diet for *P. persimilis*. Although adults were produced on it, they did not lay.

Dietary range of feeding habits of phytoseiids

Most phytoseiids have a wide range of natural foods, though preference may be shown for certain foods on which development may be faster and egg production greater than on others (McMurtry & Scriven, 1964; Elbadry & Elbenhawy, 1968; Zaher *et al.*, 1969; Knisley & Swift, 1971). In general, tetranychids are usually preferred as prey (Collyer, 1964*a*; Dosse, 1967; Croft & Jorgensen, 1969), but there are exceptions. *Amblyseius gossypii* Elbadry fed both on adults and immature stages of *Tetranychus cinnabarinus* (Bois.) and *Oligonychus mangiferus* (R. & P.) but only on immature stages of *Eutetranychus orientalis* (Klein). Development is inhibited by a pure diet of eggs of *T. cinnabarinus* and the active stages are essential for completion of development (Elbadry & Elbenhawy, 1968). Similar performances were recorded for other phytoseiids (Herbert, 1959; Knisley & Swift, 1971).

Some phytoseiids appear to be adapted to feeding only on certain tetranychid species. McMurtry & Scriven (1964) reported that neither *A. hibisci* nor *A. limonicus* could prey upon *T. cinnabarinus, Panonychus citri*, or *Oligonychus punicae* (Hirst), which develop dense colonies and produce copious amounts of webbing, because the predators became trapped in the webs.

The total number of prey consumed is proportional to the size of the predator. Gravid females consume more prey than do nymphs. Females consume more prey than do males and mated females more than do unmated females. The predation rate is relatively low in ageing females that have ceased laying eggs. Furthermore, there is a direct relationship between the size of the predator and the size of the prey.

Tenuipalpid, eriophyid, and tydeid mites are sometimes accepted as prey. Female *A. limonicus* preying upon *Brevipalpus phoenicis* Geijskes continued egg-laying for 10 days, but the rate dropped considerably after the fourth day (Swirski & Dorzia, 1968). Larval *A. gossypii* fed on *Genopalpus pulcher* (C. & F.) but did not mature beyond the deutonymphal stage (Elbadry & Elbenhawy, 1968).

Phytoseiids have been recorded as feeding on thrips larvae, eggs and first instars (crawlers) of scale insects, moth eggs, and whitefly eggs and nymphs (Elbadry, 1968; Swirski & Drozia, 1968; Knisley & Swift, 1971).

Many photoseiids feed on non-prey foods such as pollen, honeydew, and fungi. The ability to feed on pollen varies with the species of phytoseiid and the kind of pollen (Putman, 1962; McMurtry & Scriven, 1964, 1965; Elbadry & Elbenhawy, 1968; Zaher *et al.*, 1971). *A. gossypii* fed readily on small pollen grains of corn and dates, but rejected the spiny pollen grains of cotton. The malae of *A. gossypii* (11 μm length) are shorter than the cotton pollen spines (10–11 μm) which, together with the thickness of the exine (4–5 μm), prevents their reaching the internal contents of the pollen. The importance of pollen as an alternative food to increase the effectiveness of some phytoseiids was demonstrated by McMurtry & Scriven (1966, 1968) and Elbadry & Elbenhawy (1968).

Food availability and quality affect the success or failure of competing phytoseiid species. For instance, *A. limonicus* has a greater reproductive potential when feeding on spider mites than has *A. hibisci*, its rate of development being faster and its egg production higher, though if *A. hibisci* also has access to pollen its population growth becomes comparable to that of *A. limonicus* (McMurtry & Scriven, 1965, 1971).

Evidently some phytoseiids extract plant juice even when an abundance of prey is available. Unfed, mated female *A. gossypii* survived on broad bean juice for about 10 days but produced no eggs. Similarly, *Amblyseius aleyrodis* Elbadry survived on plant juice for eight days although they produced few eggs. The ability to extract plant juice would favour predator survival in the field when prey become scarce.

Hibernation of phytoseiids
Subtropical regions

In warm climates, adult and immature phytoseiid mites are present on plants throughout the winter. Thus, females with ripe eggs, males, and all immature stages can be found at all times in Florida (Muma, 1955), India (Gupta, 1969), Israel, and the Philippine Islands (Wysoki & Swirski, 1971).

Wysoki & Swirski (1971) divided the Israeli species into four groups:

(i) species in which all postembryonic stages are active throughout the winter on the above-ground parts of the plants, oviposition decreasing only during the coldest weather;

(ii) species passing the winter mainly as non-reproducing females;

(iii) species in which the mode of overwintering depends on the temperature in that in cooler regions the over-wintering form is the non-reproductive female whereas in warmer locations both males and females occur in winter and egg-laying is common; and (iv) species in which females containing undeveloped eggs constitute the majority of the overwintering population.

Apparently little phytoseiid mortality results on sub-tropical crops from cold, although a high mortality may result from food shortage (Flaherty, 1967).

Temperate regions

In temperate climates phytoseiids overwinter on deciduous trees as fertilised females in crevices in the bark of the crown and trunk, in wounds, in empty cocoons of insects, and in empty shells of coccids (McMurtry *et al.*, 1970).

Winter mortality is high: 80–90 % in England (Chant, 1959) and 85–95 % in New Jersey, USA (Knisley & Swift, 1971). Mortality of *T. occidentalis* in British Columbia, Canada, depends upon prey density. If the preferred prey *Tetranychus mcdanieli* McGregor is present and wintering on the lower trunks of apple trees, *Typhlodromus occidentalis* will concentrate there where it is somewhat protected by snow cover from freezing. If *P. ulmi* is the main prey, this predator will spend the winter in the upper parts of the tree on twigs near the winter eggs of the red mite. These locations offer little or no protection from cold weather and such populations almost all die. Other species, however, for example *Typhlodromus caudiglans* Sch. and *Typhlodromus columbiensis* Chant, survive under such conditions (Downing & Moilliet, 1971).

Extremely low temperatures are necessary to freeze the body fluids of some phytoseiid mites. Downing & Moilliet (1971) found that at a temperature of

−35 °C there was almost 100 % survival of *T. caudiglans*, yet almost 100 % mortality of *T. occidentalis*. The cold-hardiness of *P. persimilis* was studied by Begljarov & Ushchekhov (1972). They showed that at +3 °C and 80–98 % relative humidity females could be stored for 30–45 days with mortality as low as 35 %. Rambier (1972) found that in the Mediterranean area *P. persimilis* can develop outdoors throughout the year as far north as southern France.

Studies on the hibernation of phytoseiid mites in Polish orchards (Boczek, Dabrowski & Kapala, 1970) are summarised here. In autumn, before the leaves fall, the phytoseiids tend to form aggregations of five to 12 mites on leaves. These are usually of one species; if there are two, one is dominant. The descent of the predators by leaf fall or earlier is similar in both old and young orchards, and is gradual until the temperature drops to 0 °C. On average, 5–15 % of the population present before leaf fall drop to the ground on the falling leaves.

In young orchards of other types and on plum trees which have smooth bark, overwintering occurs mainly in bark crevices on the spurs of branches. The distribution of hibernation sites depends also on the variety and state of the orchard. Usually fewest overwintering phytoseiids were found on the northern sides of the trees.

The phytoseiids occupy the overwintering crevices in aggregates, mostly of two to eight individuals with about 90 % in aggregates of more than five. Larger aggregates were often found on trees inhabited by *Typhlodromus aberrans* (Oud.), *T. rhenanus* (Oud.), and *T. finlandicus* (Oud.).

Dispersal begins in March, earlier in the upper part of the crown although *Phytoseiulus macropilis* (Banks) moves little during early spring.

Only 4–25 % of the autumn female populations survive the winter. The lowest mortality is observed in young orchards with high coccid infestations that provided good hibernation conditions. Mortality of *Typhlodromus soleiger* (Rib.) and *T. finlandicus* is high immediately after the first frosts of early winter. The second period of high mortality occurs in March and April before bud swelling. At that time many individuals of those two species, as well as *T. pyri* Scheut., die. *T. rhenanus* suffers greater mortality than does *P. macropilis*.

The hibernation sites of phytoseiid mites are usually in the vicinity of winter eggs of spider mites and/or aggregates of other mites. At the time of bud swelling in Polish orchards phytoseiids feed intensively on the larvae and nymphs of spider mites and on small insect larvae. Later in the spring they feed to a greater extent on female tetranychids. They die rapidly if there is a shortage of food during this period.

Diapause

Photoperiod is the prime factor evoking diapause of phytoseiids (Putman, 1962; Sapozhnikova, 1964; Hoy & Flaherty, 1970; Croft, 1971; Knisley & Swift, 1971; Rock, Yeargan & Rabb, 1971; Hoy, 1972). Knisley & Swift (1971) induced diapause in *Amblyseius umbraticus* Chant in New Jersey, USA, by exposure during development to a 12-h photoperiod at 22 °C.

Croft (1971) found a differential diapause response in populations of *T. occidentalis* originating from different geographical latitudes and altitudes. Populations from higher altitudes entered diapause earliest. Hoy (1972) studied intensively the diapause of this species. Temperature influenced the photoresponse, reproductive diapause being entirely averted regardless of photophase at 22, 25 and 30 °C. The critical photophase period in the field was in early October when day length was 11.5 h. While diapause would terminate spontaneously, higher temperatures and photophase interacted to terminate it more quickly.

Rock *et al.* (1971) stated that induction and duration of diapause depended on both photoperiod and temperature. There was no significant parental effect on the incidence of diapause in F_1 and F_2 progeny of adults reared under either short or long photoperiods. However, parental preconditioning did alter diapause intensity. When both parents and offspring experienced short day length, the resulting diapause duration was significantly longer than when parents experienced long photoperiods and offspring short ones.

The biology of stigmaeid mites

Whereas mites of the family Stigmaeidae frequently occur in the same habitats as spider mites, their potential importance as predators has only been recognised recently. Like so many groups of small inconspicuous mites they have been ignored by taxonomists, and have not been identified in economic literature.

Summers (1960) revised the group of stigmaeids that had previously been included in the genus *Mediolata* and defined several other genera. This facilitated further work by Gonzalez (1963, 1965) on Chilean material and by Wood (1967) on New Zealand material taken from orchards where stigmaeids were predaceous on plant-feeding species (Gonzalez, 1961; Collyer, 1964*b*).

The genera *Mediolata* G. Canestrini, *Zetzellia* Oud., and *Agistemus* Summers (Wood, 1967, includes *Agistemus* in *Zetzellia*) contain most of the foliage species found in orchard trees (15 of the 48 known species are leaf-inhabiting and occur on fruit trees). A few bark-inhabiting species occur in other genera. They are mostly predaceous in habit, and their known prey are detailed in the reviews by Huffaker *et al.* (1969, 1970), and McMurtry *et al.* (1970). While it is known that many of these mites are actively predaceous on

7-2

spider mites of economic importance, and also on eriophyids, acarids, tenui-palpids, aphid eggs and pollen, little is known of their role in the regulation of populations.

Fluctuations in numbers are great. Those species observed on deciduous fruit trees show extreme seasonal changes, with high numbers in late summer and autumn but only a few individuals in early and mid summer. It is likely that the natural host plants of many species are evergreens as in New Zealand where they can feed throughout the year, and it is therefore likely that these mites will be of more permanent value on evergreen than on deciduous host plants. Those that live permanently on deciduous trees are probably partly bark-inhabiting.

The influence of pesticides on stigmaeids is inadequately documented. *Zetzellia mali* Ewing seems to be susceptible to pesticides and is therefore most common on unsprayed trees (Lord, 1972), but Parent (1967) stated that this species is not very susceptible to ordinary insecticides with the exception of endrin. There is evidence that some species may be fairly resistant to pesticides and that organo-phosphorus materials lead to an increase in numbers (Abdel-Salam, 1967). In New Zealand high populations of *Agistemus longisetus* Gonzalez followed use of these materials. In experiments, partly reported by Collyer (1964*b*), it was found that lime sulphur delayed the seasonal population increase compared with captan or glyodin. Where captan was used in each case, with azinphosmethyl (seven summer applications) or DDT, there was greater increase which started earlier and peaked earlier. A somewhat reduced increase with demetonmethyl, and the least increase when lead arsenate, karathane or oil were used. This could be regarded as a reproductive stimulation of *A. longisetus* rather than as a response to tetrany-chid mite density, for the latter were abundant on the trees sprayed with DDT but very low on those sprayed with organophosphates.

These indications are only a start. Much research is needed before we understand the population characteristics and phenology of the various stigmaeid mites and their significance as biological control factors.

Biology and ecology of predaceous insects

During this programme significant progress was made in advancing know-ledge of the biology and ecological roles of certain coccinellids and antho-corids as well as some other predaceous insects and mites. This is exemplified by the work in Australia and California where *Stethorus* spp. are of high importance and have received major emphasis and in Poland on anthocorids.

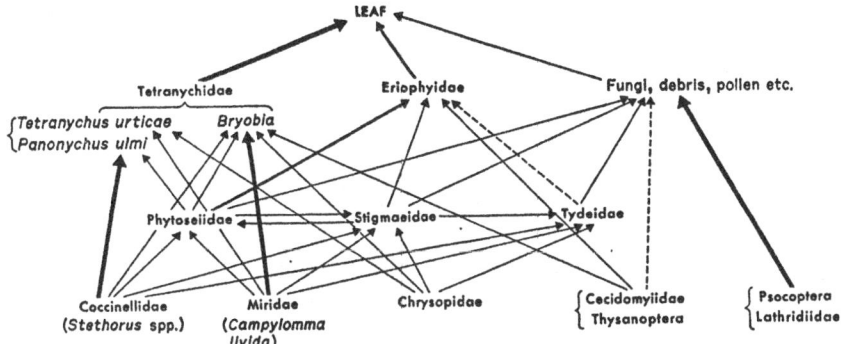

Fig. 22. Food-web of observed foliage faunal elements in unsprayed apple orchards in Australia (Readshaw, 1971).

The role of predators in the ecology and control of spider mites in Australia

The control of mites in Australian pome fruit and peach orchards is becoming more difficult and expensive every year, largely because of resistance problems. The intensive use of insecticides to control codling moth and oriental fruit moth, and the rapid development of resistance by mites to acaricides, together with favourable climate, have created optimal conditions for mite increase, frequently resulting in serious defoliation. The main species are *Tetranychus urticae* on the mainland, where the summers are hot and dry, and *Panonychus ulmi* on the more temperate island state of Tasmania. *Bryobia rubrioculus* (Scheut.) was important before the advent of OP-insecticides to which it easily succumbs. It still predominates at low densities in orchards that have not been treated with OP-insecticides for several years. It can be particularly trouble-some on trees treated repeatedly with DDT or related compounds.

The purpose of this investigation, begun in 1968, was to determine the basic cause of the mite problem in Australian orchards in order to develop improved methods of control.

Figure 22 summarises the food web of observed foliage organisms in un-sprayed orchards. The tetranychids are only one component of a complex involving many different groups of insects and mites. The more important connections are shown by the heavier lines.

In this system *Bryobia* is normally controlled at low densities by the mirid *Campylomma livida* Reut. *T. urticae*, and probably also *P. ulmi*, are controlled chiefly by predatory coccinellids of the genus *Stethorus*. In contrast to much experience in other world areas, phytoseiid mites, although frequently present, do not appear to be very important.

Variations in numbers of *T. urticae* were governed by three native species of *Stethorus* that are widely distributed throughout southeastern and western

197

Australia (Britton & Lee, 1972). *S. vagans* Blackb. and *S. nigripes* Kapur prevailed in the cooler regions, and the third species, *S. loxtoni* Britton & Lee, was more frequent in the hotter inland areas. Their biologies are similar. Adults overwinter in the orchard, on trees and surrounding vegation and frequently in association with overwintering female spider mites. They move to apple foliage in spring (October–November), and, if undisturbed by pesticides, they prevent increase of *T. urticae* for the remainder of the growing season beyond levels of one or two mites per leaf. At such low densities of prey, the beetles are very scarce and are difficult to locate other than by searching entire trees. This might explain why some investigations concluded that they are often relatively unimportant.

The beetles are long-lived and very mobile. They are also extremely efficient predators for a variety of reasons which cumulatively cause their percentage effect to increase as prey density increases. In fact, their effects on *T. urticae* can be predicted quite accurately from a simple Lotka–Volterra type model of the form $\mathrm{d}M/\mathrm{d}t = a+bM+cS$ and $\mathrm{d}S/\mathrm{d}t = d+eM+fs$, where $\mathrm{d}M/\mathrm{d}t$ and $\mathrm{d}S/\mathrm{d}t$ are rates of increase, M and S log numbers of mites and of *Stethorus* per 100 leaves, and a, b, c, etc. are positive or negative constants. Given initial ratios of from 10:1 to 1000:1 mites:predators, the model generates a series of population curves with *T. urticae* peaking at from 10 to 100 mites per leaf, then slowly declining to an equilibrium level of less than 1 mite and 0.1 predator per leaf, which is near the natural level found in undisturbed orchards represented in Fig. 22. In practice, the declining part of the curve is much steeper than the model predicts because of changes in age distributions of the mites and predators.

Any significant external disturbance of the system can lead to wide oscillations in the numbers of both predator and prey. In fact, this appears to be the basic cause of the Australian spider mite problem, as commercial spray programmes using broad-spectrum insecticides completely disrupt this delicately balanced system of natural control by eliminating *Stethorus* and thus create a serious and continuing mite problem requiring repeated applications of acaricides.

These findings suggest two ways of improving mite control in Australian orchards. One is to use modified spray schedules designed to control the major pests with minimum impact on the mite predator *Stethorus*. This approach has been successful in peach orchards in South Australia (Richardson, 1972) where the *T. urticae* problem was drastically reduced by *Stethorus* if the oriental fruit moth was controlled with parathion or malathion early in the season (September) before either the mite or the predator had emerged from their overwintering shelters. No further spraying with insecticides was usually necessary thereafter and the density of mites remained very low throughout the season. However, in apple orchards, the use of modified spray schedules has not been successful because the control of codling moth, in particular, demands the repeated use of potent broad-spectrum insecticides such as azinphosmethyl which virtually eliminates *Stethorus* from the orchard for the greater part of the summer season (Readshaw, 1971, 1972).

The other approach being examined is the introduction of insecticide-resistant natural enemies (Readshaw, personal communication). A strain of the phytoseiid *Typhlodromus occidentalis* that is resistant to azinphosmethyl and other commonly used insecticides was recently introduced from North America and is now established in commercial orchards in Canberra. The best results were obtained where the predators were released in autumn (February 1972). They increased naturally on the mite-infested trees, overwintered in bark crevices, and emerged in spring (November 1972), and have kept the levels of *T. urticae* at less than five per leaf despite use of four commercial sprays of azinphos (0.05 %) and other chemicals. *T. occidentalis* occurs naturally in Australia but it is rare, and it may be that the introduced strain could fulfill, in sprayed orchards, the role of *Stethorus* in unsprayed ones. If so, there would be not only extremely important benefits in the field of pest control but there would be a greatly reduced need to apply environmentally disturbing pesticides.

Studies on Stethorus *in California*

The general biology and ecology of *Stethorus* spp. was summarised by McMurtry *et al.* (1970).

Because numbers of *Stethorus picipes* Casey in field situations in southern California have not always increased in response to high densities of spider mites, studies were initiated to determine if reproduction was affected by photoperiod (McMurtry, Scriven & Malone, 1973). Nearly 100 % of beetles reared and maintained in a 16-h photoperiod at 21 °C oviposited, whereas only about 30 % of those reared and maintained in a 10-h photoperiod oviposited during a 40-day period. Nearly all beetles reared at a 16-h photoperiod and then transferred to a 10-h regime one week after emergence oviposited during the first few days after being transferred but then oviposition declined to about 30 %. The reciprocal treatment (reared at 10 h and transferred to 16 h) resulted in only a low percentage ovipositing at first but an increase to nearly 100 % after about two weeks. Nearly all beetles held in continuous light or continuous darkness produced eggs. The results indicated that a proportion of the population of *S. picipes* in southern California are induced into diapause by short photoperiods. Field observations in winter revealed diapausing individuals (as indicated by lack of ovarian development) on oak trees in the absence of mites, and non-diapausing females associated with mite infestations on such plants as *Ricinus communis* L.

In a 16-h photoperiod at 21 °C, females laid an average of 577 eggs each during an average oviposition period of 130 days. Eggs and larvae of *Tetranychus pacificus* supplied at the rate of about 100 mg every three days were

199

more favourable for oviposition than were other developmental stages or the adults.

A life table was developed and the rate of increase (r) calculated for *S. picipes* feeding on *O. punicae*. The r value obtained was 0.12 compared to 0.22 for *O. punicae*.

One first instar *S. picipes* larva per leaf on avocado seedlings in the laboratory suppressed prey populations of *O. punicae* averaging 70 per leaf within one week. They completed development, but then left the seedlings as larvae and died even though the prey population had again started to increase (Tanigoshi, 1973). This demonstrated the potential of *S. picipes* to suppress high and increasing spider mite populations rapidly, and also the necessity of a continuing high or adequate prey density for continuing reproduction, the latter being apparently contrary to Readshaw's results for other species (above).

Estimates of the prey densities required by *Stethorus* species range from four to five per cm² of leaf for *S. punctillum* Weise females to reproduce (Begljarov, 1970) to 10 eggs per cm² for larvae of *S. japonicus* Kamiya to mature (Tanaka, 1966).

Laboratory tests on the tolerance of *S. punctum* LeConte to various pesticides indicated that most materials tested could be used in integrated control programs with little harm to the predator (Colburn & Asquith, 1971).

The role of anthocorids in the ecology and control of spider mites

Anthocorids are rather general predators and only a few species are frequently associated with spider mites.

Orius minutus (L.) is one of the principal predators of *P. ulmi* in Western Europe (Günthart, 1945; Collyer, 1953; Berker, 1958; Post, 1962). *Orius vicinus* (Ribaut) is the most important predator of several mite species in Southern France (Fauvel, 1971). *O. vicinus* maintains *P. ulmi* populations below the level of economic importance in apple orchards sprayed according to an integrated programme. *Orius insidiosus* (Say) and/or *O. tristicolor* (White) are common predators of tetranychids on several crops in North America (Quayle, 1912; Newcomer & Yothers, 1929; Iglinsky & Rainwater, 1950; Michelbacher, Middlekauff & Bacon, 1952; Allen, 1959; Oatman & McMurtry, 1966; Putman & Herne, 1966; Horsburgh & Asquith, 1968).

Anthocoris musculus Say may play an important role in controlling populations of *P. ulmi* in apple orchards in Nova Scotia (Lord, 1956; Lord, Herbert & MacPhee, 1958; Sanford & Herbert, 1970). *Anthocoris nemorum* L. is one of the main insect predators of *P. ulmi* on deciduous fruits in Europe (Kuenen, 1942; Günthart, 1945; Berker, 1958; Böhm, 1960; Collyer, 1953; Post, 1962; Niemczyk, 1968). It also feeds upon *T. urticae* on vegetables (Fritzsche, 1958).

The role of *A. nemorum* in controlling *P. ulmi* has been investigated by

Niemczyk (1973). Under natural conditions each bark bug female attacks an average of about 150, and a maximum of about 500 winter eggs of *P. ulmi* daily. In the laboratory, a larva devours 50 to 550 mite larvae or from three to 75 adult females daily. In high mite populations this predator kills about twice as many females as it consumes.

A. nemorum can survive on plant juices alone; first instar larvae live an average of five days, and fifth instar 21 days. During development anthocorid larvae may consume daily, according to the instar, from one to 12 *P. ulmi* females or from 15 to 200 *P. ulmi* larvae.

P. ulmi females are found by the predator more easily than their eggs or larvae and even when summer eggs are abundant few are eaten.

On isolated apple twigs in the laboratory *A. nemorum* larvae matured on the economically unimportant prey density of only one to two females per leaf although each consumed at least 350 females during its larval life. The majority of the predators leave the trees at such a low prey density in the field. Apparently about five mites per apple leaf is the minimum density to maintain the predators on the trees and to attract them from surrounding areas.

In sufficient numbers (one per 50 leaves or less), *A. nemorum* can rapidly reduce very high populations of *P. ulmi* on apples to the low level of 0.5–3 mites per leaf and keep the mites at a low level throughout the season (Niemczyk, personal communication). In the field, if mites are their only available prey, such high predator density would be maintained only for a few days because they would leave. However, high predator numbers and hence effective control may be maintained over a long period, if other suitable food sources are available. The number of *A. nemorum* in the environment surrounding the orchard is important.

The biology of *Orius* and of *Anthocoris* is similar. Active stages occur from early spring to late autumn. The number of generations usually ranges from one to three per annum depending on the species and climate. *A. nemorum* has only one generation in Scotland (Hill, 1957), two generations in much of Europe (Peska, 1931; Collyer, 1953; Böhm, 1960; Niemczyk, 1966), and three in central France (Bonnemaison & Missonnier, 1956). Fertilised females overwinter under bark scales, in litter and in other protected places. Eggs are inserted into the soft tissues of plants, with only the cap being visible on the surface. The number laid varies with the species, with the food consumed, and between individuals (usually about 50 to 100). Developmental time of the five larval instars varies from several days to several weeks, depending upon the species and the temperature.

Most but not all anthocorids occur on a wide range of plants (Anderson, 1962). Some, for instance, *A. gallarum-ulmi* DeG. and *A. sarothamni* Douglas & Scott, inhabit mainly a single plant species and feed on a narrow range of prey. Anthocorids appear in higher numbers on plants that harbour large prey populations, including spider mites. As mite density increases, the numbers of anthocorids increase (Lord *et al.*, 1958; Putman & Herne, 1966). They thus appear to act as a density dependent factor (Lord, 1956; Van de Vrie, 1965), their role in mite control increasing with increasing mite density. If only mites and anthocorids were present these predators would be unable to maintain the pest at an economically unimporant level because they require a higher total prey density and would leave.

The biologies of the following species of anthocorid predators of spider mites have been studied: *Orius pilosus* (Jak) (Korcz, 1968), *O. minutus* (Collyer, 1953), *O. vicinus* (Fauvel, 1971), *O. insidiosus* (Marshall, 1930; Iglinsky & Rainwater, 1950), *Anthocoris confusus* Reut. (Anderson, 1962), *A. sarothamni* and *A. confusus* (Anderson, 1962; Hill, 1961, 1965, 1968), *A. gallarum-ulmi* (Anderson, 1962; Piasecka, 1969), *A. nemoralis* Fab. and *A. minki* Dohrn (Anderson, 1962), *A. nemorum* (Peska, 1931; Collyer, 1953, 1967; Hill, 1957; Anderson, 1962), and *Montandoniola maraquesi* Puton (Tawfik & Nagui, 1965; Funusaki, 1966). The possibilities of utilising anthocorids in biological control were discussed by Carayon (1961).

Effect of pesticides on spider mites and their natural enemies

Effect of pesticides on spider mites through changes in host plant physiology

Under natural conditions, the population dynamics of a phytophagous pest is the resultant of two antagonistic processes: its reproductive potential and the limiting effect of its natural enemies.

The reproductive potential depends not only on genetic factors and general ecological conditions, but also very distinctly on the nature of the mite's nutrition. Indeed, the physiological condition of the plant, and consequently the nutrition of the mite (or insect) is dependent on many factors such as composition and structure of the soil, nature of the root stock, fertilisation, age, and, as well, on effects of pesticides on the plant's metabolism.

It is suggested that the increase in mites following pesticide application is caused by improved nutritional factors. This possibility was first suggested by Huffaker & Spitzer (1950) for DDT, and other workers have found similar indications (Fleschner, 1952; Saini & Cutkomp, 1966). In tests on grape and apple made by Chaboussou (1969), DDT nearly always produced an increase in the fecundity and survival of *P. ulmi*, although Putman (1963) failed to find such a relationship. With *T. urticae*, parathion, carbaryl, and certain fungicides had similar effects (Chaboussou, 1969).

Through their influences on the plant's physiology, various chemicals, fungicides, and insecticides increase, through nutritional effects, the fecundity and longevity of tetranychids. In many cases, this augmented power of increase may explain the failure of natural enemies to limit tetranychids' populations.

Many authors have shown a relationship between reproduction of *T. urticae* and the percentage of nitrogen and other mineral elements in the leaves of various host plants (see above, pp. 182–4). However, the relationship is by no means constant, and is confounded by the occurrence of other elements such as phosphorus, potassium and other elements (Van de Vrie, 1974).

Reducing glucosides seem to play an important role, and it is known that certain pesticides can influence the levels of these compounds in the leaves.

There are few data on the requirements of *P. ulmi*, but it seems that, for different species of mites, or different plant hosts there is an optimal equilibrium between nitrogen and glucosides for maximum rates of spider mite increase. This optimal balance between soluble nitrogen and soluble glucosides is dependent upon the cationic equilibrium within the plant (Rodriguez, Chen & Smith, 1960; Fritzsche, Wolfgang & Opel, 1957; Henneberry, 1961; Watson, 1964; Markkula & Tiittanen, 1969), and this balance may be affected by pesticides. Some chemicals enrich the plant with metallic ions – copper, iron, zinc, phosphorus etc. – others affect osmosis (?), e.g. DDT, and some affect the main physiological processes of the plant such as respiration and photosynthesis (Pickett, Fish & Shan, 1951; Heinricke & Foot, 1966).

Similar effects as described for nitrogen are recorded for phosphorus. Thus, Rodriguez *et al.* (1960) found a significant positive correlation between the per cent of phosphorus and populations of *T. urticae* on apple at a level of phosphorus below 0.20 %, although the correlation was negative above this level. Similar results on pear were obtained by Chaboussou (1969). Fritzsche *et al.* (1957) and Watson (1964), showed that a deficiency of potassium on *Phaseolus* increases the amount of reducing glucosides.

Rodriguez *et al.* (1960) found that, on Black Valentine beans, the addition of DDT to soil at doses of 800 to 1600 pounds/acre (about 1000 to 2000 kg/ha) increased the concentrations in the leaves of nitrogen and total and reducing sugars and resulted in an increase in populations of *T. urticae*.

At normal doses on apple, fermate, 2–4D and chlordane stimulate photosynthesis, whilst parathion inhibits it (Pickett *et al.*, 1951). Heinricke & Foot (1966) also noted negative effects on the photosynthetic potential of apple following use of diazinon and gusathion.

Similar phenomena can occur on vines. On plants treated with carbaryl, populations of *P. ulmi* increase in numbers and those of *Eotetranychus carpini* Oud. decline, but the opposite occurs with parathion (Chaboussou, 1969). Analysis of foliage following chemical applications illustrate significant effects in the amount of potassium, calcium, magnesium and phosphorus. The significant correlation between the increase in potassium/calcium with use of carbaryl and the multiplication of *P. ulmi* was established. Conversely, the regression of this same relation (potassium/calcium) with parathion use and the increase in *E. carpini* and the decrease in *P. ulmi* are significant.

In conclusion, it is stressed that the action of pesticides is not by any means totally one of effect on predators. Various biochemical effects are known to occur and these can have dramatic effects both positive and negative on the reproduction and development of the mites themselves. However, Van de Vrie & Boersma (1970) demonstrated that the phytoseiid *Typhlodromus potentillae* Garm., when not inhibited by pesticides, coped with a near maximum fecundity in *P. ulmi* produced by favourable fertilisation, and prevented economic densities.

The effects of pesticides on behaviour of phytoseiids

The comprehensive reviews of McMurtry *et al.* (1970) and Huffaker *et al.* (1970), include considerations of effects of pesticides on the predator/spider mite system, together with information on the more pronounced effects of chemicals on the behaviour of predaceous mites. Until recently it was thought that any integrated control programme for red spider mites would require occasional sprays that have acaricidal action, and so workers have looked at the effect of many such pesticides on the predators as well as on the population dynamics of the spider mites (Smith, Henneberry & Boswell, 1963; Herne & Chant, 1965; Legowski, 1966; Begljarov, 1967; Krambias, 1968; McMurtry *et al.*, 1970; Huffaker *et al.*, 1970). While such studies have relevance in open field environments, in enclosed environments, such as glasshouses, control of the main arthropod pests may seem likely to be achieved with limited use of pesticides (Hussey, 1965; Hussey & Bravenboer, 1971). Thus the emphasis here is on fungicides. Fungicides also pose a problem in orchards and field crops, for although there is good evidence to suggest that some fungicides, such as wettable sulphur, captan, and glyodin, have little effect on predatory mites, there is mounting evidence that other compounds, such as zineb, maneb, ziram, karathane, ferbam, and selbar, can be detrimental (McMurtry *et al.*, 1970). Some fungicides also have strong acaricidal effects, e.g. sulphur, karathane, binapacryl and oxythioquinox, but others probably have detrimental effects that are more subtle. Smith *et al.* (1963) showed that *Phytoseiulus persimilis* can tolerate residues of zineb and maneb, and Bartlett (1964) showed that zineb and ferbam did not kill predators. Ferbam and ziram act as repellents to *Popillia japonica* Newm., but there is no evidence that the adverse effects on phytoseiid mites exhibited by these fungicides in the field are caused by a similar action.

Little work has been done to check whether spraying spider mites, their predators, and the environment, either with non-toxic materials or toxic materials at non-toxic levels, leads to the prey becoming either less or more readily available and/or acceptable to the predator through possible behavioural changes in the groups of organisms. If it could be shown that chemical deposits affect the number of prey eaten, either by rendering the prey repellent or no longer attractive or by making the location of prey more difficult, there could be important effects on the resultant numbers of predator and prey. This would be particularly important in glasshouses where a delay in the build-up of predators could be critical.

Chant (1961) underestimated the contribution that *Tetranychus urticae* eggs make to the diet of *P. persimilis*, but subsequently Bravenboer & Dosse (1962), Smith *et. al.* (1963), Mori & Chant (1966), and Laing (1968) showed that many eggs are eaten. Jackson (1969), using malathion and the two fungicides captan and dinocap, investigated some aspects of the feeding behaviour of *P. persimilis*. Normal *T. urticae* eggs alternating with eggs that had been sprayed were presented in rows to the predators. Consumed eggs were replaced with a similar type and each experiment was allowed to continue until 100 eggs had been eaten. A control experiment comparing eggs washed in distilled water with unwashed eggs showed a significantly higher proportion of normal eggs eaten. Halstead (1970) was unable to confirm this, but he showed that *Phytoseiulus persimilis* selectively preyed upon normal 'turgid' eggs in preference to 'flaccid' eggs, and that the number of 'flaccid' eggs eaten were increased by spraying with distilled water and subsequent drying. Initial contact with the prey appeared to be a matter of chance. Prey were located with the anterior legs as was also observed by Mori & Chant (1966), but in the case of the washed eggs the subsequent palpating movements of the pedipalps were continued for a longer time before feeding began and the chances of later rejection was higher (Jackson, 1969). Malathion-treated eggs were almost totally rejected by *P. persimilis* within a few seconds of the first contact; significantly fewer eggs sprayed with captan were eaten, and those treated with dinocap were eaten as readily as those washed in distilled water. Halstead (1970) confirmed that malathion-sprayed eggs were strongly repellent, that captan rendered the eggs less acceptable than unsprayed eggs, and that eggs sprayed with triarimol were not repellent. He also considered the effect on palatability of spraying *T. urticae* larvae with fungicides. He was able to make use of the fact that no difference was shown to exist between an albino strain and a 'wild' strain of *T. urticae* when unsprayed mixed populations of larvae were preyed upon by *P. persimilis*. Individuals of one strain were sprayed with a fungicide, dried, and allowed to mix with similar aged mites of the remaining strain which acted as the untreated control. No effect was shown on predation using benomyl, dimethiramol, or captan, but treatment with triamirol rendered larvae significantly less palatable.

The food-chain accumulation of pesticides is now a well-known phenomenon in the animal kingdom, but little is known about it in relation to mites. Systemic root drenches of thionazin and dimethoate proved less toxic to *T. urticae* on the treated plants than to *P. persimilis* feeding upon them (McClanahan, 1967). Krambias (1968) observed no effect on a predator fed *Panonychus ulmi* that had been sprayed with 0.01 % malathion, but a solution five times as concentrated produced 100 % mortality in the predator within 24 h. Similar experiments with chlorfenson had no effect after 24 h but produced a

25 % kill after 48 h whereas dinocap had no lethal effect after the same time.

The distribution, as well as the concentration of spray deposits on crop surfaces, affects the behaviour and susceptibility of pest organisms. Fisher & Hansell (1964) suggested that *Tetranychus telarius* (L.) may avoid discontinuous deposits of dicofol, particularly where these are widely spaced, and Fisher & Morgan (1968) confirmed this. They showed that when a constant volume and quantity of dicofol was applied to standard leaf discs as increasing numbers of droplets (droplets becoming appropriately smaller) spider mites laid eggs in inverse relationship to the number of droplets. Mite mortality was directly related to the number of deposits even though the total amount of pesticide per disc was constant. When the droplets were restricted to 13 (covering roughly 19 % of the surface), mites were repelled by the deposits irrespective of the concentration, and eggs were laid on the 'in-between' clean areas. Increasing the number of deposits to 50 (covering approximately 25 % of the surface) increased contact with the pesticide and resulted in a significant reduction in egg production, and in increased mite mortality. Kirkby (1969) showed that high mortality occurred in both *T. urticae* and *P. persimilis* if 5 to 10 % of leaf disc surfaces were covered with metasystox, all the predators dying within 1 h. Dinocap caused some mortality and reduced fecundity in both predator and prey at the highest concentrations used, but the predation rate seemed relatively unaffected. Captan had no such effect. During these experiments, spider mites at first avoided all dried deposits, the response diminishing after 24 h, but by three days some egg-laying occurred on the lower concentrations of dinocap and captan. *P. persimilis* appeared to move indiscriminately over all chemicals though no eggs were recorded on the deposits.

Many questions remain unanswered, but sufficient is known to suggest that pesticide applications at non-lethal rates may have important effects on predator/prey relationships through behavioural changes in the presence of chemical deposits. As a simple barrier effect such chemicals may simply retard satisfactory spread of the predators, whereas broken spray deposits may hinder searching ability. However, it is highly probable that in many instances the effect is complex, altering in subtle ways both the environment and behaviour of the interacting organisms.

Toxicity of pesticides to phytoseiids

Pesticides in integrated control should be chosen to allow maximum phytoseiid survival. More than 70 papers have been published on the effects of pesticides on phytoseiids, most of them dealing with orchard situations.

Any characterisation of pesticide toxicity upon phytoseiid populations should consider the following:

(i) laboratory tests on direct and residual toxicity (Ristich, 1956; Mathys, 1958; Bravenboer, 1959; Van de Vrie, 1962; Daneschwar, 1963; Smith *et al.*, 1963; Bartlett, 1964; Dabrowski, 1969a; Krambias, 1969; Rock & Yeargan, 1972);

(ii) field trials on direct and residual toxicity (Huffaker & Kennett, 1953; Clancy & McAlister, 1956a, b; MacPhee & Sanford, 1956, 1961; Collyer & Kirby, 1959; Thill, 1957; Van de Vrie & Fluiter, 1958; Van de Vrie, 1962; Bartlett, 1964; Böhm, 1966; Swift, 1968; Steiner & Baggiolini, 1968; Dabrowski, 1969b; Hoyt, 1969a; Herbert & Sanford, 1969; Herne, Simpson & Putman, 1969; Karg, 1970; Madsen, 1970);

(iii) changes in phytoseiid and tetranychid populations caused by pesticide application under natural conditions (most of the authors mentioned above; Downing, 1966; Sanford, 1967; Sanford & Herbert, 1967, 1970; Specht, 1968; Caltagirone, 1969; Collyer, 1969; Hamstead, 1970; Dabrowski, 1970a, b, c; Rock & Yeargan, 1971; Westigard, 1971; Knisley & Swift, 1972);

(iv) factors determining increase in numbers of predaceous mites and spider mites during the growing season during which the pesticide was used (MacPhee & Sanford, 1961; Lord, 1962; Putman & Herne, 1966; Parent, 1967; Sanford & Herbert, 1967, 1970; Dabrowski, 1970d; Karg, 1970); and

(v) long-term effects resulting from the use of a pesticide in earlier years on the density of phytoseiid and tetranychid populations (Specht, 1968; Dabrowski & Dabrowska, 1972; Dabrowski, Dabrowska & Labanowski, 1972).

It should be emphasised that the density of the prey left on the treated crop must be evaluated when the effect of pesticides on phytoseiids is considered.

If the toxicity of pesticides is classified into four groups, where O means harmless to phytoseiids (0–5 % mortality); L – low toxicity (6–35 % mortality); M – moderate toxicity (36–75 % mortality); H – high toxicity (76–100 % mortality), there is considerable variation in the results obtained by different workers.

Most fungicides have shown low or moderate direct toxicity to phytoseiids under laboratory conditions (although as detailed on pages 205–6 the more subtle effects on behaviour can be equally damaging). The following materials have been rated according to the direct mortality caused: binapacryl (L–M), captan (O–M), copper oxychloride (L–M), dinocap (M–H), dodine (O–L), ferbam (O–L), maneb (M), Morestan (L–M), Nirosan (L), TMTD (O–M), Wepsyn (O), wettable sulphur (L–H), zineb (O–H), ziram (L). Solbar and lime-sulphur are always toxic. Except for ziram similar ratings have been determined under field conditions: binapacryl (M–H), captan (O–L), copper oxychloride (O–H), dichlone (O–M), dinocap (M–H), dikar (H), dodine (O–M), Euparen (L–M), ferbam (M–H), Glyodin (O–L), maneb (M), Morestan (M–H), Nirosan (O), Ryania (O–L), TMTD (L), Solbar (O), Wepsyn (O–L), wettable sulphur (L–H), zineb (O–H), and ziram (H).

Under laboratory conditions the toxicity to phytoseiids of chlorinated hydrocarbon insecticides varies from O to H, DDT (L–H), methoxychlor (L–H), lindane (L–H), aldrin (O–L), dieldrin (O–M), endrin (O–H), chlordane (L), endosulfan (M), heptachlor (L–M), Rothane (H), toxaphene (K–H). Under field conditions a similar classification is obtained: DDT (L–H), methoxychlor (O–M), lindane (M), aldrin (L–M), dieldrin (O–M), endrin (O–L), chlordane (M–H), endosulfan (O–H), heptachlor (L), Rothane (M–H), and toxaphene (O–M).

The carbamate insecticide carbaryl has always been considered extremely destructive to phytoseiids in the field, but in laboratory tests toxicity has varied from low to high. Methomyl has shown high toxicity and aldicarb low to high toxicity while Isolan had almost no direct harmful effect on phytoseiids.

Most organophosphorous insecticides are extremely toxic to most predatory mites, with a few exceptions. In the laboratory: azinphosmethyl (M–H), carbophenothion (M), demeton-O (M–H), diazinon (L–H), dichlorvos (H), dimethoate (H), trichlorphon (L–H), ethion (H), malathion (H), Metasystox (H) paratin (M–H), Phosdrin (M–H), phosphamidon (H), schradan (L) and TEPP (M–H). In the field: azinphosmethyl (M–H), Bidrin (H), carbophenothion (H), chlorthion (H), demeton-O (M–H), diazinon (M–H), dimethoate (H), dichlorvos (L–H), trichlorphon (L), ethion (M–H), Gardona (M), Imidan (M–H), menazon (L), malathion (M–H), Metasystox (H), parathion (L–H), parathion-ethyl (H), Phosdrin (H), phosalone (L), phosphamidon (H), and TEPP (H).

Most acaricides tested in Poland caused almost no harm to predatory mites but some decreased egg production. Acaricides were classified by laboratory tests as follows: Aramite (O–M), chlorobenzilate (O–M), chlorfenson (O–M), Eradex (H), fenson (O), Genite 923 (L), dicofol (O–H), phencapton (H), and tetradifon (O–M). The toxicity of some acaricides was lower if tested under field or greenhouse conditions and for others apparently higher: Aramite (O–M), chlorobenzilate (L–M), chlorobenside (L–M), chlorfenson (O–L), chlorocide (O), chloropropylat (M), Dessin (M), Eradex (M–H), fenson (O–L), formetanate (H), Galecron (H), Genite 923 (O–M), dicofol (M), Omite (L–H), phencapton (H), Fundal (H), Plictran (M), tetradifon (O–H), and tetrasul (L).

Ryania, a material of plant origin, and lead arsenate were largely innocuous to most species of predaceous mites and insects. Emulsifiable oils, for instance, Niagara Superior 70 oil or Volk Supreme oil, applied at the time of bud break, were not detrimental to natural predator populations in orchards. If it is to be practised, the preventative treatment of orchards is therefore recommended in the early season when few predators are present.

The wide variation in rated effects of these pesticides are due to: methods of testing, variation in species susceptibility between different stages and strains, concentrations, and timing of the spray. Comparing the direct dip and deposit toxicity of pesticides to *Typhlodromus finlandicus* and *Phytoseiulus macropilis*, Dabrowski (1969 a) divided pesticides into four groups:

(i) Chemicals with a considerably stronger residual than direct action upon dipped phytoseiids. For *T. finlandicus*: dinocap, chlorobenzilate, chloropropylat, methoxychlor, Sumithion; and for *P. macropilis*: phencapton, metoxychlor, lindane, Sumithion and Anthio.

(ii) Compounds with a stronger direct than residual effect. For *T. finlandicus*: maneb, Miltox, lime-sulphur and Ekatin; and for *P. macropilis*: binapacryl, lime-sulphur, Morestan, diazinon, dichlorvos, Metasystox, phosphamidon, and parathion-ethyl.

(iii) Highly toxic compounds which, regardless of the method of application, kill 80–100 % of the predators. For *T. finlandicus*: carbaryl, malathion, Folithion, dichlorvos, Intration (thiometon), Metasystox, parathion-ethyl, dimethoate, demeton-O, and vamidithion; and for *P. macropilis*: malathion, Intration, and dimethoate.

(iv) Non-toxic compounds which do not kill dipped mites and which have no residual effects. For *T. finlandicus*: TMTD, dodine, wettable sulphur, Milbex, dicofol, chlorfenson, lindane; and for *P. macropilis*: all fungicides tested (above) except binapacryl, lime-sulphur, and Morestan; all acaricides tested (above) except phencapton and chlorobenzilate.

Several pesticides have displayed a repellent action upon *T. finlandicus*, *P. macropilis* and *T. pyri* (Dabrowski, 1969 a), but this was not confirmed by Hobart *et al.* (1970) for *P. persimilis*.

Field studies in various countries have also shown significant variations depending upon:

(i) *The predominant developmental stage.* A majority of eggs in the population diminishes the destructive effect of pesticides, with the exception of ovicidal and larvicidal acaricides. Nypmhs are usually more sensitive than adult mites.

(ii) *The species structure of the phytoseiid community.* Dominant species in untreated orchards are usually replaced by others less susceptible after repeated pesticide applications. This phenomenon was noted in orchards in England, the Netherlands, Canada, USA, Poland and Germany.

(iii) *The behaviour of predatory mites.* After treatment some phytoseiids migrate from treated leaves into crevices on the branches. This may cause rapid fluctuations in the number of phytoseiids present on leaves during the first two weeks after pesticide application.

Aside from development of resistance, some changes in the predatory mite complex in successive growing seasons have modified the apparent effects of pesticides. These have involved the different levels of sensitivity in successively dominant species of phytoseiids, and consequently, increase in phytoseiid numbers compared with untreated orchards after some pesticide applications, furthermore changes conditioned by the presence or lack of sufficient prey as food, and changes related to decreased numbers of predatory insects in orchards treated with pesticides. Contact insecticides (e.g. methoxychlor) have caused large decreases in predatory insect populations and usually decreases in phytoseiids as well.

Observations on the long-term effects of 23 pesticides in an experimental apple orchard in Poland have shown that six compounds resulted in significantly decreased numbers of phytoseiids during the second year following cessation of treatments. Only parathion-ethyl and schradan, however, significantly reduced phytoseiid densities over three consecutive years after the last application. Phytoseiid densities on trees treated with all other pesticides in previous years were the same as those on untreated trees. Other observations in untreated plots within commercial orchards have shown that the biological balance of arthropod populations has re-established subsequently, sometimes only after three to five years.

In conclusion, it should be stressed that earlier general statements that pesticides completely eliminate all phytoseiids from orchards are untrue. Some compounds are sufficiently safe to be used for integrated control in appropriate situations. They include: emulsifiable oils, dicofol, tetradifon, tetrasul, Plictran, chlorofenson, chlorobenzilate, Milbex, Omite, chlorobenside, Ryania, endosulfan, Imidan, diazinon, fenthion, Isolan, Gardona, phosalone, ethion, azinphosmethyl, captan, ziram, dodine, Eurapen, Morestan, wettable sulphur, Wepsyn, Glyodin. In the case of other compounds regarded as more harmful to phytoseiids, it was found that the time of application, variations in mite behaviour, and differential sensitivity between developmental stages can contribute to survival of predaceous mites.

Pesticide resistance in phytoseiids

T. occidentalis was reported to be resistant to, or tolerant of, parathion by 1952 (Morgan & Anderson, 1958), but studies by Huffaker & Kennett (1953) and Croft & Jeppson (1970) suggest that this was a natural tolerance. A strain of *T. fallacis* (Garm.) which occurred in laboratory colonies of spider mites was reported apparently resistant to DDT (Smith *et al.*, 1963) while populations of *T. caudiglans* from a peach orchard were also apparently resistant to DDT (Herne & Putman, 1966), but these populations were not subjected to comparative laboratory studies. Hoyt (1965) indicated that *T. occidentalis* from apple in Washington was resistant to some chlorinated hydrocarbon and several organophosphorous compounds. Resistance in this strain and in a strain from Utah to several compounds was confirmed by Croft & Jeppson (1970). Three other cases of resistance in phytoseiids have been confirmed by laboratory studies. These are the resistance of *T. fallacis* from apple to azin-

Table 22. *Comparison of resistance to azinphosmethyl in three species of Phytoseiidae*

Species	Azinphosmethyl formulation used	LC$_{50}$ value in ppm		Ratio R:S
		S strain	R strain	
*T. occidentalis**	Technical in acetone	20	2030	101:5
A. fallacis†	50% WP	43	4300	100
T. pyri‡	50% WP	72	700	9.7

* Modified from Croft & Jeppson (1970).
† Modified from Motoyama *et al.* (1970).
‡ From Hoyt (1972).
WP = weight per cent.

phosmethyl and parathion (Motoyama, Rock & Dauterman, 1970), of *A. hibisci* from citrus to parathion (Kennett, 1970) and of *T. pyri* from apple in New Zealand to azinphosmethyl (Hoyt, 1972).

More detailed studies on the potential use of three of these species have been made.

T. pyri is effective in regulating the density of *Panonychus ulmi* in New Zealand when not destroyed by pesticides (Collyer, 1964*b*). But resistance to azinphosmethyl (Table 22) was apparently too low to allow adequate survival for the regulation of *P. ulmi*. This predator could only be of value in an integrated programme based on azinphosmethyl if additional selection induced higher levels of resistance.

A. fallacis occurred in commercial orchards in eastern and southern states (USA) in a few years before the existence of resistant strains was confirmed in North Carolina in 1969 (Motoyama *et al.*, 1970). Table 22 shows a very high level of resistance to azinphosmethyl, and cross-resistance to Gardona® to Imidan®, a slight increase in tolerance to Omite®, but it was more susceptible to Plictran® than the S strain (Rock & Yeargan, 1971). That *A. fallacis* has a high level of resistance to azinophsmethyl was shown in Michigan (Croft & Nelson, 1972), and also cross-resistance to several organophosphorous compounds but not to phosalone and diazinon. No negative correlation in resistance to azinphosmethyl and susceptibility to Plictran® was observed.

Much interest has been expressed in using resistant strains to control *T. urticae* and *P. ulmi* on fruit trees in the USA. Commercial programmes involving *A. fallacis* are now in use in Michigan and New Jersey. It was suspected that *T. occidentalis* was resistant to some organophosphorous compounds and to DDT before 1960 in Washington State, but resistance was masked by extensive use of other compounds (carbaryl, dicofol, azinphosmethyl) which were toxic to that species. By 1965 predators apparently resistant to some insecticides were observed in many orchards in the Washington State (Hoyt, 1966) and a commercial programme of integrated control using *T. occidentalis* was initiated in 1966 (Hoyt, 1969 *a, b*). This programme is used successfully on most of the apple acreage in Washington, and to some extent in other western states and in British Columbia. *T. occidentalis* are also being used on pears in Oregon, and in grapes, peaches and strawberries in California.

Croft & Jeppson (1970) showed the Washington strain of *T. occidentalis* and a strain from Utah to have a high level of resistance to azinphosmethyl (Table 22) and cross-resistance to Gardona®. The Washington strain also had general cross-resistance factors or a general vigour tolerance as it exhibited slightly higher tolerance to carbaryl, dicofol, oxythioquinox, and Omite® than the susceptible strains or the Utah strain. Resistant *T. occidentalis* from Washington and Utah and a susceptible strain were released into sprayed apple orchards in California in 1969 (Croft & Barnes, 1971). The resistant predators increased and regulated populations of *Tetranychus mcdanieli* at lower levels than did the susceptible strain. By 1970 the resistant predators had migrated up to nine orchard rows from the release sites and had probably hybridised with the native population (Croft & Barnes, 1972). Croft (1972) indicated that researchers in Michigan are attempting to hybridise resistant *Typhlodromus occidentalis* with *T. longipilus* Nesbitt (these forms may be conspecific) to develop a resistant strain of *T. longipilus* which is adapted to Europe and eastern North America.

The recorded cases of pesticide resistance in phytoseiids are from fruit trees. This is probably due to the extensive selection pressure induced by repeated applications of pesticides. It seems probable that other resistant strains might develop more readily if they were not destroyed by other pesticides to which they are not resistant, or because their prey are kept at low levels by the repeated use of acaricides.

Croft & Barnes (1971, 1972) indicate that resistant phytoseiids can be successfully introduced to other areas and further studies on the ability of resistant strains of phytoseiids to adapt to new areas and habitats seem warranted.

Roles of natural enemies in the control of spider mites on various crops

Glasshouse crops

Tetranychus urticae has become increasingly difficult to control with acaricides in glasshouses over the past 20 years and so the demonstration of efficient predation on it by *Phytoseiulus persimilis* by Dosse (1958) stimulated interest in its biological control. Workers in many countries confirmed the efficiency of this predator although some questioned whether it could be used for control under commercial conditions. In the United Kingdom, Switzerland, Finland and Holland (Bravenboer, 1972), however, it is now mass-produced by private companies for commercial use, and in 1972 about ⅓ of the cucumber crop was treated with this predator in those countries. A smaller but increasing acreage of tomatoes is also under treatment.

Much of this acreage was treated by introducing the predator uniformly throughout the crop when spider mite damage was first detected. Although control is inevitably achieved by *P. persimilis* some damage may occur when the predator fails to spread sufficiently rapidly. Whereas crop reduction undoubtedly occurs in severely damaged patches some growers are content to accept this loss for the benefits from eliminating acaricide applications. In the United Kingdom, however, a more comprehensive approach has been undertaken at the Glasshouse Crops Research Institute.

Since the problem of control under commercial conditions is greatly influenced by the different cultural systems used, the various crops are considered separately.

Cucumbers

The first step was to seek basic information to develop techniques to reduce the selection pressure from acaricides. Work on damage assessment revealed that cucumbers could lose almost 30 % of their photosynthetic area without crop loss and that, not surprisingly, mite damage increased at a predictable rate under the uniform environmental conditions. The economic threshold was established as a leaf-damage index that could easily be determined by the grower or adviser. Where use of *Phytoseiulus* was contemplated, it was realised that if control was to succeed then the pest/predator interaction must proceed predictably and uniformly throughout the crop. Thus, as a prelude to control, the pest is uniformly distributed throughout the crop (10–20 mites per plant), either on the propagating benches or within a few days after planting. When the leaf-damage has reached a relatively low level (0.1 in a scale of 0 to 5) two predators are released on to alternate plants. Control is achieved in three to four weeks without exceeding the economic damage threshold of 1.9 providing that no further spider mites migrate into the crop. Normally, *T. urticae* emerges from hibernation in glasshouses over a period of about three months after planting and the gradual increments of prey ensures predator survivals and provide active control for that period. Otherwise, the predators would have to be sustained by artificial re-introductions of spider mites. Throughout this time *Phytoseiulus* kills the invading mites before any feeding centres exceed 1 cm in diameter. It is important to realise that the predator will eventually eliminate the spider mite and so it should be possible to eradicate the pest from a nursery. Biological control may not, therefore, be permanently necessary and should enable acaricides to be brought back again into effective use against spasmodic outbreaks when, in the absence of continued selection pressure by pesticides, any resistance should have disappeared.

In commercial greenhouses, successful biological control must operate while other pests and diseases are controlled, either chemically, biologically or culturally. Attempts are made to develop biological techniques for pests which are difficult to control chemically either because of widespread occurrence of resistant strains or because of difficulties with chemical residues or with phytotoxicity.

Hussey & Bravenboer (1971) and Anonymous (1972) described control procedures for other pests of glasshouse cucumbers which do not interfere with use of *Phytoseiulus* for mite control.

Development of an integrated control programme of this kind (Gould, 1970, 1971) depends on the availability and selection of suitable chemicals for all the pests and this has created difficulties in laboratory screening. Sometimes the more subtle, indirect rather than direct, toxicity effects are of greatest importance. Work by Parr (1971) and Binns (1971) revealed that benomyl can have dramatic effects on *P. persimilis* through the predator consuming contaminated prey and hence becoming sterile. Gould (personal communication) demonstrated that control of spider mites is unaffected when cucumbers are drenched by benomyl but that severe damage followed the loss of predator control where the fungicide was applied to the foliage as a high volume spray. Because of these and other results showing altered behaviour of various natural enemies exposed to certain pesticides, no chemical can be safely evaluated for integrated control by laboratory toxicity assay alone. Practical field trials furnish the only safe criteria.

Tomatoes

Pre-establishment of spider mites on tomatoes, as with cucumbers, is normally rendered uneconomic by dense planting (13 000 plants/acre or 32 500/ha, as compared with 3500/acre or 8 750/ha for cucumbers). But some method of achieving a uniform distribution of both pest and predator is essential. Excellent results were recently achieved by placing both organisms on every fifth plant. Because of the large number of plants involved and of the problem of mites emerging from diapause, it is essential to achieve this pre-establishment as early as possible. It is convenient to introduce the mites while the plants are still in pots on the propagating benches. A fifth of the plants are therefore segregated on the benches for infestation by the methods previously outlined for cucumbers and then planted out uniformly at a rate of 1:5. This degree of dispersion ensures that control on the infested plants is followed by rapid dispersion of the predators over the other plants. In trials the predator was found throughout the crop 21 days after planting.

Again, other pests and diseases must be contained in a manner not inimical to *Phytoseiulus*. Whitefly is controlled by *Encarsia formosa* Gah., the aphid *Myzus persicae* by either pirimor or the parasite *Aphidius matricariae*. Promising control of the leaf-miner *Liriomyza bryoniae* Kalt. has been achieved by dimethoate granules applied to the potting compost at 2 ppm. As far as fungal diseases are concerned benomyl drenches or dichlofluanid sprays can be used satisfactorily.

Ornamental crops

Pest control on chrysanthemums will probably be achieved by soil treatment with aldicarb in the future, but, as strains of *T. urticae* resistant to this material have already been reported from Scandinavia (Markkula & Tiittanen, 1970), a method was developed for controlling mites with *Phytoseiulus* and aphids with *Aphidius* by treating boxes of cuttings with both hosts and natural enemies before planting. Forty *M. persicae*, 15 % of which are parasitised by *Aphidius*, are released on every 400 cuttings, together with 400 spider mites and eight *Phytoseiulus*.

Successful control of mites by use of *P. persimilis* on roses (Simmonds, 1972), dahlias (Harris, 1971), and other ornamental crops (Gould & Light, 1971) has also been reported.

The predator *Phytoseiulus* has therefore proved an effective control for red spider mite in glasshouses and its use seems likely to become an accepted commercial control technique.

Strawberry

In southern California, commercial strawberries are primarily grown as an annual crop consisting largely of summer plantings (planted in August) with a few winter plantings (planted in October or November). Harvest usually starts in February and continues through June, after which the plants are disked under. Before planting, the soil is treated with a 2:1 mixture of methyl bromide and chloropicrin (Oatman & Voth, 1972).

In the central coastal areas of California, strawberries are grown as a two-year crop, with plantings made in the fall and winter. Harvest begins in April and continues throughout the summer for two years before the plants are disked under. As in the south, the soil is treated with methyl bromide and/or chloropicrin prior to planting. In the central coastal area, the dormant plants are sprayed in winter with dinitrol and oil to aid mechanical pruning and to encourage new growth. Under these cultural conditions, insect populations seldom reach injurious levels, although the strawberry aphid, *Chaetosiphon fragaefolii* (Ckll.), occasionally requires treatment. The cyclamen mite, *Steneotarsonemus pallidus* (Banks), is not a problem in southern California, but some growers spray routinely to control this species although it can be

215

kept under good biological control by *Typhlodromus cucumeris* (Oud.) and/or *T. reticulatus* Oud. in older fields which are not heavily treated for other pests.

The principal pest under the systems in both areas is *Tetranychus urticae*, which often reaches 100 to 500 active stages per leaflet, seriously affecting plant growth and vigour and occasionally killing plants (Oatman & McMurtry, 1966; Oatman *et al.*, 1967).

In California, eight insect species and nine phytoseiid species are predaceous on *T. urticae* on strawberry (Oatman *et al.*, 1967; McMurtry, Oatman & Fleschner, 1971; Oatman & Voth, 1972).

The Phytoseiidae *Typhlodromus arboreus* Chant, *T. occidentalis*, *Amblyseius californicus* (McGregor), *A. brevispinus* (Kennett), *A. hibisci*, *A. lindquisti* (Shust. & Pritch.), *A. aurescens* Athias-Henriot, *T. cucumeris*, and *A. similoides* Buch. & Pritch. (McMurtry *et al.*, 1971) occur on strawberries in California. *T. occidentalis* and *T. arboreus* in this order are usually the most common species. The predaceous mites are most abundant during the spring months when their prey are common.

Two species of thrips are also present, and one of these, *Scolothrips sexmaculatus* (Pergande), is the commonest predator in southern California in terms of numbers. It comprises more than 80 % of the total predators (Oatman *et al.*, 1967; Oatman & Voth, 1972). However, in the central coastal area, *T. occidentalis* is more common due to its abundance in second-year fields (Laing, personal communication). Both *Tetranychus urticae* and *S. sexmaculatus* are present throughout the time that the plants are available.

The coccinellid *Stethorus picipes* is the second most common insect predator in the spring months and the minor predator beetle *Oligota oviformis* (Casey) is also present.

The Hemipterans *Orius tristicolor* and *Geocoris punctipes* (Stål) are present in relatively low numbers from September to November and again in the spring.

An unidentified species of hemerobiid is the third most common insect predator and is present primarily during the spring.

The cecidomyiid fly *Mycodiplosis acarivora* (Felt) is similar to *Oligota oviformis* in numbers and in distribution in time except that it also occurs in low numbers in September and October.

These native predators, especially the phytoseiids, are adversely affected by prevailing spray practices. Soil fumigation (Oatman *et al.*, 1967; Oatman & Voth, 1972) and, in the central area, use of dinitrol and oil and severe prunings affect the overwintering predators (Laing, personal communication). Because of these frequent disturbances only the more mobile insect predators can readily migrate to the strawberry fields as their prey populations increase. However, in spite of this early entrance of predators, *T. urticae* usually reaches extremely high populations in March, April, or May in southern California and then abruptly declines. In the central coastal area, the pest reaches its highest populations in May, June, or July, then also rapidly declines. The population crash is caused primarily by physiological changes in the plants induced by high numbers of the spider mite (Oatman & Voth, 1972). In the central coastal area. *Typhlodromus occidentalis* can prevent high densities of *Tetranychus urticae* if it remains in synchrony with its prey and is not killed by insecticides.

To help the native predators suppress *T. urticae* to below damaging numbers, the introduced phytoseiid, *P. persimilis* has been released in strawberry fields in southern California shortly after a plastic mulch has been installed, usually in late January for summer plantings. Weekly releases over six to eight weeks at rates equivalent to a total of 300000 to 320000 predators/acre (750000 to 800000/ha) have resulted in successful control (Oatman, McMurtry & Voth, 1968).

P. persimilis overwinters on untreated strawberries in the central coastal area. Experimental releases of two to four predators on every alternate plant in the fall resulted in good control of *T. urticae* throughout the following growing season. However, this control of the pest was assisted by an increase of the native predator *Typhlodromus occidentalis* during the following season (Laing, personal communication).

Apple

Phytophagous mites are major pests wherever apples are grown. They increase rapidly when pesticides detrimental to their natural enemies are used. Important species of spider mites include the European red mite, *Panonychus ulmi*, the two-spotted mite, *Tetranychus urticae* and the brown mite, *B. rubrioculus*, which are widely distributed. *T. mcdanieli* is mainly restricted to western North America. Species of lesser importance are *T. viennensis* Zacher in Japan (Mori, 1967) and *T. cinnabarinus* and *T. lobosus* Boudr. in Australia (Gibson, 1966).

The apple rust mite, *Aculus schlechtendali* (Nal.), has attracted attention and is included here because it is an important food source for predators of spider mites (Hoyt, 1969a; Herbert & Sanford, 1969). Although this species can be injurious, relatively high populations are tolerable on apple and, in the absence of tetranychids, provide food to maintain predators at favourable population levels. It is found in all apple growing areas of North America, Europe, and Australia, and probably elsewhere.

The natural enemies of spider mites are similar in all apple-growing areas, but the prevalence and importance of each species varies. Most of the important predacious mites are of the family Phytoseiidae; members of the Stigmaeidae, Bdellidae, Trombidiidae, Anystidae, and Chyletidae are of lesser importance. Among the predaceous insects the family Miridae is perhaps the most important although their prey is not restricted to mites. Other sometimes important predators include *Stethorus* spp. and several species of Anthocoridae, the latter being noteworthy for their relative tolerance to pesticides. Predaceous thrips (*Thysanoptera*) are part of the predator complex in several areas and often are useful at high prey density.

Gilliatt (1935) was one of the first to recognise the importance of predation in the control of phytophagous mites on apple. Later, Pickett *et al.* (1946) and Lord (1949) laid the foundations of integrated control for orchard pests in general and for phytophagous mites in particular. Recent investigations are summarised below.

North America

Under natural conditions in Nova Scotia predators are numerous and diverse and provide adequate control of phytophagous mites so that chemical control is necessary only when natural enemies have been inhibited artificially (Sanford & Herbert, 1967, 1970). Only two species of phytophagous mites are economically important, *P. ulmi* and *Aculus schlechtendali*, both of which are susceptible to natural control. The periods of activity for predaceous mirids and anthocorids are as follows. Early season: *Anthocoris musculus*, *Atractotomus mali* (Meyer), *Campylomma verbasci* (Meyer), *Plagiognathus obscurus* (Uhl.), *Deraeocoris fasciolus* Knight, and *Diaphnocoris* spp. Mid season: *Blepharidopterus angulatus* (Fall.), *Hyaliodes harti* Knight, *Coniortodes salicellus* (H. & S.), *Pilophorus perplexus* (D. & S.), and *Phytocoris* spp. Late season: *Deraeocoris nebulosus* (Uhl.), *O. insidiosus*, and second generations of *C. verbasci*, *Anthocoris musculus* and *Diaphnocoris* spp. The phytoseiid *Z. mali* and *Haplothrips faurei* Hood, and *Leptothrips mali* (Fitch) are present throughout the season. Fourteen species of phytoseiids have been recorded of which *Typhlodromus pyri* is most prevalent. Under certain conditions *T. rhenanus*, *T. fallacis*, *T. finlandicus* and *Phytoseilus macropilis* are also numerous. Less commonly found are the predaceous mites *Anystis agilis* Banks and *Atomus* spp., and species of chrysopids, coccinellids, clerids and nabids.

It is evident that many species form the predator complex on apple. Those such as *H. faurei*, *L. mali*, *Diaphnocoris* spp. and *A. musculus*, which have more than one generation, are more responsive to seasonal changes in mite density than are single-generation species and thus regulate better. Phytoseiids do not become numerous in Nova Scotia even when prey populations are high, probably because they are eaten by other predators, chiefly mirids. It appears also that the predaceous mites maintain phytophagous mites at population levels that are too low to support many mirids. Recently, with the use of some of the newer pesticides in particular the phosphates, the complex of predators has changed. More susceptible species, such as the mirids, are removed but control may still occur because the more resistant species, such as *Z. mali,* increase in importance and can be assisted by the judicious use of selective acaricides.

In apple-growing areas of Quebec and Ontario mite predators exist but they are suppressed by the harsh spray practices followed. Phytoseiids and *Z. mali* are the main predaceous species (Parent, 1967; Parent & Lord, 1971). Swift (1968, 1970) considered *T. fallacis* to be the only important predator and suggested that its removal by insecticides caused *P. ulmi* and *Tetranychus urticae* to increase, implying that it has a regulating effect.

Horsburgh & Asquith (1968) surveyed the predator complex in Pennsylvania apple orchards and reported the presence of 32 species: one anthocorid, six coccinellids, eleven mirids, one thripid, two phleothripids, nine phytoseiids, one bdellid, and one stigmaeid. *O. insidiosus, S. punctum, Typhlodromus pomi* (Parrott), *T. fallacis,* and *Z. mali* were found most frequently in sprayed orchards. *S. punctum* was the only predator that occurred in sufficient numbers under their integrated spray programme to be useful (Asquith & Colburn, 1971). These authors point out that the overall population of *S. punctum* has risen dramatically as the acreage under integrated control has increased, making it easier for additional growers to implement successful integrated programmes.

Holdsworth (1968, 1972*a*) developed an integrated control programme involving *Panonychus ulmi* in an experimental apple orchard in Ohio, and listed the major useful predators: *L. mali, O. insidiosus, Plagiognathus politus* Uhl., *Chrysopa rufilabris* Burm., *C. carnea* Stephens, and *S. punctum.* Other predators present were *H. harti, Hyaliodes vitripennis* (Say), *D. nebulosus, Hemerobius humulinus* L., *Z. mali* and *Agistemus fleschneri* Summers (Holdsworth, 1972*b, c*).

Thirteen species of phytoseiids have been reported in Missouri apple orchards (Poe &·Enns, 1969) but only two in sufficient numbers to offer promise as biological control agents. One, *Galendromus longipilis* (Nesbitt), is readily eliminated by pesticides, but re-invades orchards after cessation of spraying. Only the other, *T. fallacis,* is numerous throughout the year.

Effective natural control of *Tetranychus mcdanieli,* the most important species in British Columbia, Washington, and Oregon, is provided by *Typhlodromus occidentalis.* Because of the widespread use of pesticides for codling moth this species has developed resistance to organophosphates and other pesticides. It can survive most pest control programmes. Unfortunately, it does not adequately control *P. ulmi.* In British Columbia and Washington, *A. schlechtendali* is also present and serves as an alternative food until *T. mcdanieli* populations become troublesome (Hoyt, 1969*a, b*; Madsen, 1968). In Oregon *Aculus schlechtendali* is not present and *T. occidentalis* is unable to increase sufficiently quickly to suppress *Tetranychus mcdanieli* without assistance from selective acaricides (Zwick, 1972).

Other predators reported from British Columbia (Madsen, 1968) and

Washington (Hoyt, 1969*b*, Hoyt & Caltagirone, 1971) include *S. picipes*, anthocorids, the mirid *Deraeocoris brevis* Uhl., and *Z. mali*, but none occurs in sufficient numbers to be important.

In the Pacific Northwest the main predator is *Typhlodromus occidentalis*. Over the years this species has developed resistance to many pesticides, particularly the organophosphates, thus allowing their use to control other pests without affecting it. Croft & Barnes (1971) released resistant and non-resistant strains of *T. occidentalis* on apple trees in southern California and Croft & McMurtry (1972) concluded that such release programmes could be both practical and economical.

Other apple-growing areas

Solomon (1972) studied the predators in English apple orchards and found that many *Anthocoris* spp., *O. minutus*, and *B. angulatus* occurred with *P. ulmi*. Reviewing earlier work, Dicker (1967) stated that in addition to the species listed above, *Typhlodromus* spp., *Psallus ambiguus* (Fall.), *Atractotomus mali*, *Orthotylus marginalis* Reut. and *Phytocoris* spp. are predatory on *P. ulmi* in England.

In Holland, Van de Vrie & Kropczynska (1967) demonstrated that *T. potentillae* Gar. could reduce high densities of *P. ulmi*, even ones possessing a maximum fecundity under a high nitrogen fertilisation programme.

In Poland, *C. verbasci*, *Atractotomus mali*, *Anthocoris nemoralis*, *A. nemorum*, *Orius minutus*, *Psallus ambiguus*, *Himacerus apterus* (Fab.), *O. majusculus* (Reut.) and *Phytocoris dimidiatus* Kbm. were important (Korcz, 1967, 1969).

Karg (1970) mentions 11 species of predatory mites in East German orchards, *T. tiliarum* (Oud.), *T. finlandicus*, *Phytoseiulus macropilis*, and *Z. mali* being dominant.

Near Minsk, USSR (Sidlyarevich, 1965), *Panonychus ulmi* was preyed on by the following phytoseiids: *T. pyri*, *T. soleiger*, *T. rhenanus*, *T. subtilisetosus* Begl., *T. finlandicus* and *P. macropilis*. *A. nemorum* was especially effective in spring and autumn while phytoseiids and other predators were important in the summer. The predators observed were *Pilophorus perplexus*, *Atractotomus mali*, *Globiceps sphaegiformes* (Rossi), *Malacarcoris chlorezins* (Panz.), *Campylomma verbasci*, *O. pilosus* and *O. niger* (Wolff.). More recently, Bondarenko & Emelyanov (1970) found that *Typhlodromus subsolidus* Begl. was numerous and widespread on apples in the Leningrad area. Populations of *P. ulmi* rarely exceeded the economic threshold in orchards where these predators were present.

A complex of predators consisting of phytoseiids, anthocorids, chrysopids, thrips, and coccinellids is considered important in the control of the phytophagous mite complex in apple orchards in Japan (Hukusima, 1968, 1969).

In Australia *P. ulmi* and *Tetranychus urticae* are largely controlled in apple by predatory beetles, *Stethorus* spp. (Readshaw, 1971).

The more complex predaceous faunas are found in orchards close to wooded areas or to hedgerows that act as their natural reservoirs. Few attempts have been made to propagate and release or to introduce predators for mite control in commercial orchards. Effective control of mites has been achieved in some areas by a single species of phytoseiid; in others a complex of predators seems necessary. However, a diverse predator population that includes predaceous insects (anthocorids, mirids, thrips, coccinellids and chrysopids) in addition to phytoseiids and some other species of predaceous mites, will provide a more durable and dependable regulation, with less fluctuations in the phytophagous mite complex than one or a few species. Within a complex of this nature the predaceous insects are, in general, more mobile and have a greater prey consumption capacity than the predaceous mites. They also require more prey to maintain their own numbers.

Successful natural regulation of mites requires faunal management of commercial apple orchards. Pesticides detrimental to predators must not be used. Food for the predators must be provided either by the prey species at numbers below the economic threshold or by alternate prey that are of little or no economic importance.

Peach

Information on the management of natural enemies to control spider mites on peaches is meagre and not yet clearly demonstrated. Nevertheless, the importance of natural enemies in the regulation of mites on peaches is generally accepted. In their lists of promising predators and pathogens of spider mites, McMurtry *et al.* (1970) cited two North American references to predators of spider mites on peaches and Van de Vrie *et al.* (1972) listed two, while Boczek *et al.* (1970) did not mention any specific natural enemies for eastern Europe, South America, and Japan, respectively. Kazimi & Ghani (1970) cited two predators as important in controlling tetranychids on peaches in Pakistan.

Putman & Herne (1966) stated 'Earlier in these studies the usual scarcity of predators in orchards where *P. ulmi* was also scarce lead us to doubt whether endemic densities of the mite were actually regulated by these agents. Further consideration based on later work however has removed these doubts.' But they acknowledge that they 'have not in any sense "proved" the effectiveness of the predaceous species either individually or collectively'.

The phytophagous mite species of most importance are *T. urticae*, *P. ulmi*, and *B. rubrioculus*. To these must be added the eriophid *Aculus cornutus* (Banks) not so much because it is a pest but because it plays an important role in natural control of the spider mites.

The phytoseiids appear to be the most important predators of spider mites

on peaches. In Ontario, *Typhlodromus caudiglans* is the most important predator of *P. ulmi* and *B. rubrioculus* (Putman & Herne, 1966). *T. occidentalis* is the
dominant predator of mite pests on peach in parts of California (Hoyt &
Caltagirone, 1971) and of *Tetranychus urticae* in Washington (Tamaki &
Powell, 1972). The introduced *Phytoseiulus persimilis* is an important predator
of *T. urticae* on indoor peaches in the Netherlands (Bravenboer & Dosse,
1962). *S. punctillum* is an important predator of *Panonychus ulmi* in Ontario
peaches (Putman & Herne, 1966), while *S. gilvifrons* Muls. and *S. tetranychi*
Kapur are useful predators of *E. orientalis* Klein on peaches in Pakistan
(Kamizi & Ghani, 1970). The phloeothripid *H. faurei* and the chrysopid *C.
carnea* are considered useful predators of phytophagous mites on peaches in
Ontario, whereas the anthocorid *Orius* sp. is not very important because it
commonly occurs in very low numbers although it sometimes shows numerical
response to the fluctuations of mite densities (Putman & Herne, 1966). The
role of spiders as regulatory agents has not been demonstrated or appreciably
claimed. They have been reported as preying on tetranychids on peaches
(Putman, 1967; Putman & Herne, 1966).

Putman & Herne (1966) reported a disease, later proved to be a virus (Bird,
1967), of unknown etiology in natural populations of *P. ulmi* on peach in
Ontario. Field epizootics may occur from early August to late September.
Apparently the virus does not reduce mite populations early enough to prevent foliage damage, but it can restrict production of winter eggs (Putman,
1970).

Putman & Herne (1966) noticed that mirids are virtually absent from peach
orchards. The same authors concluded after some 25 years of study that
'endemic densities of *P. ulmi* are maintained in that state by predators,
chiefly *Typhlodromus caudiglans*, that subsist to a considerable extent on other
sources of food, whereas epidemics are reduced largely by other predators,
chiefly *H. faurei* and *S. punctillum*, that increase by feeding on the mite during
its period of rapid population growth but exert their greatest effect through
later destruction of winter eggs'.

Alternative prey seem to play an important role in the control of mites by
phytoseiids. Putman & Herne (1966) indicate that *T. caudiglans* increases most
rapidly where the peach silver mite, *A. cornutus*, is abundant and is then most
likely to prey heavily on *P. ulmi*. A similar dependency on *A. cornutus* exists
in California with the control of tetranychids by *T. occidentalis* (Hoyt &
Caltagirone, 1971).

Tactics to maximise the control achieved by mite predators in peach
orchards have not been adequately developed. One estimate needed as a basis
is the size of the overwintering population of phytoseiids. It seems that in
temperate regions phytoseiids overwinter as mated females (McMurtry *et al.*,
1970). Putman (1959) studied the overwintering sites for several phytoseiids in
peach orchards in Ontario. He found that *T. caudiglans* (as *T. rhenanus*) over-

winters in cankers on limbs, *T. pyri* in cankers and trunks, *T. andersoni* Chant in trees and ground litter, and *T. fallacis* and *T. cucumeris* in litter or soil. Caltagirone (1969) reported that *T. occidentalis* overwinters in the fruit stalks left on the trees when the fruits are picked. They congregate under loose bark at the base and in the spongy tissue at the tip of the stalk. Apparently, two-year-old stalks or older are preferred. No phytoseiids were found over-wintering under bud scales or in bark crevices on the branches.

Populations of *T. occidentalis* in California are comparatively resistant to organophosphorous compounds commonly used in orchard pest control. This facilitates the manipulation of this predator. When overwintering populations reach some 1.2 mites per fruit stalk in January there are sufficient predators to keep tetranychids under control during the season provided they are not too greatly reduced by chemicals. Two practices damaging to them are the early spring application of organophosphorous compounds, and sulphur treat-ments to control the silver mite, *A. cornutus*. Applications of organophos-phorous compounds early in the spring kill so many overwintered *T. occi-dentalis* that it is unable to prevent mite damage.

Applications of sulphur to control brown, red and/or silver mite greatly reduce *T. occidentalis* because its alternative prey, the silver mite, is almost eliminated. This is especially noticeable in varieties on which sulphur is applied late in the season.

Tamaki & Powell (1972) reported that the defoliant sodium chlorate used to control green peach aphid, *Myzus persicae*, on peach and applied early in the season (September), causes many *T. occidentalis* to fall with the leaves. Presumably, these predators do not attain a successful diapausing stage. Fewer predators are affected if the defoliant is applied later.

Abundant peach silver mite is essential to ensure sufficiently high *T. occi-dentalis* densities to control tetranychids on peaches. Consequently it is im-portant to avoid treatments with chemicals that eliminate or seriously reduce peach silver mite populations. The economic injury level of this eriophyid is not known, but it is believed to be of minor importance (Putman & Herne, 1966). Hoyt & Caltagirone (1971) proposed a pest management programme for peaches in California which consists of the following:

(i)   control of peach twig borer, San José scale, and mites with dormant spray;

(ii) elimination of the early spring sprays;

(iii) control of oriental fruit moth, timing the treatments very carefully, and for which sprays during the season may be sufficient;

(iv) reduction or elimination of the sprays to control two-spotted spider mite and European red mite, with treatments not necessary, if there are enough predators; and

(v)   limitation of treatments against peach silver mite to the absolute minimum.

Grape

Despite a large complex of natural enemies on San Joaquin Valley grapes the phytoseiid *T. occidentalis* is the essential element regulating the spider mites *T. pacificus* and *Eotetranychus willamettei* Ewing at sub-economic levels. Several insect predators, such as *Scolothrips sexmaculatus* and *O. tristicolor* may occasionally be important in controlling outbreaks but they do not appear to be efficient because of their large food requirements, low numerical responses and, particularly, their sporadic occurrence and distribution in vineyards (Flaherty & Huffaker, 1970).

T. occidentalis has a sufficiently large numerical and functional response (Laing & Huffaker, 1969), and sufficiently low food requirements to be capable of reducing high densities of the *T. pacificus* and maintaining it at low numbers for indefinite periods provided the predators remain well distributed throughout the vineyard (Flaherty & Huffaker, 1970; Kinn & Doutt, 1972). Flaherty & Huffaker (1970) considered that for such persistent regulation the predator must remain well dispersed and an alternate food source must be available during the late summer and early autumn months, the first condition arising from the second. This reliance on alternate food is necessitated by the lack of synchrony of the life histories of *T. occidentalis* and *T. pacificus*. Adult female *T. pacificus* begin to seek overwintering sites under the heavy bark of the vine and enter diapause as early as June. This behaviour continues through July and August. By late September and October only a few scattered individual *T. pacificus* remain on the grape leaves. *T. occidentalis* overwinters in different sites which they enter much later (October and November). Thus, it must have an alternative food supply during late summer and early autumn if they are to overwinter in sufficient numbers to prevent an early spring surge in *T. pacificus*.

In undisturbed vineyards the less seasonally restricted mite *E. willamettei*, several species of tydeid mites (*Pronematus* spp.), and/or occasionally *Tetranychus urticae* can support a substantial number of *Typhlodromus occidentalis* after the autumn decline of *T. pacificus*, and thus ensure that many predators enter diapause. As noted elsewhere, Hoy (1972) studied the effect of day length and temperature on diapause in this species and concluded that the critical photophase occurred during the first part of October when the day length was approximately 11.5 h. By early November 100 % of the females in the field were in diapause. By early February all field-collected females had terminated diapause. It is noteworthy that *T. occidentalis* on grape may feed on tydeid mites overwintering with it under the leaf bud scales.

Predator survival may be severely reduced in vineyards heavily treated with pesticides, not only because of the direct toxic effect on the predators and their primary prey but also because of the scarcity of alternative prey (Flaherty & Huffaker, 1970). *T. pacificus* populations in heavily treated vineyards typically crash during the summer, either because of severe feeding injury to the vines or because of overexploitation by *T. occidentalis*. Regardless of the cause, the predator declines to low numbers, and because of lack of alternative prey (due to use of the pesticides) is unable to increase during late summer and early fall to a level that can provide a sufficiently well-distributed number of adequately nourished, successful-diapausing, overwintering predators to be able to control the ensuing spring build-up of *T. pacificus*. Yet it is this early spring activity of *T. occidentalis* which seems to be necessary to prevent serious outbreaks of *T. pacificus* (Flaherty & Huffaker, 1970).

Because of the difficulty in distinguishing between *T. pacificus* and *E. willamettei*, the latter was at one time considered to be a serious pest of grapes (Pritchard & Baker, 1955). However, Flaherty & Huffaker (1970) demonstrated that this species is not as serious a pest in the San Joaquin Valley as was formerly believed but actually can be a valuable alternative prey for *T. occidentalis* because it is available during the critical time before early September (Hoy, 1972). Kinn & Doutt (1972) found that *E. willamettei* is also an important alternate food source on grapes in the northern coastal region of California.

Flaherty (1969) demonstrated the potential value of cover crops, e.g. Johnson grass, for the maintenance of *Tetranychus urticae* as an alternate prey for *Typhlodromus occidentalis*. *T. urticae* is usually not found on grapes but occurs in weed-infested vineyards, and moves from the drying Johnson grass and other low herbage to the vines late in the summer where it provides food for *T. occidentalis*.

The tydeidds *Pronematus anconai* Baker and *P. ubiquitus* (McGregor) may serve as alternative prey for *T. occidentalis* when spider mites are absent from vines. Flaherty & Hoy (1971) successfully reared large numbers of tydeids on wind-blown pollens of cattail (*Typha* sp.), bottlebrush (*Melaleuca* sp.), and grape (*Vitis* sp.). They also demonstrated that when cattail pollen is dusted on the vines the tydeid mite population increases and so, indirectly, do the numbers of *T. occidentalis* late in the season. The importance of tydeids for increasing the overwintering populations of *T. occidentalis* was also shown by Calvert and Huffaker (personal communication). In a non-commercial vineyard where few tetranychids occurred, various densities of tydeids were achieved through applications of cattail pollen as contrasted to areas receiving none. Following applications of pollen, tydeids began to increase in July and reached peak numbers in September, followed closely by an increase in predators. This late increase is characteristically absent from heavily sprayed vineyards. Chemicals such as sulphur were detrimental to the tydeid population and resulted in a reduced number of *T. occidentalis* reaching overwintering sites.

These studies show that though *T. occidentalis* plays a dominant role in the control of spider mites on grapes in California, it is dependent upon a complex of trophic relationship involving alternative prey and/or pollen sources. It is also significant that, once this balanced complex is disturbed by regular chemical spraying, natural control may not be regained for three or four years after the treatments stop (Flaherty & Huffaker, 1970), though re-establishment of control has been demonstrated in some instances in less than three years.

Knowledge of economic injury levels is critical in the utilisation of natural enemies in any integrated pest control programme. The quantity and quality of grapes was unaffected by maximum average densities up to 250 *T. pacificus* per leaf over a three- to four-week period (Laing, Calvert & Huffaker, 1972). However, these densities approached levels that possibly have a carry-over effect on the yield in the following season. Higher densities of *T. pacificus*, averaging up to 2500 mites per leaf over the five-week period, had no effect on the weight of grapes produced in the same season but caused a significant reduction in sugar content (Laing and Preston, personal communication).

Citrus

The phytoseiid *Amblyseius hibisci* is the most common predator on California citrus and can sometimes maintain *P. citri* at sub-economic levels. Data from both southern and central California areas (McMurtry, 1969; Kennett, personal communication) indicate both a functional and a numerical response by *A. hibisci* to increasing densities of *P. citri*.

A. hibisci reproduces more rapidly on pollen than on spider mites (McMurtry & Scriven, 1964). When wind-blown pollens become available in late winter and early spring, *A. hibisci* increases to densities as high as three per leaf, sometimes in the near absence of other foods. Warm, dry weather during later winter appears to increase production and dissemination of wind-blown pollens whereas cold, wet weather retards it. This apparently delays the peak numbers of *A. hibisci* until later in the spring.

The key to effective biological control of *P. citri* in central California is achieving a low prey/predator ratio during late winter. A ratio of less than three adult female *P. citri* to one *A. hibisci* during late winter has resulted in effective control (i.e. adult female *P. citri* of less than 0.5 per leaf) during the entire mite season. Early spring ratios of 10:1 or higher have resulted in *P. citri* densities as high as eight adult females per leaf (Kennett, personal communication), with subsequent leaf and fruit drop when trees are stressed during periods of high temperatures and low soil moisture. Leaf and fruit drop are more commonly encountered in southern than in central California. In the former region, serious infestations may occur in the fall and winter when *A. hibisci* does not usually produce control.

Efforts to enhance early-season predator populations through the applications of large quantities of commercial pollens to the trees have not proved effective (Kennett, personal communication).

In southern California, *A. limonicus* occurs only near the ocean (McMurtry & Scriven, 1965) where it is the main predator responsible for the control of *P. citri* in lemon trees in San Diego County. *A. peregrinus* (Muma) is the most common phytoseiid on Florida citrus (Muma, 1955). Extensive laboratory and field studies (Muma, 1967) indicate that it possesses many attributes of a good natural control agent, including an arboreal habit, high reproductive rate, broad range of primary, secondary, and survival foods, and some tolerance to pesticides. *A. peregrinus* sometimes showed direct or lag synchronisation with populations of *P. citri* and *Eotetranychus banksi* (McGregor) with peak predator numbers occurring during winter and spring. However, long-term studies suggested that this predator has little economic effect.

Metaseiulus floridanus (Muma) is considered to be an important predator of *Eotetranychus sexmaculatus* (Riley) on Florida citrus (Muma, 1970). In laboratory studies it was found to reduce greatly tetranychid populations (*P. citri, E. sexmaculatus, E. banski*) on citrus fruits within two weeks after three or four adult predators were introduced into each fruit or citrus seedling. Muma (1970) concluded that *M. floridanus* is responsible in many instances for the control of *E. sexmaculatus* in Florida citrus but is probably ineffective against other mites.

Stethorus spp. are generally considered to be 'high density' predators. *S. japonicus* is an important predator of *P. citri* in Japan (Tanaka, 1966) and an undescribed species is an effective predator of *P. citri* in Chile and Peru (Gonzalez, personal communication). It has been noted earlier that *Stethorus* spp. may be the principal regulating agents on apple in some areas. *Saula japonica* Garm., a predaceous endomychid, is considered to be the most important natural enemy of *P. citri* on citrus in Japan (Nohara, 1970); it has two to five generations per year, with long-lived adults that begin hibernation in October. Eggs are laid in small groups on the lower leaf surface and the first instar larvae are gregarious. *S. japonica* also feeds readily on sooty mould as well as on the scale insects *U. yanonensis* and *I. purchasi*.

A non-inclusion virus which attacks *P. citri* was first reported from southern California in 1959 (Smith *et al.*, 1959). Subsequent studies (Shaw, Tashiro & Dietrick, 1968) have shown the disease to be endemic in *P. citri* populations in both southern and central California. When *P. citri* numbers are low the disease remains at a chronic level and as numbers increase during the spring months epizootics can occur and result in near annihilation of the mites within a month and the *P. citri* population may remain at near-extinction levels until the following spring (Kennett, personal communication). Studies in Ventura County indicate that population crashes of *P. citri* are more closely correlated with an increase in percentage of mites infected with virus than with the abundance of *A. hibisci* (McMurtry and Shaw, personal communication).

Field studies indicate that other factors in addition to density, such as temperature and humidity, may contribute to the onset of epizootics. For example, in central California it appears that epizootics in dense populations are less likely to occur at cooler average temperatures (15 °C) than at warmer ones (20–24 °C).

The incidence of *Entomophthora floridana* (Weiser & Muma) on the Texas citrus mite, *Eotetranychus banksi*, indicates that the fungus may frequently reduce spider mite infestations in southern Florida. The potential exists all the time but is realised only occasionally. Intensive studies have not been conducted on the east and west coasts of Florida, but the fungus is known to occur there and probably possesses some utility (Muma, 1969).

Avocado

The avocado brown mite, *Oligonychus punicae*, sometimes reaches high and damaging numbers on avocado in southern California. Normally associated as predators of this mite are the phytoseiid *A. hibisci* and the coccinellid *S. picipes* (McMurtry & Johnson, 1966). *A. hibisci*, although usually outnumbering its prey in the spring when pollen is abundant, does not prevent rapid summer increases of *O. punicae*. As greenhouse studies indicate that *A. hibisci* could control this prey if pollen is also available (McMurtry & Scriven, 1966), field applications of cattail pollen were made at bi-weekly intervals during the summer, but with little effect.

S. picipes is the key predator of *O. punicae* (McMurtry & Johnson, 1966), although in some cases its response to prey increase was too late to result in economic control. Field releases of *S. picipes* were made early in the season to determine if heavy infestations of *O. punicae* would be prevented. In 1967, releases of 400 adults per tree in 16-tree plots in each of three study orchards resulted in an increased build-up of *Stethorus*, lowered peak populations of *O. punicae*, and lowered percentages of heavily-damaged leaves, compared to control plots (McMurtry, Johnson & Scriven, 1969). Releases of 200 adults per tree produced similar successful results in a subsequent experiment. In 1969 and 1970 *O. punicae* was generally scarce and control occurred on both release and control blocks. The problem of economic mass production of *S. picipes* poses a major obstacle to such release programmes.

8. Concluding remarks

M. J. WAY

It is a salutary reflection on our understanding of classical biological control that at present 'no amount of planning and preliminary research can replace the actual empirical search for natural enemies in the field and their trial by release in the new environment' (DeBach *et al.*, 1971).

It is probably unlikely that we will ever be able to predict with accuracy the consequences of introducing a particular natural enemy or of manipulating a particular part of the environment to favour indigenous natural enemies. However, there is now abundant evidence that success in biological methods of control, whether by introducing natural enemies or encouraging indigenous natural enemy species, or, by host plant resistance or autocidal methods, will depend in large measure on *appropriate* detailed understanding at all levels from the fundamental to the immediately practical. At the immediately practical level, there are usually clear-cut questions to answer; otherwise we can sometimes point specifically to the more basic information that is needed but mostly we have faith in fundamental knowledge 'for its own sake' which eventually may become relevant to problems of practical control. There is no doubt that, in devising control measures, whether by chemicals alone or by natural enemies or by more complicated procedures of integrated control, we depend upon and take for granted much fundamental information on the overall biology of pest and natural enemy species.

The IBP work on biological control has been limited to five groups of species, two known to be readily amenable to control by biological methods especially natural enemies, and the others which are not satisfactorily controlled at present and against which biological control methods (in the widest sense) must be improved either as a sole method of control or, much more likely, used in integrated control programmes. As indicated in Chapters 1 and 2, the IBP projects have not aimed specifically at practical biological control programmes but have aimed to encourage and coordinate relevant basic work, to collate existing information and to stimulate world-wide collaboration in the study of the insects concerned thereby perhaps initiating the kinds of collaborative work which are so urgently needed for the understanding and solution of many pest problems.

The foregoing reports on the individual projects show in detail what has been done. This final discussion aims to summarise what seem to be the major contributions by each group and also to examine their findings in relation to the vital need to strengthen the contribution of biological methods of pest control.

Fruit flies

Although this work was the last to be started, much of what was done was initiated or stimulated by the working group. Emphasis was placed on background information, both in terms of literature reviews and, more especially, field studies on life tables, mortality factors (including natural enemies of the pupae), problems of dispersal, attractants, sexual behaviour and population genetics. For the particular examples studied, whilst natural enemies can sometimes cause large mortalities of pupae in the soil, biological control by natural enemies does not appear to be an important component either actually or potentially in control programmes. Much of the work therefore was aimed at understanding factors that might be important in developing autocidal methods of control based on sterile male release or other 'genetic' methods. Reference should be made to the introduction to Chapter 3 for an excellent summary of what was accomplished. It is sufficient here to point to some of the highlights, notably the work on flight behaviour, so important in relation to population isolation needed for most autocidal methods of control, also to the collaborative studies on life-table analysis, particularly on pupal mortality by natural enemies. Useful collaborative studies were also undertaken on the responses of the fruit fly to colour stimuli aimed at standardising trapping methods for forecasting pest infestations.

Aphids – Myzus persicae

This project, unlike the others, highlights a particular species of world-wide importance chosen because it exemplifies inadequacies in current control methods against several major aphid pests, particularly those that are vectors of plant virus diseases. Besides reviewing current knowledge and establishing methodologies, other objectives included re-assessment of the role of natural enemies in the dynamics and control of aphids, especially *M. persicae*, deeper understanding of aphid/host plant relationships aimed at improving host plant resistance, the problem of dispersion – particularly aerial movement – and the analysis of the biological properties of aphid species with particular reference to biotype variation as exemplified by *M. persicae* in different parts of its distributional range. The 'core' study was a specially devised collaborative project on population dynamics of *M. persicae* on potato. This aimed to quantify the relative importance in control of different naturally occurring mortality factors, notably predators, parasites and pathogens, in relation to the breeding rate of the aphid which is influenced by the nature of the host plant and by the environmental conditions in the different study sites. This ambitious study which covered a range of tropical and temperate conditions indicated that natural enemies are likely to be of little value once the viruliferous aphid has reached the crop. It highlighted however the potential for natural enemy

230

action outside the crop. *M. persicae* is a relatively low density pest so the conclusions drawn from this study are not unexpected but the technique provides an example of what should be done in studies on other pests which are more amenable to natural enemy action. Another collaborative project also helped greatly to clarify the nature of intraspecies variation in *M. persicae* throughout its distributional range, especially in terms of the life cycle and also of host specificity. That natural enemies are sometimes vitally important is exemplified by the work in glasshouses where successful biological control methods have been established and also in peach orchards where infestations of *M. persicae* on the primary host have now increased in severity, seemingly through destruction by insecticides of their natural enemies. Perhaps the most important outcome of the work was to demonstrate the need to understand more about the role of the host plant in colonisation and build-up of populations on the plant, such as the sensitivity of the aphid to the physiological state of its host plant.

Rice stem-borers

Stem-borers are important pests of rice in most regions where this crop is grown and are key pests in the study area of the IBP work, which was limited to parts of South-East Asia. However, the study area exemplified a wide range of conditions in which rice is grown, from temperate to tropical zones. The work included a review of the main pest species and valuable keys to their identity and to that of the main parasites, and also an important handbook on methodology. Work on natural enemies showed that mortalities of up to 99 % could be caused by natural enemies – mostly parasites and pathogens. This has highlighted the need to use insecticides selectively, since we do not adequately know what harm is being done by insecticides in those areas where they are now widely used. Insecticide-induced resurgences of rice pests combined with development of resistance to the insecticide could severely jeopardise rice production, so there is an urgent need to understand more about the dynamics of rice pests. The report points to agricultural practices that could be used to enhance natural enemy action and also emphasises the opportunities for intra-area transfer of natural enemies. Emphasis is also placed on integration of chemical and biological methods as the most rational approach in controlling rice stem-borers; but much imaginative work needs to be done before suitable methods can be recommended, let alone established.

It is hoped that the individual efforts that were stimulated by the rice stem-borer project will now be developed into more comprehensive and cooperative attempts to deal with the urgent problem of rice stem-borer control through a combination of basic studies aimed at understanding the dynamics of the species and practical work aimed at integrated control, including its various forms of biological control.

M. J. Way

Armoured scale insects

In the past, much work has been done on the biology and introductions of natural enemies of Diaspididae. This is a reflection of early and continued successes in classical biological control of scale insects. A major objective of this IBP project was to accumulate and analyse the formidable literature on the main pest species of Diaspididae and their natural enemies and also to examine experimental procedures for studying the scale insects and their natural enemies. A major part of the section is also devoted to a review of successes and failures in biological control of Diaspididae. Special attention is paid to species of the parasitic genus *Aphytis* which are considered to be much the most effective natural enemies of Diaspididae. Many outstanding successes are highlighted. It would have been interesting to have analysed in more detail the possible causes of success and of failure where it occurred.

A major part of the project was a special study of *Aphytis* spp. of which 80 distinct species in about six groups were identified. Several potentially valuable species for biological control were recognised, and studies made on biological characteristics such as systematics, host specificity, developmental rates and problems of hybridisation and sexual isolation. This outstanding work, undertaken in California, is not yet published except as a preliminary progress report.

Biological control work using introduced natural enemies of diaspidid scales during the period of the IBP work is fully documented and demonstrates striking successes in the USA, Peru, Australia, Oceania, Israel and South Africa. Details are also given of introductions made in many other countries, the results of which have not yet been established. The important conclusion is drawn that although much remains to be done in terms of importing natural enemies, priority must be given to work aimed at minimising harmful effects of pesticides on existing natural enemies. Indeed, Diaspididae provide classic examples of 'upset' pests.

In general, the scale insect project seems to have been a great stimulus to widespread importations of natural enemies whereas the outstanding basic studies remained limited to individuals or localised groups who have long been recognised as authorities in this field.

Spider mites

Throughout much of the world, spider mites have become major pests because pesticides have destroyed natural enemies that otherwise successfully controlled the spider mites. Such evidence has pointed to the key role of natural enemies in control of spider mites and has led to exciting developments such as the organised use of natural enemies in control of spider mites in glasshouses. The IBP project has provided an opportunity to review the available information on the basic biology of the more important tetranychids and their natural enemies; this has provided a valuable stimulus for already existing research as well as for some new developments. Whilst no collaborative programmes of research were established, important individual projects were maintained and developed on the biology of the mites and their natural enemies, the interactions between mites and their enemies and the effects of environmental conditions on spider mite/natural enemy interaction. Other work of notable interest includes reports on studies on genetics (with implications for control by genetic means), on feeding habits and nutritional physiology of spider mites (relevant to enhancement of host plant resistance), on hibernation which is crucially important to the pest status of spider mites in glasshouses, on the general ecology of the mites and their natural enemies in orchards and on detailed interactions between mites and natural enemies in laboratory conditions, including studies on how the natural enemy's behaviour is affected by pesticides. The many short reports on current projects provide a valuable summary of existing work on all aspects from the basic to the applied. An outstanding development during the period of the IBP work was the discovery of certain predacious mite species that have become resistant to some pesticides in parts of North America. This should provide a long overdue stimulus for work aimed at creating pesticide resistance in other important natural enemy species.

Biological control possibilities as exemplified by these studies

The studies conducted on the five groups of arthropods chosen for special consideration, exemplify an almost complete spectrum of approaches to biological control, using the term as defined for the overall biological control work, namely the control of biota by biota. For species at one end of the spectrum natural enemies may provide complete control, at the other end they are seemingly a relatively minor factor and biological control must be sought through other methods. Throughout, however, the emphasis must be on integrated control since most of the species chosen for study form only a part of a complex of pest species attacking a particular crop.

Scale insects and spider mites are widely recognised as 'upset' pests. They are usually relatively uncommon in their native habitats, including artificial environments such as orchards, except when control by natural enemies is upset by selective destruction of the latter by pesticides, or where the pest has been accidentally introduced to a new region without its indigenous natural enemies. In nature, a very wide range of different natural enemies may feed on red spider mite but, as with the scale insects, only a few species or even one natural enemy species may be crucially important. This is especially well demonstrated by the extraordinarily efficient biological control of spider mites in the highly artificial environment of some glasshouses, which depends on a single species of predatory mite.

The IBP work has therefore been correctly concentrated on certain important natural enemy groups notably *Aphytis* spp., parasitic on Diaspididae, and on certain biotypes or species of predacious mite important in the control of the red spider mites. The comparatively sedentary habit of scale insects and mites makes them particularly amenable to realistic experimental work such as the use of exclusion techniques to assess the impact of natural enemies of pests. It is perhaps to be regretted that opportunities were not made to use the already well-established methods in collaborative projects to assess natural enemy action in different stituations throughout the world.

With rice stem-borers there is fascinating evidence of apparent variation in the importance of natural enemies in control of rice pests in different areas. In some areas natural enemies may control stem-borers at below the economic injury threshold implying that priority should be put on assessing the importance of natural enemies, either actually or potentially, in other areas where they appear to be comparatively ineffective, perhaps due sometimes to the use of pesticides. The information summarised in the report and the methodology manual and reviews stimulated by the IBP provide the essential basis for such work.

In contrast to the perennial crop environment the annual crop seldom provides opportunities for pest control dependent solely on the action of natural enemies. This is due to the dislocating influence of cultivation practices which periodically destroy the continuity needed to establish appropriate equilibria between the more host-specific natural enemy and its host, the pest. In theory, however, rice could provide ideal opportunities for a form of stable equilibrium in places where it is grown continuously throughout the year. Here, there are opportunities for pest and natural enemy to move from crop to crop and thereby maintain the continuity within a particular area throughout the year. The question to be asked is whether and in what circumstances this situation most favours the natural enemies by providing them with a continuous supply of hosts, or, whether, in contrast, it may give greater advantage to the pest by providing a convenient succession of host plants throughout the year. No doubt, as implied in the report, there will be very many circumstances where natural enemies, however carefully preserved or augmented, will be inadequate in preventing economic damage by stem-borers but there remains the opportunity, so little studied, of developing integrated controls based on the combination of natural enemy action, cultural control, partial host plant resistance and selective use of pesticides. The outstanding importance of rice as a food crop in many parts of the world and the dangers of increasing its pest problems through misuse of pesticides and use of artificial fertilisers highlights the outstanding importance of developing integrated control methods against stem-borers and other pests of rice.

In spite of their close taxonomic affinity with scale insects, aphids are a notably different problem in terms of control. This stems from their highly dispersive behaviour based on the periodic production of vast numbers of emigrant alatae. This habit and the fact that many of the major aphid pests are on annual crops creates situations where they can multiply quickly to pest density in newly colonised crops before natural enemies arrive. *M. persicae*, the species chosen for special study in this project, presents especially unfortunate difficulties for control by natural enemies. These are a consequence of its role as a major vector of plant virus diseases even when relatively scarce. The evidence that an average of 0.5 aphids per sugar beet plant can spread damaging virus diseases does not augur well for control by specific natural enemies which characteristically depend on the maintenance of a residue of pests in the field which retains the natural enemies. Whilst *M. persicae* has many different species of natural enemy and whilst it is known that natural enemies can strikingly influence numbers of the pest either locally or seasonally, natural enemies are unlikely to make a major contribution to control, at any rate on annual crops where virus spread is a major factor. Perhaps nonspecific predators can sometimes make a larger contribution than we appreciate but there seems little doubt that many crop production practices are proving increasingly detrimental to such predators, e.g. Carabidae.

The extreme sensitivity of *M. persicae* and many other aphids to the nutritional status of the host plant indicates that host plant resistance should provide a powerful tool in the broad biological control of this pest. The success of control based on host plant resistance is, however, always threatened by the spectre of appearance of resistance-breaking biotypes of the pest; hence there has been special emphasis on biotype variation in *M. persicae* as part of the programme. Whilst this work provides a knowledge of fascinating fundamental interest, from a practical viewpoint it seems more likely that it will enable us to explain failure rather than permit us to maintain success by using resistant host plants.

The control of *M. persicae* and other aphids will no doubt depend on fully integrated methods involving selective use of insecticides, maintenance of natural enemies, not only on the crop but especially in places where the pest is present at other times, and on cultural and other methods of making the host plant more resistant or less attractive than current methods of crop production make it nowadays.

Achievements

All the projects have provided a notable stimulus to relevant basic and practical work by individual workers. Some have laid a foundation for collaborative research. Up-to-date reviews of current knowledge directly stimulated by the IBP proposals have been published on all the chosen species and are now augmented by the reports in this volume. Published methodologies have also provided bases for standardised field and laboratory techniques. This amounts to a notable achievement but perhaps of special importance is the opportunity that the IBP work has provided for active workers in the field to meet and establish long lasting opportunities for collaborative work.

References

Asterisks denote publications which are the result of IBP collaborative activities.

Abdelrahman, I. (1973). Toxicity of malathion to the natural enemies of California red scale (Hemiptera: Diaspididae). *Aust. J. agric. Res.*, **24**, 119–34.

Abdel-Salam, F. (1967). On the effect of phosphoric-acid esters on some arthropods in the apple-tree biocoenosis in relation to their density. *Z. angew. Zool.*, **54**, 233–83.

Abernathy, C. O. & Thurston, R. (1969). Plant age in relation to the resistance of *Nicotiana* to the green peach aphid. *J. econ. Ent.*, **62**, 1356–9.

Adams, J. B. (1946). Aphid resistance in potatoes. *Am. Potato J.*, **23**, 1–22.

Ahmad, R. (1970a). A new species of *Pharoscymnus* Bedel (Coleoptera: Coccinellidae) predaceous on scale insects in Pakistan. *Entomophaga*, **15**, 233–5.

(1970b). A new genus and species (*Pakencyrtus pakistanensis* gen. & sp. nov.) of the family Encyrtidae (Hymenoptera: Chalcidoidea) from Pakistan. *Entomophaga*, **15**, 237–40.

Allen, W. W. (1959). Strawberry pests in California. *Calif. agric. Exp. Stn. Circ.*, **484**, 1–39.

Anderson, N. H. (1962). Bionomics of six species of *Anthocoris* (Heteroptera: Anthocoridae) in England. *Trans. R. ent. Soc. Lond.*, **114**, 67–95.

*Annecke, D. P. (1969). Recent developments in biological and integrated control of citrus pests in South Africa. *Proceedings of the 1st international citrus symposium*, Riverside, **2**, 849–54.

*Annecke, D. P. & Insley, H. P. (1970). New and little known species of *Azotus* Howard, *Ablerus* Howard and *Physcus* Howard (Hym., Aphelinidae) from Africa and Mauritius. *Bull. ent. Res.*, **60**, 237–51.

* (1971). Catalogue of Ethiopian Encyrtidae and Aphelinidae (Hymenoptera: Chalcidoidea). *S. Afr. Dep. Agric. Tech. Services ent. Mem.*, **23**, 53 pp.

*Annecke, D. P. & Mynhardt, M. J. (1970). On some species of *Habrolepis* Foerster and *Adelencyrtus* Ashmead (Hymenoptera: Encyrtidae) in southern Africa and Mauritius. *Entomophaga*, **15**, 127–48.

Anonymous (1944). Entomological investigations. *Rep. Coun. sci. ind. Res. Organ.*, Canberra, 1942–43, pp. 15–20.

Anonymous (1972). The biological control of cucumber pests. *Glasshouse Crops Res. Inst. Growers Bull.*, **1**, 12 pp.

Applebaum, S. W., Kfir, R., Gerson, U. & Tadmor, U. (1971). Studies on the summer decline of *Chilocorus bipustulatus* in citrus groves of Israel. *Entomophaga*, **16**, 433–44.

References

Archer, T. L., Cate, R. H., Eikenbary, R. D. & Starks, K. J. (1974). Parasitoids collected from greenbugs and corn leaf aphids in Oklahoma. *Ann. ent. Soc. Am.*, **67**, 11–14.

*Argyriou, L. C. (1975). Data on the biological control of citrus scales in Greece. *Annls Inst. Phytopath. Benaki*, in press.

Argyriou, L. C. & DeBach, P. (1968). The establishment of *Metaphycus helvolus* (Compere). (Hym. Encyrtidae) on *Saissetia oleae* (Bern.) (Hom. Coccidae) in olive groves in Greece. *Entomophaga*, **13**, 223–8.

Arthur, D. R. (1944). *Aphidius granarius*, Marsh., in relation to its control of *Myzus kaltenbachi*, Schout. *Bull. ent. Res.*, **35**, 257–70.

Asquith, D. & Colburn, R. (1971). Integrated pest management in Pennsylvania apple orchards. *Bull. ent. Soc. Am.*, **17**, 89–91.

Avidov, Z., Blumberg, D. & Gerson, U. (1968). *Cheletogenes ornatus* (Acarina: Cheyletidae), a predator of the chaff scale on citrus in Israel. *Israel J. Ent.*, **3**, 77–93.

Avidov, Z. & Gerson, U. (1968). Some interactions of two hymenopterous parasites of the chaff scale. *Res. Popul. Ecol.*, **10**, 171–6.

Bailey, P. (1971). The physiological effects of ionizing radiation on *Dacus cucumis* French (Diptera: Tephritidae). *Ph.D. Thesis*, University of New South Wales.

Balduf, W. V. (1959). Obligatory and facultative insects in rose hips, their recognition and bionomics. *Ill. biol. Monogr.*, **26**, 1–194.

Baranyovits, F. (1953). Some aspects of the biology of armoured scale insects. *Endeavour*, **12**, 202–9.

Barlow, C. A. (1962). The influence of temperature on the growth of experimental populations of *Myzus persicae* (Sulz.) and *Macrosiphum euphorbiae* (Thomas) (Aphididae). *Can. J. Zool.*, **40**, 145–56.

Barnes, H. F. (1954). Gall-midge larvae as endoparasites, including the description of a species parasitising aphids in Trinidad, B.W.I. *Bull. ent. Res.*, **45**, 769–75.

Bartlett, B. R. (1953). Retentive toxicity of field-weathered insecticide residues to entomophagous insects associated with citrus pests in California. *J. econ. Ent.*, **46**, 565–9.

(1964). The toxicity of some pesticide residues to adult *Amblyseius hibisci*, with a compilation of the effects of pesticides on phytoseiid mites. *J. econ. Ent.*, **57**, 559–63.

Bateman, M. A. (1968). Determinants of abundance in a population of the Queensland fruit fly. *Symp. R. ent. Soc. Lond.*, **4**, 119–31.

* (1972). The ecology of fruit flies. *A. Rev. Ent.*, **17**, 493–518.

*Bateman, M. A. & Elliott, P. O. (1971). *Report to 2nd meeting of IBP Working Group on Fruit Flies*, 31 pp. (mimeo.).

Bateman, M. A., Friend, A. H. & Hampshire, F. (1966). Population suppression in the Queensland fruit fly, *Dacus (Strumeta) tryoni* II. Experiments on isolated populations in western New South Wales. *Aust. J. Agric. Res.*, **17**, 699–718.

Bateman, M. A., Sonleitner, F. J. (1967). The ecology of a natural population of the Queensland fruit fly, *Dacus tryoni*. I. The parameters of the pupal and adult populations during a single season. *Aust. J. Zool.*, **15**, 303–35.

Batra, H. N. (1953). Aphids infesting peach and their control. *Indian J. Ent.*, **15**, 45–51.

— (1954). Biology and control of *Dacus diversus* Coquillett and *Carpomyia vesuviana* Costa and important notes on other fruit flies in India. *Indian J. agric. Sci.*, **23**, 87–112.

Beardsley, J. W. & Gonzalez, R. H. (1975). The biology and ecology of armored scales. *A. Rev. Ent.*, **20**, 47–73.

*Bedford, E. C. G. (1973). Citrus scale insects: biological control proves successful. *Citrus Sub-Trop. Fruit J.*, no. **470**, 4, 5, 7, 9, 11.

Begljarov, G. A. (1967). Ergebnisse der Untersuchungen und Anwendung von *Phytoseiulus persimilis* Athias-Henriot (1957) als biologisches Bekämpfungsmittel gegen Spinnmilben in der Sowjetunion. *NachrBl. dt. Pfl-Schutzdienst*, **21**, 197–200.

— (1970). Report in the *Spider Mite Newsletter*, **3**, 29.

Begljarov, G. A. & Ushchekhov, A. T. (1972). Ecology of the predatory mite *Phytoseiulus persimilis* A.-H. and its effectiveness in practical application in the U.S.S.R. *Zesz. Probl. Post. Nauk. Rolniczych*, Warsaw, **129**, 93–102.

Behrendt, K. (1968). Das Abwandern parasitierter Aphiden von ihren Wirtspflanzen und eine Methode zu ihrer Erfassung. *Beitr. Ent.*, **18**, 293–8.

*Beingolea, O. (1973). Control biológico de la queresa redonda de los citricos *Selenaspidus articulatus* Morgan como un ejemplo actual de las posibilidades de este metodo de control de plagas para la fruticultura peruana. *An. Censos Nac.*, Peru, in press.

Bénassy, C. (1967). Note sur une diaspine peu connue de l'olivier au Maroc: *Quadraspidiotus maleti* (Hom. Coccoidea). *Annls Soc. ent. Fr.* (N.S.), **3**, 1133–9.

* — (1969). The biological control against coccids of citrus in the country of the French zone. *Proc. 1st int. Citrus Symp.*, Riverside, **2**, 793–9.

*Bénassy, C., Bianchi, H. & Milaire, H. G. (1971). La température, facteur prévisionnel du cycle évolutif de *Prospaltella perniciosi* Tow. *Entomophaga* **16**, 411–20.

*Bénassy, C. & Euverte, G. (1967). Perspectives nouvelles dans la lutte contre *Aonidiella aurantii* au Maroc. (Hom. Diaspididae). *Entomophaga*, **12**, 449–59.

*Ben Dov, Y. & Rosen, D. (1969). Efficacy of natural enemies of the California red scale on citrus in Israel. *J. econ. Ent.*, **62**, 1057–60.

Berker, J. (1958). Die natürlichen Feinde der Tetranychiden. *Z. angew. Ent.*, **43**, 115–72.

Berry, R. E. & Simpson, R. G. (1967). Flight activity of the green peach aphid, *Myzus persicae* (Sulz.), a natural vector of potato leafroll virus in Colorado. *Tech. Bull. Colo. agric. Exp. Stn*, **92**, 1–34.

Binns, E. S. (1971). The toxicity of some soil applied systemic insecticides to *Aphis gossypii* and *Phytoseiulus persimilis* on cucumber. *Ann. appl. Biol.*, **67**, 211–12.

References

Bird, F. T. (1967). A virus disease of the European red mite *Panonychus ulmi* (Koch). *Can. J. Microbiol.*, **113**, 1131.

Blackman, R. L. (1971*a*). Variation in the photoperiodic response within natural populations of *Myzus persicae* (Sulz.). *Bull. ent. Res.*, **60**, 533–46.

(1971*b*). Chromosomal abnormalities in an anholocyclic biotype of *Myzus persicae* (Sulzer). *Experientia*, **27**, 704–6.

(1972). The inheritance of life cycle differences in *Myzus persicae* (Sulz.). *Bull. ent. Res.*, **62**, 281–94.

Boczek, J., Dabrowski, Z. T. & Kapala, T. (1970). Studies on the hibernation of phytoseiid mites (Acarina: Phytoseiidae) in orchards. *Zesz. Probl. Post. Nauk. Rolniczych*, Warsaw, **109**, 43–64.

Bodenheimer, F. S. (1951). *Citrus Entomology in the Middle East*. Junk, The Hague, 663 pp.

Boller, E. F. (1966). Der Einfluss natürlicher Reduktionsfaktoren auf die Kirschenfliege *Rhagoletis cerasi* L. in der Nordwestschweiz, unter besonderer Berücksichtigung des Puppenstadiums. *Schweiz. Landw. Forsch.*, **5**, 153–210.

(1974). Status of the sterile insect release method against the cherry fruit fly (*Rhagoletis cerasi* L.) in Northwest Switzerland. In *The Sterile-Insect Technique and its Field Applications*. IAEA, Vienna, pp. 1–3.

Boller, E. F. & Bush, G. L. (1974). Evidence for genetic variation in populations of the European cherry fruit fly, *Rhagoletis cerasi* (Diptera: Tephritidae), based on physiological parameters and hybridization experiments. *Entomologica exp. appl.*, **17**, 279–93.

Boller, E. F., Haisch, A. & Prokopy, R. J. (1971). Sterile insect release method against *Rhagoletis cerasi* L. Preparatory ecological and behavioural studies. In *Sterility Principle for Insect Control or Eradication*. IAEA, Vienna, pp. 77–86.

*Boller, E. F. & Remund, U. (1971). Life table studies on the European cherry fruit fly in Switzerland. *Report to the 2nd Meeting of IBP Working Group on Fruit Flies*, 8 pp, mimeo.

Bondarenko, N. V. & Emelyanov, V. A. (1970). Characteristics of the biology of the predacious mite *Typhlodromus subsolidus* Begl. (Acarina, Phytoseiidae) in the Leningrad region and its role in regulating the numbers of the fruit-tree red spider mite *Panonychus ulmi* Koch (Tetranychidae). *Ent. Obozr.*, **49**, 163–7.

Bonnemaison, L. (1950). Observations biologiques sur le Puceron gris du Pêcher (*Myzus persicae* Sulz.) et le Puceron noir (*Aphis fabae* Scop.) en relation avec la transmission des maladies à virus de la Pomme de terre et de la Betterave. *C. r. Acad. Agric. Paris*, **36**, 525–7.

(1971). Observations sur les fluctuations des populations aphidiennes du Chou, de la Betterave et de la Pomme de terre. *Annls Soc. ent. Fr.* (N.S.), **7**, 505–51.

Bonnemaison, L. & Missonnier, J. (1956). Le psylle du poirier (*Psylla piri* L.). Morphologie et biologie, méthodes de lutte. *Annl. épiphyt.*, **7**, 263–331.

Borkhsenius, N. S. (1963). *Practical Keys to the Coccids* (*Coccoidea*) *of*

Cultural Plants and Forests in the USSR. USSR Academy of Science: Moscow and Leningrad, 311 pp. (in Russian).

Boudreaux, H. B. (1963). Biological aspects of some phytophagous mites. *A. Rev. Ent.*, **8**, 137–54.

Bowers, W. S., Nault, L. R., Webb, R. W. & Dutky, S. R. (1972). Aphid alarm pheromone: isolation, identification, synthesis. *Science, Wash.*, **177**, 1121.

Boyce, A. M. (1934). Bionomics of the walnut husk fly, *Rhagoletis completa*. *Hilgardia*, **8**, 363–579.

Böhm, H. (1960). Untersuchungen über Spinnmilben in Oesterreich. *Pflanzenschutz-Berichte*, **25**, 23–46.

(1966). Ein Beitrag zur biologischen Bekämpfung von Spinnmilben in Gewächshäusern. *Pflanzenschutz-Berichte*, **34**, 65–77.

Böhm, O. (1962). Kartoffelblattläuse im steirischen Ennstal. *Pflanzenschutz-Berichte*, **28**, 79–87.

Bravenboer, L. (1959). De chemische en biologische bestrijding van de spintmijt *Tetranychus urticae* Koch. Diss. Landbouwhogeschool, Wageningen, 85 pp.

(1972). Some aspects of large-scale introduction of *Phytoseiulus riegeli* in practice. *Proceeding of the OILB conference on integrated control in glasshouses* (1970), pp. 76–8.

Bravenboer, L. & Dosse, G. (1962). *Phytoseiulus riegeli* Dosse als Prädator einiger Schadmilben aus der *Tetranychus urticae*-Gruppe. *Entomologica exp. appl.*, **5**, 291–304.

Brewer, R. H. (1971). The influence of the parasite *Comperiella bifasciata* How. on the populations of two species of armoured scale insects, *Aonidiella aurantii* (Mask.) and *A. citrina* (Coq.) in South America. *Aust. J. Zool.*, **19**, 53–63.

Britton, E. B. & Lee, B. (1972). *Stethorus loxtoni* sp.n. (Coleoptera: Coccinellidae) a newly-discovered predator of the two-spotted mite. *J. Aust. ent. Soc.*, **11**, 55–60.

Broadbent, L. (1949). The grouping and overwintering of *Myzus persicae* (Sulzer) on *Prunus* species. *Ann. appl. Biol.*, **36**, 334–40.

(1964). Control of plant virus diseases. In *Plant Virology* (eds. M. K. Corbett and H. D. Sisler), pp. 330–64. University of Florida Press, Gainesville.

Broadbent, L. & Heathcote, G. D. (1955). Sources of overwintering *Myzus persicae* (Sulz.) in England. *Pl. Path.*, **4**, 135–7.

(1961). Winged aphids trapped in potato fields, 1942–1959. *Entomologia exp. appl.*, **4**, 226–37.

Brooks, F. E. (1921). Walnut husk-maggot. *U.S. Dep. Agric. Bull.*, **992**, 1–8.

Brown, H. D. (1972). Predacious behaviour of four species of Coccinellidae (Coleoptera) associated with the wheat aphid, *Schizaphis graminum* (Rondani), in South Africa. *Trans. R. ent. Soc. Lond.*, **124**, 21–36.

Burk, L. G. & Stewart, P. A. (1969). Resistance of *Nicotiana* species to the green peach aphid. *J. econ. Ent.*, **62**, 1115–17.

Burki, T. & Boller, E. F. (1969). Production and testing procedures of antiserum of *Rhagoletis cerasi* and *R. indifferens* for the recognition of pupal

References

predators. IBP Working Group on Fruit Flies, Document K. (Project manual), pp. 45–7.

Bush, G. L. (1962). The cytotaxonomy of the larvae of some Mexican fruit flies in the genus *Anastrepha* (Tephritidae: Diptera). *Psyche*, **69**, 87–101.

(1966a). Taxonomy, cytology and evolution of the genus *Rhagoletis* in North America (Diptera: Tephritidae). *Bull. Mus. comp. Zool. Harv.*, **134**, 431–562.

(1966b). Female heterogamety in the family Tephritidae (Acalyptratae: Diptera). *Am. Nat.*, **100**, 119–26.

(1969a). Sympatric host race formation and speciation in frugivorous flies of the genus *Rhagoletis* (Diptera: Tephritidae). *Evolution*, **23**, 237–51.

(1969b). Mating behaviour, host specificity and the ecological significance of sibling species in frugivorous flies of the genus *Rhagoletis* (Diptera: Tephritidae). *Am. Nat.*, **103**, 669–72.

(1974). Mechanisms of sympatric host race formation of the true fruit flies (*Tephritidae*). In *Genetic analysis of Speciation Mechanisms* (ed. M. J. D. White), pp. 3–23. Australia and New Zealand Book Co., Sydney.

Bush, G. L. & Huettel, M. D. (1970). The cytogenetics and description of a new North American species of the neotropical genus *Cecidocharella* (Diptera: Tephritidae). *Ann. ent. Soc. Am.*, **63**, 88–91.

(1972). Starch gel electrophoresis of tephritid proteins, a manual of techniques. *IBP Working Group on Fruit flies* (mimeo).

Bush, G. L. & Taylor, S. C. (1969). The cytogenetics of *Procecidochares*. I. The mitotic and polytene chromosomes of the pamakani fly, *P. utilis* Stone (Tephritidae: Diptera). *Caryologia* **22**, 311–22.

Byford, W. J., Dunning, R. A., Heathcote, G. D. & Hull, J. (1966). Aphid parasites. *Reps. Rothamsted Exp. Stn*, 1965, p. 262.

Byford, W. J. & Reeve, G. J. (1969). *Entomophthora* species attacking aphids in England 1962–1966. *Trans. Br. mycol. Soc.*, **2**, 342–6.

Byford, W. J. & Ward, L. K. (1968). Effect of the situation of the aphid host at death on the type of spore produced by *Entomophthora* sp. *Trans. Br. mycol. Soc.*, **51**, 598–600.

Caesar, L. & Spencer, G. J. (1915). Cherry fruit flies. *Ont. Dep. Agric. Bull.*, **227**, 30 pp.

Caltagirone, L. E. (1969). Report in the *Spider Mite Newsletter*, **2**, 13–14.

Calvert, D. J. & van den Bosch, R. (1972). Host range and specificity of *Monoctonus paulensis* (Hymenoptera: Braconidae), a parasite of certain dactynotine aphids. *Ann. ent. Soc. Am.*, **65**, 422–32.

*Cameron, P. J. (1971). Pupal mortality on the apple maggot *Rhagoletis pomonella* (Walsh). *Report to the IBP Working Group on Fruit Flies*, 10 pp. (mimeo).

Campbell, A. (1974). Seasonal changes in abundance of the pea aphid and its associated parasites in the southern interior of British Columbia. Ph.D. Thesis, Simon Fraser University, 282 pp.

Campbell, A., Frazer, B. D., Gilbert, N., Gutierrez, A. P. & Mackauer, M. (1974). Temperature requirements of some aphids and their parasites. *J. appl. Ecol.*, **11**, 431–8.

Campbell, A. & Mackauer, M. (1973). Some climatic effects on the spread and abundance of two parasites of the pea aphid in British Columbia (Hymenoptera: Aphidiidae–Homoptera: Aphididae). *Z. angew. Ent.*, **74**, 47–55.

Carayon, J. (1961). Quelques remarques sur les Hémipteres–Héteroptères: leur importance comme insectes auxiliaires et les possibilités de leur utilisation dans la lutte biologique. *Entomophaga*, **6**, 133–41.

Caspersson, T., Zech, L., Johansson, C. & Modest, E. J. (1970). Identification of human chromosomes by DNA-binding fluorescing agents. *Chromosoma*, **30**, 215–27.

Cate, R. H., Archer, T. L., Eikenbary, R. D., Starks, K. S. & Morrison, R. D. (1973). Parasitization of the greenbug by *Aphelinus asychis* and the effect of feeding by the parasitoid on aphid mortality. *Environ. Ent.*, **2**, 549–53.

Catling, H. D. (1971). Studies on the citrus red scale, *Aonidiella aurantii* (Mask.) and its biological control in Swaziland. *J. ent. Soc. S. Afr.*, **34**, 393–411.

Cavalloro, R. & Delrio, G. (1971). Rilievi sul comportamento sessuale di *Dacus oleae* Gmelin (Diptera: Trypetidae) in laboratorio. *Redia*, **52**, 201–30.

Chaboussou, F. (1969). Recherches sur les facteurs de pullulation des acariens phytophages de la vigne à la suite des traitements pesticides du feuillage. Thèse Fac. Sci. Paris, 238 pp.

Chambers, D. L. & O'Connell, T. B. (1969). A flight mill for studies with the Mexican fruit fly. *Ann. ent. Soc. Am.*, **62**, 917–20.

Chant, D. A. (1959). Phytoseiid Mites. *Can. Ent.*, suppl. **12**, 45–166.

(1961). An experiment in biological control of *Tetranychus telarius* (L.) (Acarina: Tetranychus) in a greenhouse, using *Phytoseiulus persimilis* Athias-Henriot (Phytoseiidae). *Can. Ent.*, **93**, 437–43.

*Chu, Yau-i (1969). On the bionomics of *Lyctocoris beneficus* (Hiura) and *Xylocoris galactinus* (Fieber) (Anthocoridae, Heteroptera). *J. Fac. Agric. Kyushu University*, Fukuoka, **15**, 1–136.

Christenson, L. D. & Foote, R. H. (1960). Biology of fruit flies. *A. Rev. Ent.*, **5**, 171–92.

Cilliers, C. J. (1971). Observations on circular purple scale, *Chrysomphalus aonidum* (L.), and two introduced parasites in Western Transvaal citrus orchards. *Entomophaga*, **16**, 269–84.

Cirio, U. (1971). Reperti sul meccanismo stimolo-risposta nell'ovideposizione del *Dacus oleae* Gmelin (Diptera: Trypetidae). *Redia*, **52**, 577–600.

(1973). Osservazioni sul comportamento di ovideposizione della *Rhagoletis completa* Cresson (Diaptera: Trypetidae) in laboratorio. *9th Congr. Ital. Ent. Soc.*, Siena, pp. 99–117.

Clancy, D. W. & McAlister, H. J. (1956a). Effects of spray practices on apple mites and their predators in West Virginia. *Proc. 10th int. Congr. Ent.*, Montreal, **4**, 597–601.

(1956b). Selective pesticides as aids to biological control of apple pests. *J. econ. Ent.*, **49**, 196–202.

References

Clausen, C. P. (1940). *Entomophagous Insects*. McGraw-Hill, New York and London, 688 pp.

(1958). Biological control of insect pests. *A. Rev. Ent.*, **3**, 291–310.

Clausen, C. P., Clancy, D. W. & Chock, Q. C. (1965). Biological control of the oriental fruit fly (*Dacus dorsalis* Hendel) and other fruit flies in Hawaii. *Tech. Bull. U.S. Dep. Agric.*, **1322**, 1–102.

*Cochereau, P. (1969). Controle biologique d'*Aspidiotus destructor* Signoret (Homoptera, Diaspinae) dans l'île Vaté (Nouvelles Hébrides) au moyen de *Rhizobius pulchellus* Montrouzier (Coleoptera Coccinellidae). *Cahiers ORSTOM*, Serie Biologie, **8**, 57–100.

Cognetti, G. (1967). Sexual genotypes and migratory tendencies in *Myzus persicae* Sulz. *Monit. zool. ital.* (N.S.), **1**, 229–34.

Colburn, R. & Asquith, D. (1971). Tolerance of the stages of *Stethorus punctum* to selected insecticides and miticides. *J. econ. Ent.*, **64**, 1072–4.

Collyer, E. (1953). The biology of some predatory insects and mites associated with the fruit tree red spider mite (*Metatetranychus ulmi* Koch) in southeastern England. II. Some important predators of the mite. *J. hort. Sci. Lond.*, **28**, 85–97.

(1964 a). The effect of alternate food supply on the relationship between two *Typhlodromus* species and *Panonychus ulmi* Koch. *Entomologia exp. appl.*, **7**, 120–4.

(1964 b). Phytophagous mites and their predators in New Zealand orchards. *N.Z. J. agric. Res.*, **7**, 551–68.

(1967). On the ecology of *Anthocoris nemorum* (L.) (Hemiptera–Heteroptera). *Proc. R. ent. Soc. Lond.*, **42**, 107–18.

(1969). Report in the *Spider Mite Newsletter*, **2**, 28–9.

Collyer, E. & Kirby, A. H. M. (1959). Further studies on the influence of fungicide sprays on the balance of phytophagous and predacious mites on apple in south-east England. *J. hort. Sci. Lond.*, **34**, 39–50.

Compere, H. (1961). The red scale and its insect enemies. *Hilgardia*, **31**, 173–278.

Costa, C. L. (1969). Occurrência, no Estado de São Paulo, de forma sexuada de *Myzus persicae*, importante vector de virus de plantas. *Rev. Soc. bras. Fitopatologia*, **3**, 59–60.

Coudriet, D. L. & Tuttle, D. M. (1963). Seasonal flights of insect vectors of several plant viruses in South Arizona. *J. econ. Ent.*, **56**, 865–8.

Croft, B. A. (1971). Comparative studies on four strains of *Typhlodromus occidentalis*. V. Photoperiodic induction of diapause. *Ann. ent. Soc. Am.*, **64**, 962–4.

(1972). Resistant natural enemies in pest management systems. *Span*, **15**, 1–4.

Croft, B. A. & Barnes, M. M. (1971). Comparative studies of four strains of *Typhlodromus occidentalis*. III. Evaluations of releases of insecticide resistant strains into an apple orchard ecosystem. *J. econ. Ent.*, **64**, 845–50.

(1972). Comparative studies of four strains of *Typhlodromus occidentalis*. IV. Persistence of insecticide-resistant strains in an apple orchard ecosystem. *J. econ. Ent.*, **65**, 211–16.

Croft, B. A. & Jeppson, L. R. (1970). Comparative studies of four strains of *Typhlodromus occidentalis*. II. Laboratory toxicity of ten compounds common to apple pest control. *J. econ. Ent.*, **63**, 1528–31.

Croft, B. A. & Jorgensen, C. D. (1969). Life history of *Typhlodromus mcgregori* (Acarina: Phytoseiidae). *Ann. ent. Soc. Am.*, **62**, 1261–8.

Croft, B. A. & McMurtry, J. A. (1972). Minimum release of *Typhlodromus occidentalis* to control *Tetranychus mcdanieli* on apple. *J. econ. Ent.*, **65**, 188–91.

Croft, B. A. & Nelson, E. E. (1972). Toxicity of apple orchard pesticides to Michigan populations of *Amblyseius fallacis*. *Environ. Ent.*, **1**, 576–9.

*Crouzel, I. S. de. (1971). Studies on the biological control of diaspidid scales on citrus in Argentina. *Proceedings of the 12th Pacific Science Congress*, Canberra, **1**, 200.

Crozier, R. H. (1970). On the potential for genetic variability in haplo-diploidy. *Genetica*, **41**, 551–6.

Dabrowski, Z. T. (1969*a*). Laboratory researches on the toxicity of pesticides to *Typhlodromus finlandicus* (Oud.) and *Phytoseiulus macropilis* (Banks) (Phytoseiidae, Acarina). *Roczn. Nauk roln.*, **95-A-3**, 337–69 (in Polish with English and Russian summaries).

(1969*b*). Toxicity of pesticides to the predatory mites (Phytoseiidae, Acarina) occurring in apple orchards. *Roczn. Nauk roln.*, **95-A-3**, 265–311 (in Polish with English and Russian summaries).

(1970*a*). Density of spider mites (*Tetranychidae*) and predatory mites (Phytoseiidae) in apple orchards treated and not treated with pesticides. *Ekol. Pol.*, **18**, 111–36.

(1970*b*). Effect of pesticides on the associations of predatory mites in apple orchards. *Ekol. Pol.*, **18**, 817–36.

(1970*c*). Investigations on successive action of pesticides on the spider mites (Tetranychidae) and on the predatory mite (Phytoseiidae) populations in the apple orchards. *Roczn. Nauk. roln.*, Ser. E, **1**, 7–26 (in Polish with English and Russian summaries).

(1970*d*). Factors determining the increase of density of predatory mites (Acarina, Phytoseiidae) in apple orchards treated with pesticides. Part I. *Bull. Ent. Pologne*, Wroclaw, **40**, 141–89 (in Polish with English and Russian summaries).

Dabrowski, Z. T. & Dabrowska, B. (1972). A comparison of estimation methods measuring the long term effect of pesticides on some Acarina populations in apple orchards. Advances in Agric. Acarology in Europe. *Zesz. Probl. Post Nauk Rolniczych*, Warsaw, **129**, 159–69.

Dabrowski, Z. T., Dabrowska, B. & Labanowski, G. S. (1972). Statistical estimation of long term effect of pesticides on predatory mites (Phytoseiidae) and spider mites (Tetranychidae) in apple orchards. *Roczn. Nauk roln.*, Ser. E, **3**.

Dadd, R. H. & Krieger, D. L. (1968). Dietary amino acid requirements of the aphid, *Myzus persicae*. *J. Ins. Physiol.*, **14**, 763–78.

Daiber, C. C. & Schöll, S. E. (1959). Further notes on the overwintering of the green peach aphid, *Myzus persicae* (Sulzer), in South Africa. *J. ent. Soc. S. Afr.*, **22**, 494–520.

References

Daneschwar, G. (1963). Einfluss einiger Pflanzenschutzmittel auf die Raub-milbe *Typhlodromus* (*Typhlodromus pyri* Scheuten) (Acarina, Phyto-seiidae). Ph.D. Thesis, Hohenheim, 74 pp.

Das, B. (1918). The Aphididae of Lahore. *Mem. Indian Mus.*, **6**, 135–274.

Davis, E. W. & Landis, B. J. (1951). Life history of the green peach aphid on peach and its relation to the aphid problem on potatoes in Washington. *J. econ. Ent.*, **44**, 586–90.

de Jong, J. K. (1929). Enkele resultaten van het onderzoek naar de biologie van de tabaksluis *Myzus persicae* Sulzer. *Bull. Deli Proefstn Medan*, **28**, 1–36.

De Lotto, G. (1971). A preliminary note on the black scales (Homoptera: Coccidae) of North and Central America. *Bull. ent. Res.*, **61**, 325–6.

De Murtas, I. D., Enkerlin, S. D. & Cirio, U. (1972). Dispersion de *Ceratitis capitata* (Wiedemann) dans l'île de Procida (Italy). *Bull. Eur. Med. Pl. prot. Organ.*, **6**, 69–76.

De Santis, L. (1967). *Catálogo de los himenópteros argentinos de la serie para-sítica, incluyendo Bethyloidea.* Comision de Investigación Cientifica, Provincia de Bueños Aires, La Plata, 337 pp.

de Villiers, J. F. (1970). The establishment of *Aphytis lepidosaphes* Compere, a parasite of citrus mussel scale, *Lepidosaphes beckii. S. Afr. Citrus J.*, no. **436**, 7–14.

Dean, G. J. W. & Wilding, N. (1971). *Entomophthora* infecting cereal aphids *Metopolophium dirhodum* and *Sitobion avenae. J. Invertebr. Path.*, **18**, 169–76.

(1973). Infection of cereal aphids by the fungus *Entomophthora. Ann. appl. Biol.*, **74**, 133–8.

DeBach, P. (1943). The importance of host-feeding by adult parasites in the reduction of host populations. *J. econ. Ent.*, **36**, 647–58.

(1964). Successes, trends and future possibilities. In *Biological Control of Insect Pests and Weeds* (ed. P. DeBach), pp. 673–713. Chapman and Hall, London.

(1966). The competitive displacement and coexistence principles. *A. Rev. Ent.*, **11**, 183–212.

* (1969). Uniparental, sibling and semi-species in relation to taxonomy and biological control. *Israel J. Ent.*, **4**, 11–28.

* (1971a). Fortuitous biological control from ecesis of natural enemies. In *Entomological Essays to Commemorate the Retirement of Professor K. Yasumatsu.* Shukosha Printing Co. Ltd., Fukuoka, pp. 293–301.

(1971b). The use of imported natural enemies in insect pest management ecology. *Proceedings of the tall timbers conference on ecological animal control by habitat management*, 3, 211–33.

DeBach, P. & Argyriou, L. C. (1967). The colonization and success in Greece of some imported *Aphytis* spp. (Hym. Aphelinidae) parasitic on citrus scale insects (Hom. Diaspididae). *Entomophaga*, **12**, 325–42.

*DeBach, P. & Huffaker, C. B. (1971). Experimental techniques for evalua-tion of the effectiveness of natural enemies. In *Biological Control* (ed. C. B. Huffaker), Chap. 5, pp. 113–40. Plenum Press, New York and London.

DeBach, P. & Landi, J. (1961). The introduced purple scale parasite, *Aphytis lepidosaphes* Compere, and a method of integrating chemical with biological control. *Hilgardia*, **31**, 459–97.

*DeBach, P., Rosen, D. & Kennett, C. E. (1971). Biological control of coccids by introduced natural enemies. In *Biological Control* (ed. C. B. Huffaker), Chapter 7, pp. 165–94. Plenum Press, New York and London.

DeBach, P. & Sundby, R. A. (1963). Competitive displacement between ecological homologues. *Hilgardia*, **34**, 105–66.

*DeBach, P. & Warner, S. C. (1969). Research on biological control of whiteflies. *Citrograph*, **54**, 301–3.

Dicker, G. H. L. (1967). Integrated control of apple pests. *Proceedings of the 4th British Conference on Insects and Fungi, Brighton*, **1**, 1–7.

Dickson, R. C. & Laird, E. F., Jr (1962). Green peach aphid populations on desert sugar beets. *J. econ. Ent.*, **55**, 501–4.

Dieleman, J. & Overmeer, W. P. J. (1972). Preferential mating hampering the possibility to apply a genetic control method against a population of *Tetranychus urticae* Koch. *Z. angew. Ent.*, **71**, 156–61.

Dixon, A. F. G. & Russel, R. J. (1972). The effectiveness of *Anthocoris nemorum* and *A. confusus* (Hemiptera: Anthocoridae) as predators of the sycamore aphid, *Drepanosiphum platanoides*. II. Searching behaviour and the incidence of predation in the field. *Entomologia exp. appl.*, **15**, 35–50.

Dosse, G. (1958). Ueber einige neue Raubmilbenarten (Acar. Phytoseiidae). *Pflanzenschutz-Berichte*, **21**, 44–61.

 (1967). Schadmilben des Libanons und ihre Prädatoren. *Z. angew. Ent.*, **59**, 16–48.

Downing, R. S. (1966). The effect of certain miticides on the predaceous mite *Neoseilus caudiglans* (Acarina: Phytoseiidae). *Can. J. Pl. Sci.*, **46**, 521–4.

Downing, R. S. & Moilliet, T. K. (1971). Occurrence of phytoseiid mites (Acarina: Phytoseiidae) in apple orchards in South Central British Columbia. *J. ent. Soc. B. C.*, **68**, 33–6.

Dubynina, T. S. (1965). Onset of diapause and reactivation in *T. urticae*. *Ent. Obozr.*, **44**, 288–92.

Dunn, J. A. & Kempton, D. P. H. (1969). Resistance of rape (*Brassica napus*) to attack by the cabbage aphid (*Brevicoryne brassicae* L.). *Ann. appl. Biol.*, **64**, 203–12.

 (1971). Seasonal changes in aphid populations on Brussels sprouts. *Ann. appl. Biol.*, **68**, 233–44.

 (1972). Resistance to attack by *Brevicoryne brassicae* among plants of Brussels sprouts. *Ann. appl. Biol.*, **72**, 1–11.

Dunne, R. M. (1971). Overwintering of *Myzus persicae* and other aphids infesting sugar beet in Ireland. *Ir. J. agric. Res.*, **10**, 59–69.

Eastop, V. F. (1973). Biotypes of aphids. In *Perspectives in Aphid Biology* (ed. A. D. Lowe), pp. 40–51. *Bull. ent. Soc. N.Z.*, **2**.

Eastop, V. F. & Russell, G. E. (1967). Morphological and physiological distinction between two populations of the peach-potato aphid. *Nature, Lond.*, **215**, 514–15.

References

Economopoulos, A. P. (1972). Sexual competitiveness of γ-ray sterilized males of *Dacus oleae*. Mating frequency of artificially reared and wild females. *Environ. Ent.*, **1**, 490–7.

Economopoulos, A. P., Giannakakis, A., Tzanakakis, M. E. & Voyadjoglou, A. V. (1971). Reproductive behaviour and physiology of the olive fruit fly. I. Anatomy of the adult rectum and odours emitted by adults. *Ann. ent. Soc. Am.*, **64**, 1112–16.

Ekka, I., Rodriguez, J. G. & Davis, D. L. (1971). Influence of dietary improvement on oviposition and egg viability of the mite, *Tetranychus urticae*. *J. Insect Physiol.*, **17**, 1393–99.

Elbadry, E. A. (1968). Biological studies on *Amblyseius aleyrodis* a predator of the cotton whitefly (Acarina: Phytoseiidae). *Entomophaga*, **13**, 323–9.

Elbadry, E. A. & Elbenhawy, E. M. (1968). The effect of non-prey food, mainly pollen, on the development survival, and fecundity of *Amblyseius gossypi* (Acarina: Phytoseiidae). *Entomologia exp. appl.*, **11**, 269–72.

Elliott, W. M. (1968). Migration of the green peach aphid from peach in Essex County. *Proc. ent. Soc. Ont.*, **99**, 69–72.

Emmart, E. W. (1935). Studies of the chromosomes of *Anastrepha* (Diptera: Trypetidae). I. The chromosomes of the fruit fly, *Anastrepha ludens* Loew. *Proc. ent. Soc. Wash.*, **37**, 119–35.

Espul, J. C. & Mansur, P. S. (1968). Reproducción sexual del pulgón verde del duraznero *Myzus persicae* (Sulz.) en Mendoza (Argentina). *Rev. invest. agropecu.*, **5**, 63–71.

Essig, E. O. (1948). The most important species of aphids attacking cruciferous crops in California. *Hilgardia*, **18**, 405–22.

Euverte, G. (1975). Etude de l'efficacite d'*Aphytis melinus* au Maroc. *Annls Inst. Phytopath. Benaki*, in press.

Evenhuis, H. H. (1968). The natural control of the apple-grass aphid, *Rhopalosiphum insertum*, with remarks on the control of apple aphids in The Netherlands in general. *Neth. J. Pl. Path.*, **74**, 106–17.

*Fabres, G. (1971). Natural biological control of *Lepidosaphes beckii* (Coccoidea, Diaspididae) in New Caledonian shadow habitats. *Proceedings of the 12th Pacific Science Congress*, Canberra, **1**, 195.

Fauvel, G. (1971). Influence de l'alimentation sur la biologie d'*Orius* (*Heterorius*) *vicinus* Ribaut (Heteroptera, Anthocoridae). *Annl. zool.-ecol. anim.*, **3**, 31–42.

Fenjves, P. (1945). Beiträge zur Kenntnis der Blattlaus *Myzus persicae* Sulz., Überträgerin der Blattrollkrankheit der Kartoffel. *Mitt. schweiz. ent. Ges.*, **19**, 489–611.

Feron, M. (1962). L'instinct de reproduction chez la mouche méditerranéenne des fruits *Ceratitis capitata* Wied (Diptera: Trypetidae) Comportement sexuel. Comportement de ponte. *Rev. Path. veg. Ent. agric. Fr.*, **41**, 1–129.

Feron, M. & Andrieu, A. J. (1962). Etude des signaux acoustiques du male dans le comportement sexuel de *Dacus oleae* Gmel. (Diptera: Trypetidae). *Annls épiphyt.*, **13**, 269–76.

Ferrière, C. (1965). Hymenoptera Aphelinidae de l'Europe et du Bassin

Méditerranéen. *Faune de l'Europe et du Bassin Méditerranéen, 1.* Masson et Cie., Paris, 206 pp.

Ferris, G. F. (1942). *Atlas of the Scale Insects of North America,* part 4. Stanford University Press, California, plates 384–448.

Fiestas Ros de Ursinos, J. A., Constante, E. G., Duran, R. M. & Roncero, A. V. (1972). Etude d'un attractif naturel pour *Dacus oleae. Ann. Soc. ent. Fr.,* **8,** 179–88.

Fisher, R. W. & Hansell, R. I. C. (1964). Effect of pre- and post-treatment temperatures, age of deposit, and repellency on the toxicity of Kelthane to the two-spotted mite, *Tetranychus telarius* (L.) (Acarina: Tetranychidae). *Can. Ent.,* **96,** 1307–12.

Fisher, R. W. & Morgan, N. G. (1968). The effect on the two-spotted spider mite, *Tetranychus urticae,* of dicofol concentration and deposit distribution on the leaf surface. *Can. Ent.,* **100,** 777–81.

Fisken, A. G. (1959). Factors affecting the spread of aphid-borne viruses in potato in eastern Scotland. 1. Overwintering of potato aphids, particularly *Myzus persicae* (Sulzer). *Ann. appl. Biol.,* **47,** 264–73.

Flaherty, D. L. (1967). The ecology and importance of spider mites on grapevine in the southern San Joaquin Valley, with emphasis on the role of *Metaseiulus occidentalis* (Nesbitt). Ph.D. Thesis, University of California, Berkeley.

(1969). Vineyard trophic complexity and densities of the Willamette mite, *Eotetranychus willamettei* Ewing (Acarina: Tetranychidae). *Ecology,* **50,** 911–16.

Flaherty, D. L. & Hoy, M. A. (1971). Biological control of Pacific mites and Willamette mites in San Joaquin Valley vineyards. III. Role of tydeid mites. *Res. Popul. Ecol.,* **13,** 80–96.

Flaherty, D. L. & Huffaker, C. B. (1970). Biological control of Pacific mites and Willamette mites in San Joaquin Valley vineyards. I. Role of *Metaseiulus occidentalis.* II. Influence of dispersion patterns of *Metaseiulus occidentalis. Hilgardia,* **40,** 267–330.

Flanders, S. E. (1953). Predatism by the adult hymenopterous parasite and its role in biological control. *J. econ. Ent.,* **46,** 541–4.

(1959). Differential host relations of the sexes in parasitic Hymenoptera. *Entomologia exp. appl.,* **2,** 125–42.

Flanders, S. E., Gressitt, J. L. & Fisher, T. W. (1958). *Casca chinensis,* an internal parasite of California red scale. *Hilgardia,* **28,** 65–91.

Fleschner, C. A. (1952). Host–plant resistance as a factor influencing population density of citrus red mites on orchard trees. *J. econ. Ent.,* **45,** 637–95.

Fletcher, B. S. (1969). Structure and function of the sex pheromone glands of the male Queensland fruit fly, *Dacus tryoni. J. Insect Physiol.,* **15,** 1309–22.

(1973). The ecology of a natural population of the Queensland fruit fly, *Dacus tryoni.* IV. The immigration and emigration of adults. *Aust. J. Zool.,* **21,** 541–65.

(1974a). The ecology of a natural population of the Queensland fruit fly, *Dacus tryoni.* V. The dispersal of adults. *Aust. J. Zool.,* **22,** 189–202.

References

(1974b). The ecology of a natural population of the Queensland fruit fly, *Dacus tryoni*. VI. Seasonal changes in fruit fly numbers in the areas surrounding the orchard. *Aust. J. Zool.*, **22**, 353–63.

Fletcher, B. S. & Giannakakis, A. (1973). Factors limiting the response of females of the Queensland fruit fly, *Dacus tryoni*, to the sex pheromone of the male. *J. Insect Physiol.*, **19**, 1147–55.

Flitters, N. E. (1964). The effect of photoperiod, light intensity and temperature on copulation, oviposition and fertility, in the Mexican fruit fly. *J. econ. Ent.*, **57**, 811–13.

Foott, W. H. (1968). The importance of *Solanum carolinense* L. as a host of the pepper maggot, *Zonosemata electa* (Say) (Diptera: Tephritidae) in south-west Ontario. *Proc. ent. Soc. Ont.*, **98**, 16–18.

Ford, E. B. (1964). *Ecological Genetics*. Methuen, London, 335 pp.

Foster, G. N. (1972). The population dynamics of aphids infesting potato. Ph.D. Thesis, University of Newcastle-upon-Tyne.

Fowler, R. (1934). Green peach aphis (*Myzus persicae* Sulz.) and its control. *J. Dep. Agric. S. Aust.*, **38**, 376–82.

Fox, P. M., Pass, B. C. & Thurston, R. (1967). Laboratory studies on the rearing of *Aphidius smithi* (Hymenoptera: Braconidae) and its parasitism of *Acyrthosiphon pisum* (Homoptera: Aphididae). *Ann. ent. Soc. Am.*, **60**, 1083–7.

Frazer, B. D. (1972). Population dynamics and recognition of biotypes in the pea aphid (Homoptera: Aphididae). *Can. Ent.*, **104**, 1729–33.

Frazer, B. D. & van den Bosch, R. (1973). Biological control of the walnut aphid in California: the interrelationship of the aphid and its parasite. *Environ. Ent.*, **2**, 561–8.

Fritzsche, R. (1958). Zur Kenntnis der Raubinsekten von *Tetranychus urticae* Koch. *Beitr. Ent.*, **8**, 716–24.

Fritzsche, R., Wolfgang, H. & Opel, H. (1957). Untersuchungen über die Abhängigkeit der Spinnmilbenvermehrung von dem Ernährungzustand der Wirtspflanzen. *Z. Pflernähr.*, **78**, 13–27.

Frizzi, G. & Springhetti, A. (1953). Prime ricerche citogenetiche sul *Dacus oleae* Gmel. *Ricerca Scientifica*, **23**, 1612–20.

Funusaki, G. Y. (1966). Studies on the life cycle and propagation techniques of *Montandoniola moraguesi* Puton. (Heteroptera Anthocoridae). *Proc. Hawaii ent. Soc.*, **19**, 209–11.

Fusco, R. A., & Thurston, R. (1968). Anholocyclic overwintering of the green peach aphid in Kentucky. *J. econ. Ent.*, **61**, 1383–5.

Galecka, B. & Kajak, A. (1971). Studies on ecological mechanisms reducing population of *Myzus persicae* (Sulz.) (Hom., Aphididae). *Ekol. Pol.*, **19**, 789–806.

Gambrell, F. L. (1931). The fruit flies of New York. *J. econ. Ent.*, **24**, 226–32.

Gerling, D. & Bar, D. (1971a). Biological studies of *Pteroptrix smithi* (Hymenoptera: Aphelinidae). *Entomophaga*, **16**, 19–36.

(1971b). Reciprocal host–parasite relations as exemplified by *Chrysomphalus aonidum* (Homoptera: Diaspididae) and *Pteroptrix smithi* (Hymenoptera: Aphelinidae). *Entomophaga*, **16**, 37–44.

References

Gersdorf, E. (1955). Beiträge zur holozyklischen Überwinterung von *Myzodes persicae* Sulzer im Bereich des Pflanzenschutzamtes Hannover im Winterhalbjahr 1953/54. *Z. PflKrankh. PflPath. PflSchutz*, **62**, 1–11.

Gerson, U. (1967a). Studies on the chaff scale on citrus in Israel. *J. econ. Ent.*, **60**, 1145–51.

(1967b). Interrelationships of two scale insects on citrus. *Ecology*, **48**, 872–3.

(1968). The comparative biologies of two hymenopterous parasites of the chaff scale, *Parlatoria pergandii*. *Entomophaga*, **13**, 163–73.

(1971). The mites associated with armored scale insects. *Proceedings of the 3rd international congress of acarology*, Prague, pp. 653–4.

Geyspits, K. F., Sapozhinikova, F. D. & Taranets, M. N. (1971). Seasonal changes in the photoperiodic reaction and physiological state of the spider mite (*T. urticae* Koch). *Ent. Obozr.*, **50**, 156–62.

Gibson, F. A. (1966). A survey of the spider mites of deciduous fruit trees in New South Wales (*Tetranychus telarius*, *T. cinnabarinus*, *T. lobosus*, *T. urticae*). *Aust. J. Sci.*, **29**, 21–2.

Gibson, R. W. (1971a). The resistance of three *Solanum* species to *Myzus persicae*, *Macrosiphum euphorbiae* and *Aulacorthum solani* (Aphididae: Homoptera). *Ann. appl. Biol.*, **68**, 245–51.

(1971b). Glandular hairs providing resistance to aphids in certain wild potato species. *Ann. appl. Biol.*, **68**, 113–19.

(1974). Aphid-trapping glandular hairs on hybrids of *Solanum tuberosum* and *S. berthaultii*. *Potato Res.*, **17**, 152–4.

Gilbert, N. & Gutierrez, A. P. (1973). A plant–aphid–parasite relationship. *J. Anim. Ecol.*, **42**, 323–40.

Gilbert, N. & Hughes, R. D. (1971). A model of an aphid population – three adventures. *J. Anim. Ecol.*, **40**, 525–34.

Gilliatt, F. C. (1935). Some predators of the European red mite, *Paratetranychus pilosus* C. & F., in Nova Scotia. *Can. J. Res.*, **13**, 19–38.

Goeden, R. D. & Ricker, D. W. (1971). Biology of *Zonosemata vittigera* relative to silverleaf nightshade. *J. econ. Ent.*, **64**, 417–21.

Gonzalez, R. H. (1961). Contribución al conocimiento de los ácaros del manzano en Chile central. *Tech. Bull. Agric. Stn. Univ. Chile*, **11**, 59 pp.

(1963). Four new mites of the genus *Agistemus* Summers, 1960 (Acarina: Stigmaeidae). *Acarologia*, **5**, 342–50.

(1965). A taxonomic study of the genera *Mediolata*, *Zetzellia*, and *Agistemus*. *Calif. Univ. Publ. Ent.*, **41**, 1–64.

Gould, H. J. (1970). Preliminary studies of an integrated control programme for cucumber pests and an evaluation of methods of introducing *Phytoseiulus persimilis* Athias-Henriot for the control of *Tetranychus urticae* Koch. *Ann. appl. Biol.*, **66**, 505–13.

(1971). Large-scale trials of an integrated control programme for cucumber pests on commercial nurseries. *Pl. Path.*, **20**, 149–56.

Gould, H. J. & Light, W. I. St G. (1971). Biological control of *Tetranychus urticae* on stock plants of ornamental ivy. *Pl. Path.*, **20**, 18–20.

Greany, P. D. & Oatman, E. R. (1972). Analysis of host discrimination in the

251

References

parasite *Orgilus lepidus* (Hymenoptera, Braconidae). *Ann. ent. Soc. Am.*, **65**, 377–83.

Greathead, D. J. (1972). Dispersal of the sugar-cane scale *Aulacaspis tegalensis* (Zhnt.) (Hem., Diaspididae) by air currents. *Bull. ent. Res.*, **61**, 547–58.

Griffiths, D. C. (1960). The behaviour and specificity of *Monoctonus paludum* Marshall (Hym., Braconidae), a parasite of *Nasonovia ribis-nigri* (Mosley) on lettuce. *Bull. ent. Res.*, **51**, 303–19.

— (1961). The development of *Monoctonus paludum* Marshall (Hym., Braconidae) in *Nasonovia ribis-nigri* on lettuce, and immunity reactions in other lettuce aphids. *Bull. ent. Res.*, **52**, 147–63.

Grist, D. H. (1959). *Rice*. Longmans Green, London, 466 pp.

Günthart, E. (1945). Ueber Spinnmilben und deren natürliche Feinde. *Mitt. schweiz. ent. Ges.*, **19**, 279–308.

Gupta, S. K. (1969). Three new species of the genus *Phytoseius* (Acarina: Phytoseiidae) from India. *Israel J. agric. Res.*, **19**, 115–20.

Gustafsson, M. (1965a). On species of the genus *Entomophthora* Fres. in Sweden. I. Classification and distribution. *Lantbrukshögskolans Ann.*, **31**, 103–212.

— (1965b). On species of the genus *Entomophthora* Fres. in Sweden. II. Cultivation and physiology. *LantbrHögsk. Annlr*, **31**, 405–57.

— (1969). On species of the genus *Entomophthora* Fres. in Sweden. III. Possibility of usage in biological control. *LantbrHögsk. Annlr*, **35**, 235–74.

Guthrie, F. E., Cambell, W. V. & Baron, R. L. (1962). Feeding sites of the green peach aphid with respect to its adaptation to tobacco. *Ann. ent. Soc. Am.*, **55**, 42–6.

Gutierrez, A. P., Morgan, D. J. & Havenstein, D. E. (1971). The ecology of *Aphis craccivora* and subterranean clover stunt virus. I. The phenology of aphid populations and the epidemiology of virus in pastures in south-east Australia. *J. appl. Ecol.*, **8**, 699–721.

Gutierrez, J. & Van Zon, A. Q. (1973). A comparative study of several strains of the *Tetranychus neocaledonicus* complex and sterilization tests on males by X-rays. *Entomologia exp. appl.*, **16**, 123–34.

Hafez, M. (1961). Seasonal fluctuations of the population density of the cabbage aphid (*Brevicoryne brassicae* (L.)) in the Netherlands, and the role of its parasite *Aphidius* (*Diaeretiella*) *rapae* (Curtis). *Tijdschr. PlZiekt.*, **67**, 445–548.

Häfliger, E. (1953). Das Auswahlvermögen der Kirschenfliege bei der Eiablage. Eine statistische Studie. *Mitt. schweiz. ent. Ges.*, **26**, 258–64.

Hagen, K. S. (1962). Biology and ecology of predaceous Coccinellidae. *Ann. Rev. Ent.*, **7**, 289–326.

Hagen, K. S., Sawall, E. F., Jr & Tassan, R. L. (1971). The use of food sprays to increase effectiveness of entomophagous insects. *Proceedings tall timbers conference on ecological animal control by habitat management*, Tallahassee, Fla (1970), **2**, 59–81.

*Hagen, K. S. & van den Bosch, R. (1968). Impact of pathogens, parasites, and predators on aphids. *Ann. Rev. Ent.*, **13**, 325–84.

Haine, E. (1950). Zur Frage der Überwinterung von *Myzodes persicae* Sulz. an Sekundärwirten. II. *Anz. SchädlingsKd*, **23**, 81–6.

Halfhill, J. E. & Featherston, P. E. (1967). Propagation of braconid parasites of the pea aphid. *J. econ. Ent.*, **60**, 1756.

Halimie, M. A. & Ford, J. B. (1972). Feeding and uptake of phosphamidon by two strains of *Myzus persicae* (Sulz.) on radish plants. *Ann. appl. Biol.*, **70**, 169–74.

Hall, I. M. & Dunn, P. H. (1957*a*). Fungi on spotted alfalfa aphid. *Calif. Agric.*, **11**, 2, 5, 14.

(1957*b*). Entomophthorous fungi on the spotted alfalfa aphid. *Hilgardia*, **27**, 159–81.

Halstead, A. J. (1970). The effect of some systemic fungicides and other pesticides on the feeding preference of *Phytoseiulus persimilis* Athias-Henriot (Acarina: Phytoseiidae). Ph.D. Thesis, University of Wales, 47 pp.

Hamstead, E. O. (1970). Greenhouse integrated control studies of the two-spotted spider mite and lima beans with a predaceous mite, *Typhlodromus fallacis*, and insecticides. *J. econ. Ent.*, **63**, 1027–8.

Haniotakis, G. E. (1974). Sexual attraction in the olive fruit fly, *Dacus oleae* (Gmelin). *Environ. Ent.*, **3**, 82–6.

Hardman, J. A. & Wheatley, G. A. (1970). Lettuce root aphid. *Rep. natn. Veg. Res. Stn, 1969*, pp. 99–100.

*Harpaz, I. & Rosen, D. (1971). Development of integrated control programs for crop pests in Israel. In *Biological Control* (ed. C. B. Huffaker), Chapter 20, pp. 458–68. Plenum Press, New York and London.

Harris, K. M. (1971). A new approach to pest control in the Dahlia Trial at Wisley. *J. R. hort. Soc.*, **96**, 200–6.

Hartley, E. A. (1922). Some bionomics of *Aphelinus semiflavus* (Howard) chalcid parasite of aphids. *Ohio J. Sci.*, **22**, 209–36.

Harvey, T. L. & Hackerott, H. L. (1969). Recognition of greenbug biotype injurious to sorghum. *J. econ. Ent.*, **62**, 776–9.

Heathcote, G. D. & Cockbain, A. J. (1966). Aphids from mangold clamps and their importance as vectors of beet viruses. *Ann. appl. Biol.*, **57**, 321–36.

Heathcote, G. D., Palmer, J. M. P. & Taylor, L. R. (1969). Sampling for aphids by traps and by crop inspection. *Ann. appl. Biol.*, **63**, 155–66.

Heathcote, G. D. & Ward, J. (1958). The preference shown by *Myzus persicae* (Sulz.) for *Brassica* plants sprayed with wetting agents. *Bull. ent. Res.*, **49**, 235–7.

Heie, O. E. (1954). Studies of the overwintering of *Myzus persicae* Sulzer in Denmark and the occurrence of this aphid in beet fields. *Trans. Dan. Acad. tech. Sci.*, **1**, 1–34.

Heie, O. & Petersen, B. (1961). Investigations on *Myzus persicae* Sulz., *Aphis fabae* Scop. and virus yellows of beet (*Beta* virus 4) in Denmark. Condensed reports from the virus committee. *Trans. Dan. Acad. tech. Sci.*, 1961, pp. 7–13.

Heikinheimo, O. (1959). On the occurrence of virus vector aphids in Finland. *Valt. Maatalouskoet. Julk*, **178**, 20–40 (original in Finnish).

References

Heinricke, R. & Foot, U. W. (1966). The effect of several phosphate insecticides on photosynthesis of red delicious apple leaves. *Can. J. Pl. Sci.*, **46**, 589–91.

Heinze, K. (1939). Zur Biologie und Systematik der virusübertragenden Blattläuse. *Mitt. biol. Reichsanst. Ld-u Forstw.*, **59**, 35–48.

— (1948). Die Überwinterung der grünen Pfirsichlaus *Myzodes persicae* (Sulzer) und die Auswirkung der Überwinterungsquellen auf den Massenwechsel im Sommer. *NachrBl. dt. PflSchutzdienst*, **2**, 105–12.

Helle, W. (1969 a). Genetics and cytogenetics of spider mite species. *Proc. 2nd int. Congr. Acarol.*, Sutton Bonington, England (1967), pp. 479–83.

— (1969 b). New developments towards biological control of the two spotted spider mite by incompatible genes. *Publ. OEPP.*, Sér. A, **52**, 7–15.

Helle, W. & Bolland, H. R. (1967). Karyotypes and sex determination in spider mites. *Genetica*, **38**, 43–55.

*Helle, W., Gutierrez, J. & Bolland, H. R. (1970). A study on sex determination and karyotype evolution in Tetranychidae. *Genetica*, **41**, 21–32.

*Helle, W. & Overmeer, W. P. J. (1973). Variability in Tetranychid mites. *A. Rev. Ent.*, **18**, 97–120.

*Helle, W. & Pieterse, A. H. (1965). Genetic affinities between adjacent populations of spider mites. *Entomologia exp. appl.*, **5**, 159–62.

*Helle, W. & Van Zon, A. Q. (1970). Linkage in the pacific spider mite *Tetranychus pacificus*. II. Genes for white eye, lemon and flamingo. *Entomologia exp. appl.*, **13**, 300–6.

Hely, P. C. (1968). The entomology of citrus in New South Wales. *Aust. ent. Soc. Misc. Publ.*, **1**, 20 pp.

Henneberry, T. J. (1961). The effect of plant nutrition on the reproductive rate and susceptibility to malathion of two strains of the two-spotted spider mite *Tetranychus telarius* L. *Diss. Abstr.*, **21**, 10, 2840.

Herbert, H. J. (1959). Note on feeding ranges of six species of predaceous mites (Acarina: Phytoseiidae) in the laboratory. *Can. Ent.*, **91**, 812.

Herbert, H. J. & Sanford, K. H. (1969). The influence of spray programs of the fauna on apple orchards in Nova Scotia. XIX. Apple rust mite, *Vasates schlechtendali*, a food source for predators. *Can. Ent.*, **101**, 62–7.

Herne, D. C., Simpson, C. M. & Putman, W. L. (1969). Report in *Spider Mite Newsletter*, **2**, 11–14.

Herne, D. H. C. & Putman, W. L. (1966). Toxicity of some pesticides to predaceous arthropods in Ontario peach orchards. *Can. Ent.*, **98**, 936–42.

Herne, D. H. C. & Chant, D. A. (1965). Relative toxicity of parathion and kelthane to the predacious mite *Phytoseiulus persimilis* Athias-Henriot and its prey, *Tetranychus urticae* Koch (Acarina: Phytoseiidae, Tetranychidae) in the laboratory. *Can. Ent.*, **97**, 171–6.

Hill, A. R. (1957). The biology of *Anthocoris nemorum* (L.) in Scotland (Hemiptera: Anthocoridae). *Trans. R. ent. Soc. Lond.*, **109**, 379–94.

— (1961). The biology of *Anthocoris sarothamni* Douglas and Scott in Scotland (Hemiptera: Anthocoridae). *Trans. R. ent. Soc. Lond.*, **113**, 41–54.

— (1965). The bionomics and ecology of *Anthocoris confusus* Reuter in Scotland. I. The adult and egg production. *Trans. Soc. Br. Ent.*, **16**, 245–56.

(1968). The bionomics and ecology of *Anthocoris confusus* Reuter in Scotland. II. Life history and population changes. *Trans. Soc. Br. Ent.*, **18**, 35–48.

Hille Ris Lambers, D. (1955). Potato aphids and virus diseases in the Netherlands. *Ann. appl. Biol.*, **42**, 355–60.

Hobart, J., Ford, J., Kirby, R. & Jackson, G. J. (1970). Report of the work carried out on *Phytoseiulus*, 1968/69. *Spider Mite Newsletter*, **4**, 10–12.

Hodek, I. (1967). Bionomics and ecology of predaceous Coccinellidae. *A. Rev. Ent.*, **12**, 79–104.

(1973). *Biology of Coccinellidae*. Dr W. Junk, The Hague and Academia Press, Prague. 260 pp.

*Hodek, I., Hagen, K. S. & van Emden, H. F. (1972). Methods for studying effectiveness of natural enemies. In *Aphid Technology* (ed. H. F. van Emden), pp. 147–88. Academic Press, London and New York.

Hodek, I., Holman, J., Starý, P. & Štys, P. (1959). Natural enemies of the bean aphid (*Aphis fabae* Scop.) in Czechoslovakia. *Trans. 1st int. Conf. Insect Path. biol. Control*, Prague, 1958, pp. 553–7.

Hoffer, A. (1969). Beschreibungen der neuen Arten der Familie Encyrtidae (Hym., Chalcidoidea) aus der Tschechoslowakei. IV. *Acta entomol. Bohemoslov.*, **66**, 165–80.

Hoffman, W. E. (1937). Kwangtung Aphididae including host plants and distribution. *Lingnan Sci. J.*, **16**, 267–302.

Holdsworth, R. P. (1968). Integrated control: effect on European red mite and its more important predators. *J. econ. Ent.*, **61**, 1602–7.

(1972*a*). Major predators of the European red mite on apples in Ohio. *Ohio agric. Res. Circ.*, **192**, 17 pp.

(1972*b*). *Zetzellia mali* and *Agistemus fleschneri*: difference in spatial distribution. *Environ. Ent.*, **1**, 532–3.

(1972*c*). European red mite and its major predators: effects of sulfur. *J. econ. Ent.*, **65**, 1098–9.

Holling, C. S. (1961). Principles of insect predation. *A. Rev. Ent.*, **6**, 163–82.

Hormchong, T. & Wood, E. A. (1963). Evaluations of barley varieties for resistance to the corn leaf aphid. *J. econ. Ent.*, **56**, 113–14.

Horsburgh, R. L. & Asquith, D. (1968). Initial survey of arthropod predators of the European red mite in South Central Pennsylvania. *J. econ. Ent.*, **61**, 1752–4.

Horsfall, J. L. (1924). Life history studies of *Myzus persicae* Sulzer. *Bull. Pa agric. Exp. Stn*, **185**, 1–16.

Howard, L. O. (1929). *Aphelinus mali* and its travels. *Ann. ent. Soc. Am.*, **22**, 341–68.

Hoy, M. A. (1972). Diapause in the predaceous mite *Metaseiulus occidentalis* (Nesbitt) (Acarina: Phytoseiidae). Ph.D. Thesis, University of California, Berkeley.

Hoy, M. A. & Flaherty, D. (1970). Photoperiodic induction of diapause in a predaceous mite *Metaseiulus occidentalis*. *Ann. ent. Soc. Am.*, **63**, 959–63.

Hoyt, S. C. (1965). A possible new approach to mite control on apples. *Proc. Wash. State hort. Ass.*, **61**, 127–8.

References

(1966). The development of an integrated mite control program. *Proc. Wash. State hort. Ass.*, **62**, 71–2.

(1969*a*). Integrated chemical control of insects and biological control of mites on apple in Washington. *J. econ. Ent.*, **62**, 74–86.

(1969*b*). Population studies of five mite species on apples in Washington. *Proceedings of the 2nd international congress on acarology*, Sutton Bonnington, England (1967), 117–33.

(1972). Resistance to azinphosmethyl of *Typhlodromus pyri* (Acarina: Phytoseiidae) from New Zealand. *N.Z. J. Sci.*, **15**, 16–21.

Hoyt, S. C. & Caltagirone, L. E. (1971). The developing programs of integrated control of pests of apples in Washington and peaches in California. In *Biological Control* (ed. C. B. Huffaker), pp. 395–421. Plenum Press, New York and London.

Huettel, M. D. (1972). Comparative biology, host relationships and population genetics of sibling species in *Procecidochares* (Diptera: Tephritidae) Ph.D. Thesis, Department of Zoology, University of Texas, 212 pp.

Huettel, M. D. & Bush, G. L. (1972). The genetics of host selection and its bearing on sympatric speciation in *Procecidochares* (Diptera: Tephritidae). *Entomologia exp. appl.*, **15**, 465–80.

(1975). Enzyme polymorphism and the differentiation of sibling species. *2nd int. Symp. biol. Control of Weeds*, Rome.

Huffaker, C. B. & Kennett, C. E. (1953). Differential tolerance to parathion in two *Typhlodromus* predators on cyclamen mite. *J. econ. Ent.*, **46**, 707–8.

(1966). Studies of two parasites of olive scale, *Parlatoria oleae* (Colvée). IV. Biological control of *Parlatoria oleae* (Colvée) through the compensatory action of two introduced parasites. *Hilgardia*, **37**, 283–335.

(1969). Some aspects of assessing efficiency of natural enemies. *Can. Ent.*, **101**, 425–47.

Huffaker, C. B. & Spitzer, C. (1950). Some factors affecting red spider mite populations on pears in California. *J. econ. Ent.*, **43**, 819–31.

Huffaker, C. B., Messenger, P. S. & DeBach, P. (1971). The natural enemy component in natural control and the theory of biological control. In *Biological Control* (ed. C. B. Huffaker), pp. 16–67. Plenum Press, New York and London.

Huffaker, C. B., Van de Vrie, M. & McMurtry, J. A. (1969). The ecology of tetranychid mites and their natural control. *A. Rev. Ent.*, **14**, 125–74.

(1970). Ecology of tetranychid mites and their natural enemies: a review. II. Tetranychid populations and their possible control by predators: an evaluation. *Hilgardia*, **40**, 391–458.

Hughes, R. D. (1963). Population dynamics of the cabbage aphid, *Brevicoryne brassicae* (L.). *J. Anim. Ecol.*, **32**, 393–424.

(1972). Population dynamics. In *Aphid Technology* (ed. H. F. van Emden), pp. 275–93. Academic Press, London and New York.

Hughes, R. D. & Gilbert, N. (1968). A model of an aphid population – a general statement. *J. Anim. Ecol.*, **37**, 553–63.

Hukusima, S. (1968). Integrating arthropod predator releases and chemical manipulations for pest control in apple orchards. *Res. Bull. Fac. Agric. Gifu-ken Prefect. Univ.*, **26**, 40–63.

(1969). Ecological implications in population trends of common arthropod predators and major pests in apple orchards under different control programs. *Res. Bull. Fac. Agr. Gifu-ken Prefect. Univ.*, **28**, 64–87.

Hull, R. (1967). Aphid borne viruses of sugar beet: a retrospective exercise in integrated control. *Proceedings 4th British conference on insects and fungi*, **2**, 472–7.

Hussey, N. W. (1965). Possibilities for integrated control of some glasshouse pests. *Ann. appl. Biol.*, **56**, 347–50.

(1972). Diapause in *Tetranychus urticae* Koch and its implications in glasshouse culture. *Acarologia*, **13**, 344–50.

Hussey, N. W. & Bravenboer, L. (1971). Control of pests in glasshouse culture by the introduction of natural enemies. In *Biological Control* (ed. C. B. Huffaker), pp. 195–216. Plenum Press, New York and London.

Iglinsky, W. & Rainwater, C. F. (1950). *Orius insidiosus* as enemy of a spider mite on cotton. *J. econ. Ent.*, **43**, 567–8.

Inaizumi, M. (1968). Factors affecting the infestation of aphids on potato plants. *Jap. J. appl. Ent. Zool.*, **12**, 10–17.

Iwahashi, O. (1972). Movement of the oriental fruit fly *Dacus dorsalis* (Diptera: Tephritidae) adults among the islets of the Ogasawara Islands. *Environ. Ent.*, **1**, 176–9.

Jackson, G. J. (1969). The effect of residues of some common glasshouse pesticides on the feeding behaviour of *Phytoseiulus persimilis* Athias-Henriot (Acarina: Phytoseiidae). Ph.D. Thesis, University of Wales, 42 pp.

Jackson, H. B., Rogers, C. E. & Eikenbary, R. D. (1971). Colonization and release of *Aphelinus asychis*, an imported parasite of the greenbug. *J. econ. Ent.*, **66**, 173–6.

Jacob, F. H. (1941). The overwintering of *Myzus persicae* Sulz. on Brassicae in North Wales. *Ann. appl. Biol.*, **28**, 119–24.

Jacobson, M., Ohinata, K., Chambers, D. L., Jones, W. A. & Fujimoto, M. S. (1973). Insect sex attractants. XIII. Isolation, identification and synthesis of sex pheromones of the male Mediterranean fruit fly. *J. med. Chem.*, **16**, 248–51.

Jarraya, A. (1973). Observations bioécologiques sur une cochenille citricole dans la region de Tunis, *Saissetia oleae* (Bernard) (Homoptera, Coccoidea, Coccidae). *Annls Inst. Phytopath. Benaki*, in press.

Jarvis, J. L. (1969). Differential reaction of introductions of crambe to the turnip aphid and the green peach aphid. *J. econ. Ent.*, **62**, 697–8.

Jepson, W. F. (1954). A critical review of the world literature on the Lepidopterous stalk borers of tropical graminaceous crops. *Commonw. Inst. Ent. Lond.*, 127 pp.

Johnson, C. G. (1969). *Migration and Dispersal of Insects by Flight*. Methuen, London, 763 pp.

References

Jones, S. C. (1937). The currant and gooseberry maggot or yellow currant fly, *Epochra canadensis* (Loew). *Oreg. agric. Exp. Stn Circ.*, **121**, 11 pp.

Kamburov, S. S. (1966). Methods of rearing and transporting predaceous mites. *J. econ. Ent.*, **59**, 875–7.

Kamran, M. A. & Raros, E. S. (1969a). Seasonal fluctuations in abundance of various rice stem-borers on Luzon Island, Philippines. *J. econ. Ent.*, **61**, 650–6.

(1969b). Insect parasites in the control of species of rice stem-borers on Luzon Island, Philippines. *J. econ. Ent.*, **61**, 797–801.

Karg, W. (1970). Ueber die Möglichkeiten von integrierten Pflanzenschutz-massnahmen bei der Spinnmilbenbekämpfung im Obstbau. *NachrBl. dt. PflSchutzdienst*, **24**, 166–70.

Kawasaki, M. (1940). Some observations on the life after the fall migration to the peach tree in *Myzus persicae* Sulz. in Manchuria. *Insect World*, **44**, 101–3, 132–5.

Kazimi, S. K. & Ghani, M. A. (1970). Tentative 'Key' predators of tetranychids. 6. Pakistan. *Spider Mite Newsletter*, no. **3** (mimeo.).

Kehat, M. (1967). Survey and distribution of common lady beetles (Col. Coccinellidae) on date palm trees in Israel. *Entomophaga*, **12**, 119–25.

(1968a). The phenology of *Pharoscymnus* spp. and *Chilocorus bipustulatus* L. in date palm plantations in Israel. *Annls épiphyt.*, **19**, 605–14.

(1968b). The feeding behaviour of *Pharoscymnus numidicus* (Coccinellidae), predator of the date palm scale *Parlatoria blanchardi*. *Entomologia exp. appl.*, **11**, 30–42.

Kehat, M. & Greenberg, S. (1970). Survey and distribution of lady beetles (Coccinellidae) in citrus groves in Israel. *Entomophaga*, **15**, 275–80.

Kehat, M., Greenberg, S. & Gordon, D. (1970). Factors causing seasonal decline in *Chilocorus bipustulatus* (Coccinellidae) in citrus groves in Israel. *Entomophaga*, **15**, 337–45.

Kennedy, J. S., Day, M. F. & Eastop, V. F. (1962). *A Conspectus of Aphids as Vectors of Plant Viruses*. Commonw. Inst. Ent., London, 114 pp.

Kennett, C. E. (1970). Resistance to parathion in the phytoseiid mite *Amblyseius hibisci*. *J. econ. Ent.*, **63**, 1999–2000.

Kennett, C. E. & Caltagirone, L. E. (1968). Biosystematics of *Phytoseiulus persimilis* Athias-Henriot (Acarina-Phytoseiidae). *Acarologia*, **10**, 563–77.

Keuneke, W. (1924). Uber die Spermatogenese einiger Dipteren. *Z. f. Zellen-u. Gewebelehre*, **1**, 357–412.

Kinn, D. N. & Doutt, R. L. (1972). Initial survey of arthropods found in north coast vineyards of California. *Environ. Ent.*, **1**, 508–13.

Kirkby, R. A. (1969). Studies on the predatory efficiency of *Phytoseiulus persimilis* Athias-Henriot on the red spider mite *Tetranychus urticae* Koch in the presence of discontinuous deposits of selected pesticides. Ph.D. Thesis, University of Wales, 46 pp.

Kiritani, K. & Dempster, J. P. (1973). Different approaches to the quantitative evaluation of natural enemies. *J. appl. Ecol.*, **10**, 323–30.

Klassen, W., Knipling, E. F. & McGuire, J. U., Jr (1970). The potential for insect population suppression by dominant conditional lethal traits. *Ann. ent. Soc. Am.*, **63**, 238–55.

Klingauf, F. (1967). Abwehr- und Meidereaktionen von Blattläusen (Aphididae) bei Bedrohung durch Räuber und Parasiten. *Z. angew. Ent.*, **60**, 269–317.

Klingauf, F. & Sengonca, C. (1970). Koloniebildung von Röhrenblattläusen (Aphididae) unter Feindeinwirkung. *Entomophaga*, **15**, 359–77.

Knipling, E. F. (1970). Suppression of pest Leopidoptera by releasing partially sterile males. A theoretical appraisal. *BioScience*, **20**, 415–70.

Knisley, C. B. & Swift, F. C. (1971). Biological studies of *Amblyseius umbraticus* (Acarina: Phytoseiidae). *Ann. ent. Soc. Am.*, **64**, 813–22.

(1972). Qualitative study of mite fauna associated with apple foliage in New Jersey. *J. econ. Ent.*, **65**, 445–8.

Kono, T. & Sugino, I. (1958). On the estimation of the density of rice stems infested by the rice stem-borer. *Jap. J. appl. Zool.*, **2**, 184–8.

Korcz, A. (1967). Fauna of the predatory bugs (Hemiptera–Heteroptera) occurring on apple trees in the district of Poznan. *Pol. Pismo Entomol.*, **37**, 581–6.

(1968). Wstepne badania nad biologia *Anthocoris pilosus* (Jak.) (Hemiptera, Anthocoridae). *Pol. Pismo Entomol.*, **38**, 559–64.

(1969). The predator bug fauna of apple trees in the vicinity of Poznan. *Pol. Pismo Entomol.*, **39**, 581–6 (*Abstr. Rev. appl. Ent.*, **59**, 724).

Kovachevski, I. K. (1942). Die Viruskrankheiten der Paprikapflanze. *Arch. bulg. landw. Ges.*, **1**, 25–102.

Krambias, A. (1968). A study on the effects of three acaricidal compounds on the predatory mite *Phytoseiulus persimilis* Athias-Henriot. Ph.D. Thesis, University of Wales, 54 pp.

Krambias, K. (1969). Effect on some pesticides on *Phytoseiulus*. *Spider Mite Newsletter no.* **2**, 14.

Krejzová, R. (1972). Experimental infections of several species of aphids by species of the genus *Entomophthora*. *Vestn. česk. zool. Spol. Prague*, **36**, 17–22 (original in Czech).

Krimbas, C. B. (1963). A contribution to the cytogenetics of *Dacus oleae* (Gmel). (Diptera: Trypetidae) the salivary gland and mitotic chromosomes. *Caryologia*, **16**, 371–5.

Krimbas, C. B. & Tsakas, S. (1971). The genetics of *Dacus oleae*. v. Changes of esterase polymorphism in a natural population following insecticide control – selection or drift? *Evolution*, **25**, 454–60.

Krombein, K. V. & Burks, B. D. (1967). *Hymenoptera of America North of Mexico*. Synoptic Catalog (Agricultural Monograph No. 2). Second Supplement. U.S. Department of Agriculture, Washington, D.C., 584 pp.

Kuenen, D. J. (1942). Onderzoek over de biologie en bestrijding van het fruitspint, *Metatetranychus ulmi* Koch. (Acari, Tetranychidae). *Tijdschr. Ent.*, **88**, 303–12.

References

Lagace, C. F. (1969). Observations on the biology of *Mesidia nigra,* a parasite of *Iziphya punctata. Ann. ent. Soc. Am.,* **62,** 532–6.
Laing, J. E. (1968). Life history and life table of *Phytoseiulus persimilis* Athias-Henriot. *Acarology,* **10,** 578–88.
(1969). Life history and life table of *Tetranychus urticae* Koch. *Acarology,* **11,** 32–42.
Laing, J. E., Calvert, D. L. & Huffaker, C. B. (1972). Preliminary studies of effects of *Tetranychus pacificus* on yield and quality of grapes in the San Joaquin Valley, California. *Environ. Ent.,* **1,** 658–63.
Laing, J. E. & Huffaker, C. B. (1969). Comparative studies of predation by *Phytoseiulus persimilis* Athias-Henriot and *Metaseiulus occidentalis* (Nesbitt) (Acarina: Tetranychidae) on populations of *Tetranychus urticae* Koch (Acarina: Tetranychidae). *Res. Popul. Ecol.,* **11,** 105–26.
Laing, J. E. & Osborn, J. A. L. (1974). The effect of prey density on the functional and numerical responses of three species of predatory mites. *Entomophaga,* **19,** 267–77.
Lamb, K. P. (1953). Field trials of five swede varieties with special reference to aphid resistance. *N.Z. J. Sci. Tech.,* (A), **35,** 135–45.
Lammerink, J. (1968*a*). Rangi: new rape that resists aphids. *N.Z. J. Agric.,* **117,** 61.
(1968*b*). A new biotype of cabbage aphid (*Brevicoryne brassicae* (L.)) on aphid resistant rape (*Brassica napus* L.). *N.Z. J. agric. Res.,* **11,** 341–4.
Lampel, G. (1968). *Die Biologie des Blattlaus-Generationswechsels.* G. Fischer, Jena, 264 pp.
Lawson, F. R. & Chamberlin, F. S. (1957). Aphids on tobacco, how to control them. *U.S. Dep. Agric. Leaflet,* **405,** 1–8.
Leatherdale, D. (1970). The arthropod hosts of entomogenous fungi in Britain. *Entomophaga,* **15,** 419–35.
Leclant, F. & Remaudière, G. (1970). Eléments pour la prise en considération des Aphides dans la lutte intégrée en vergers de Pêchers. *Entomophaga,* **15,** 53–81.
Lee, M. S. & Davis, D. W. (1968). Life history and behaviour of the predatory mite *Typhlodromus occidentalis* in Utah. *Ann. ent. Soc. Am.,* **61,** 151–5.
Legowski, T. J. (1966). Experiments on predator control of the glasshouse red spider mite of cucumbers. *Pl. Path.,* **15,** 34–41.
Le Roux, E. J. & Mukerji, M. K. (1963). Notes on the distribution of immature stages of the apple maggot, *Rhagoletis pomonella* (Walsh) (Diptera: Trypetidae) on apples. *Ann. ent. Soc. Que.,* **8,** 60–9.
*Li, C. S. (1970). Some aspects of the conservation of natural enemies of rice stem-borer and the feasibility of harmonizing chemical and biological control of these pests in Australia. *Mushi,* **44,** 15–23.
(1972). Integrated control of the white rice borer, *Tryporyza innotata* (Walker) (Lepidoptera: Pyralidae), in Northern Australia. *Mushi,* **45,** *Suppl.,* 51–9.
*Lim, G. S. (1970). Some aspects of the conservation of natural enemies of rice stem-borers and the feasibility of harmonizing chemical and biological control of these pests in Malaysia. *Mushi,* **43,** 127–35.

Lim Sook Ming. (1971). The biology and ecology of some aphidophagous Coccinellidae. Ph.D. Thesis, University of Malaya, Kuala Lumpur.

Lord, F. T. (1949). Influence of spray programs on the fauna of apple orchards in Nova Scotia. III. Mites and their predators. *Can. Ent.*, **81**, 202–30.

(1956). The influence of spray programs on the fauna of apple orchards in Nova Scotia. IX. Studies on means of altering predator populations. *Can. Ent.*, **88**, 129–37.

(1962). The influence of spray programs on the fauna of apple orchards in Nova Scotia. XI. Effects of low dosages of DDT on predator populations. *Can. Ent.*, **94**, 204–16.

(1972). Comparisons of the abundance of the species composing the foliage inhabiting fauna of apple trees. *Can. Ent.*, **104**, 731–49.

Lord, F. T., Herbert, H. J. & MacPhee, A. W. (1958). The natural control of phytophagous mites in apple trees in Nova Scotia. *Proceedings of the 10th international congress on entomology*, Montreal, **4**, 617–22.

Lowe, A. D. (1962). The overwintering of *Myzus persicae* Sulz. in Canterbury (a note). *N.Z. J. agric. Res.*, **5**, 364–7.

(ed.). (1973). *Perspectives in Aphid Biology. Bull. ent. Soc. N.Z.*, **2**, 1–123.

Lowe, H. J. B. (1972a). The role of varieties in controlling pests and diseases. *J. Inst. int. Rech. better.*, **5**, 224–31.

(1972b). Resistance to aphids in sugar beet. *Rep. Pl. Breed. Inst.*, *1971*, pp. 146–7.

(1973a). Variation in *Myzus persicae* (Sulz.) (Hemiptera, Aphididae) reared on different host plants. *Bull. ent. Res.*, **62**, 549–56.

(1973b). Resistance to aphids in sugar beet. *Rep. Pl. Breed. Inst.*, *1972*, pp. 150–1.

(1974a). A method for testing for resistance to aphids in the glasshouse and some factors influencing aphid numbers on sugar beet under test. *J. agric. Sci. Camb.*, in press.

(1974b). Intraspecific variation in *Myzus, persicae* (Sulz.) (Hemiptera: Aphididae) on sugar beet (*Beta vulgaris* L.). *Ann. appl. Biol.*, in press.

Lowe, H. J. B. & Russell, G. E. (1969). Inherited resistance of sugar beet to aphid colonisation. *Ann. appl. Biol.*, **63**, 337–44.

Lundie, A. E. (1924). A biological study of *Aphelinus mali* Hald., a parasite of the woolly apple aphid *Eriosoma lanigerum* Haus. *Mem. agr. Exp. Stn Cornell Univ.*, **79**, 1–27.

MacGillivray, M. E. (1972). The sexuality of *Myzus persicae* (Sulzer), the green peach aphid, in New Brunswick. *Can. J. Zool.*, **50**, 469–71.

Mackauer, M. (1961). Zur Frage der Wirtsbindung der Blattlausschlupfwespen (Hymenoptera: Aphidiidae). *Zentbl. Bakt. ParasitKde Abt.*, **20**, 576–91.

(1967). Wirtsbindung und parallele Evolution parasitischer Hymenopteren. I. Allgemeines und Parasiten der Homopteren, 1. Teil. *Angew. Parasitol.*, **8**, 21–39.

* (1968). Insect parasites of the green peach aphid, *Myzus persicae* Sulz., and their control potential. *Entomophaga*, **13**, 91–106.

(1971). *Acyrthosiphon pisum* (Harris), pea aphid (Homoptera: Aphididae.

References

In *Biological Control Programmes against Insects and Weeds in Canada 1959–1968*, pp. 3–10. *Tech. Commun. Commonw. Inst. biol. Control*, **4**.

(1972). Genetic aspects of insect production. *Entomophaga*, **17**, 27–48.

(1973a). The population growth of the pea aphid biotype R1 on broad bean and pea (Homoptera: Aphididae). *Z. angew. Ent.*, **74**, 343–51.

(1973b). Host selection and host suitability in *Aphidius smithi* (Hymenoptera: Aphidiidae). In *Perspectives in Aphid Biology* (ed. A. D. Lowe), pp. 20–9. *Bull. ent. Soc. N.Z.*, **2**.

Mackauer, M. & Albright, L. J. (1973). The susceptibility of the pea aphid to intrahemocoelic infection by *Serratia marcescens*. *J. Invertebr. Path.*, **22**, 418–23.

Mackauer, M. & Starý, P. (1967). World Aphidiidae (Hym. Ichneumonoidea). In *Index of Entomophagous Insects* (ed. V. L. Delucchi and G. Remaudière). Le François, Paris, 195 pp.

Mackauer, M. & van den Bosch, R. (1973). General applicability of evaluation results. *J. appl. Ecol.*, **10**, 330–5.

MacLeod, D. M., Cameron, J. W. M. & Soper, R. S. (1966). The influence of environmental conditions on epizootics caused by entomogenous fungi. *Revue roum. Biol.*, Sér. Bot., **11**, 125–34.

MacLeod, D. M. & Müller-Kögler, E. (1973). Entomogenous fungi: *Entomophthora* species with pear-shaped to almost spherical conidia (Entomophthorales: Entomophthoraceae). *Mycologia*, **65**, 823–93.

MacPhee, A. W. & Sanford, K. H. (1956). The influence of spray programs on the fauna of apple orchards in Nova Scotia. x. Supplement to vii. Effects on some beneficial arthropods. *Can. Ent.*, **88**, 631–4.

(1961). The influence of spray programs on the fauna of apple orchards in Nova Scotia. xii. Second Supplement to vii. Effects on beneficial arthropods. *Can. Ent.*, **93**, 671–3.

Madsen, H. F. (1968). Integrated control of deciduous tree fruit pests. *World Crops Lond.*, **20** (4), 20–3.

(1970). Insecticides for codling moth control and their effect on other insects and mites of apple in British Columbia. *J. econ. Ent.*, **63**, 1521–3.

Maltby, H. L., Jiménez-Jiménez, E. & DeBach, P. (1968). Biological control of armored scale insects in Mexico. *J. econ. Ent.*, **61**, 1086–8.

Markkula, M. & Roukka, K. (1970). Resistance of plants to the pea aphid *Acyrthosiphon pisum* (Harris) (Hom., Aphididae). i. Fecundity of the biotypes on different host plants. *Annls agric. fenn.*, **9**, 127–32.

Markkula, M., Roukka, K. & Tiittanen, K. (1968). Reproduction of *Myzus persicae* (Sulz.) and *Tetranychus telarius* (L.) on different chrysanthemum cultivars. *Annls agric. fenn.*, **8**, 175–83.

Markkula, M. & Tiittanen, K. (1969). Effect of fertilizers on the reproduction of *Tetranychus telarius* L., *Myzus persicae* Sulz. and *Acyrtosiphon pisum* Harris. *Annls agric. fenn.*, **8**, 9–14.

(1970). Two-spotted spider mite, *Tetranychus telarius* (L.) resistant to aldicarb, Temik 1 OG. *Annls ent. fenn.*, **36**, 191–2.

Marshall, C. E. (1930). Biology of *Orius insidiosus*. *J. Kans. ent. Soc.*, **3**, 29–32.

Martin, H. (1948). Observations biologiques et essais de traitements contre la mouche de l'olive *Dacus oleae* (Rossi) dans la province de Tarragone (Espagne) de 1946 à 1948. *Mitt. schweiz. ent. Ges.*, **21**, 361–402.

Mason, A. C. (1922). Life history studies of some Florida aphids. *Fla Ent.*, **5**, 53–9, 62–5.

Mathys, G. (1958). The control of phytophagous mites in Swiss vineyards by *Typhlodromus* mites. *Proceedings of the 10th international congress on entomology*, Montreal, **4**, 607–10.

Matsuka, M., Shimotori, D., Senzaki, T. & Okada, T. (1972). Rearing some coccinellids on pulverized drone honeybee brood. *Bull. Fac. Agric. Tamagawa Univ.*, **12**, 28–38.

Mayo, Z. B. & Starks, K. J. (1972). Chromosome comparisons of biotypes of *Schizaphis graminum* to one another. *Ann. ent. Soc. Am.*, **65**, 925–8.

Mayr, L. (1973). Möglichkeiten und Grenzen des Einsatzes von *Aphidoletes aphidimyza* (Rond.) (Diptera, Cecidomyidae) gegen Blattläuse im Gewächshaus. *Z. angew. Ent.*, **73**, 255–60.

Maxwell, C. W. (1968). Apple maggot adult dispersion in a New Brunswick orchard. *J. econ. Ent.*, **61**, 103–6.

Maxwell, C. W. & Parsons, C. E. (1968). The recapture of marked apple maggot adults in several orchards from one release point. *J. econ. Ent.*, **61**, 1157–9.

MacArthur, R. H. & Wilson, E. O. (1967). *The Theory of Island Biogeography*. Princeton University Press, New Jersey, 203 pp.

McClanahan, H. J. (1967). Food chain toxicity of systemic acaricides to predacious mites. *Nature, Lond.*, **215**, 1001.

McKechnie, S. W. (1972). Differences in specific proteins within populations of the *Dacus tryoni-neohumeralis* complex. Ph.D. Thesis, University of Sydney, Australia.

McKenzie, H. L. (1956). The armored scale insects of California. *Bull. Calif. Insect Survey*, **5**, 208 pp.

McLaren, I. W. (1971). A comparison of the population growth potential in California red scale, *Aonidiella aurantii* (Maskell), and yellow scale, *A. citrina* (Coquillett), on citrus. *Aust. J. Zool.*, **19**, 189–204.

  (1972). Integrated control of insect pests on citrus fruits in Victoria, Australia. *Proceedings of the 14th international congress on entomology*, Canberra, p. 248 (abstract).

McLeod, J. H. (1937). Some factors in the control of the common greenhouse aphid, *Myzus persicae* Sulzer, by the parasite *Aphidius phorodontis* Ashm. *Rep. ent. Soc. Ont.*, **67**, 62–70.

  (1940). Biological control of greenhouse insect pests. *Rep. ent. Soc. Ont.*, **70**, 62–8.

McMurtry, J. A. (1969). Biological control of citrus red mite in California. *Proc. 1st int. Citrus Symp.*, Riverside, **2**, 855–62.

McMurtry, J. A. & Johnson, H. G. (1966). An ecological study of the spider mite *Oligonychus punicae* (Hirst) and its natural enemies. *Hilgardia*, **37**, 363–402.

McMurtry, J. A., Johnson, H. G. & Scriven, G. T. (1969). Experiments to

References

determine effects of mass releases of *Stethorus picipes* on the level of infestation of the avocado brown mite. *J. econ. Ent.*, **62**, 1216–21.

McMurtry, J. A. & Huffaker, C. B. & Van de Vrie, M. (1970). Ecology of tetranychid mites and their natural enemies: a review. I. Tetranychid enemies: their biological characters and the impact of spray practices. *Hilgardia*, **40**, 331–90.

McMurtry, J. A., Oatman, E. R. & Fleschner, C. A. (1971). Phytoseiid mites on some tree and row crops and adjacent wild plants in southern California. *J. econ. Ent.*, **64**, 405–8.

McMurtry, J. A. & Scriven, G. T. (1962). The use of agar media in transporting and rearing phytoseiid mites. *J. econ. Ent.*, **55**, 412–14.

McMurtry, J. A. & Scriven, G. T. (1964). Studies on the feeding, reproduction, and development of *Amblyseius hibisci* (Acarina: Phytoseiidae) on various food substances. *Ann. ent. Soc. Am.*, **57**, 649–55.

(1965). Life-history of *Amblyseius limonicus*, with comparative observations on *Amblyseius hibisci* (Acarina: Phytoseiidae). *Ann. ent. Soc. Am.*, **58**, 106–11.

(1966). Studies on predator–prey interactions. between *Amblyseius hibisci* and *Oligonychus punicae* (Acarina: Phytoseiidae, Tetranychidae) under greenhouse conditions. *Ann. ent. Soc. Am.*, **59**, 793–800.

(1968). Studies of predator–prey interactions between *Amblyseius hibisci* and *Oligonychus punicae*: Effects of host-plant conditioning and limited quantities of an alternate food. *J. econ. Ent.*, **61**, 393–7.

(1971). Predation by *Amblyseius limonicus* on *Oligonychus punicae* (Acarina): Effects on initial predator–prey ratios and prey distribution. *J. econ. Ent.*, **64**, 21–4.

McMurtry, J. A., Scriven, G. T. & Malone, R. S. (1973). Factors affecting oviposition of *Stethorus picipes*. (In press.)

Mendes, L. O. T. (1958). Observacões citológicas em 'moscas das frutas'. *Bragantia*, **17**, 29–39.

Messenger, P. S. (1970). Bioclimatic inputs to biological control and pest management programmes. In *Concepts of Pest Management* (eds. R. L. Rabb and F. E. Guthrie), pp. 84–99. North Carolina State University, Raleigh.

Messenger, P. S. & Force, D. C. (1963). An experimental host–parasite system: *Therioaphis maculata* (Buckton) – *Praon palitans* Muesebeck (Homoptera:Aphididae–Hymenoptera:Braconidae). *Ecology*, **44**, 532–40.

Michel, M. F. (1967). Importance écologique du comportement prédateur d'*Aphelinus asychis* Walker (Hym. Aphelinidae) endoparasite de pucerons (Hom. Aphididae). *C. r. Acad. Sci. Paris*, **264**, 936–9.

(1970). L'adaptation d'*Aphelinus asychis* Walker (Hym. Aphelinidae) à certains de ses hôtes naturels (Hom. Aphididae). *C. r. Acad. Sci.*, Paris, **271**, 2339–41.

(1971). Aphélinides, parasites de pucerons (Hym. Chalcidoidea). *Parasitica*, **27**, 127–34.

Michelbacher, A. E. & Middlekauff, W. W. & Bacon, O. G. (1952). Mites on melons in northern California. *J. econ. Ent.*, **45**, 365–70.

Miles, P. W. (1968). Insect secretions in plants. *A. Rev. Phytopath.*, **6**, 137–64.

Minoranskii, V. A. (1967). Über die Faktoren, die die Massenvermehrung der Rübenblattlaus (*Aphis fabae* Scop.) im Süden der UdSSR verhindern. *Arch. PflSchutz*, **3**, 101–14.

Missonnier, J., Robert, Y. & Thoizon, G. (1970). Circonstances épidémiologiques semblant favoriser le développement des mycoses à Entomophthorales chez trois aphides, *Aphis fabae* Scop., *Capitophorus horni* Börner et *Myzus persicae* Sulz. *Entomophaga*, **15**, 169–90.

Mitchell, R. (1972). The sex ratio of spider mite, *Tetranychus urticae*. *Entomologia exp. appl.*, **15**, 293–8.

Mittler, T. E. (1958). Studies on the feeding and nutrition of *Tuberolachnus salignus* (Gmelin) (Homoptera, Aphididae). II. The nitrogen and sugar composition of ingested phloem sap and excreted honeydew. *J. exp. Biol.*, **35**, 74–84.

(1970). Effects of dietary amino acids on the feeding rate of the aphid, *Myzus persicae*. *Entomologia exp. appl.*, **13**, 432–7.

Miyake, T. (1919). Studies on the fruit flies of Japan. *Imp. Cent. agric. Exp. Stn Bull.*, **2**, 86–160.

Moericke, V. (1950). Wo entstehen Gynoparen und Männchen der Pfirsichblattlaus (*Myzodes persicae* Sulz.)? *NachrBl. dt. PflSchutzdienst*, **2**, 99–102.

(1963). Über 'Virusartige Körper' in Organen von *Myzus persicae* (Sulz.). *Z. PflKrankh. PflPath. PflSchutz*, **70**, 464–70.

Monadjemi, N. (1972). Sur les variations de la fécondité d'*Aphelinus asychis* (Hym. Aphelinidae) en fonction des espèces aphidiennes mises à sa disposition à différentes périodes de sa vie imaginale. *Annls Soc. ent. Fr.* (N.S.), **8**, 451–60.

Monastero, S. (1967). La prima grande applicazione di lotta biologica artificiale contro la mosca delle olive (*Dacus oleae* Gmel.). *Boll. Ist. Entomol. Agrar. Oss. Fitopatol. Palermo*, **7**, 63–100.

(1969). I risultati della lotta biologica contro il *Dacus oleae* nel 1968 e nuove acquisizioni tecniche nell'allevamento della *Ceratitis capitata* W. *Boll. Ist. Entomol. Agrar. Oss. Fitopatol. Palermo*, **7**, 165–70.

Monro, J. (1953). Stridulation in the Queensland fruit fly, *Dacus (Strumeta) tryoni* Frogg. *Aust. J. Sci.*, **16**, 60–2.

(1967). The exploitation and conservation of resources by populations of insects. *J. Anim. Ecol.*, **36**, 531–47.

Montes, F. O. (1970). Biologia y morfologia de *Eriopis connexa* Germar 1824 y de *Adalia bipunctata* Linnaeus 1758 (Coleoptera). *Publs Centro Estud. ent. Univ. Chile*, **10**, 43–54.

Moore, I. (1960). Contribution to the ecology of the olive fly, *Dacus oleae* Gmel. *Israel Min. Agric. Res. Stn spec. Bull.*, **26**, 53 pp.

Moreno, D. S., Rice, R. E. & Carman, G. E. (1972). Specificity of the sex pheromones of female yellow scales and California red scales. *J. econ. Ent.*, **65**, 698–701.

Morgan, C. V. G. & Anderson, N. H. (1958). Notes on parathion-resistant

References

strains of two phytophagous mites and a predaceous mite in British Columbia. *Can. Ent.*, **90**, 92–7.

Mori, H. (1967). A review of biology on spider mites and their predators in Japan. *Mushi*, **40**, 47–65.

Mori, H. & Chant, D. A. (1966). The influence of prey density, relative humidity, and starvation on the predacious behaviour of *Phytoseiulus persimilis* Athias-Henriot (Acarina: Phytoseiidae). *Can. J. Zool.*, **44**, 483–91.

Morris, R. F. (1957). The interpretation of mortality data in studies on population dynamics. *Can. Ent.*, **89**, 49–69.

Motoyama, N., Rock, G. C. & Dauterman, W. C. (1970). Organophosphorus resistance in an apple orchard population of *Typhlodromus* (*Amblyseius*) *fallacis*. *J. econ. Ent.*, **63**, 1439–42.

Müller, F. P. (1958). Bionomische Rassen der grünen Pfirsichblattlaus *Myzus persicae* (Sulz.). *Arch. Freunde NatGesch. Mecklenb.*, **4**, 200–33.

(1971). Isolationsmechanismen zwischen sympatrischen bionomischen Rassen am Beispiel der Erbsenblattlaus *Acyrthosiphon pisum* (Harris) (Homoptera, Aphididae). *Zool. Jb.* (*Syst.*), **98**, 131–52.

Muma, M. H. (1955). Phytoseiidae (Acarina) associated with citrus in Florida. *Ann. ent. Soc. Am.*, **48**, 262–72.

(1967). *Typhlodromus peregrinus* (Muma) (Acarina: Phytoseiidae) on Florida citrus. *Proceedings of the 2nd international congress on acanology*, **1**, 135–48.

(1969). Coincidence and incidence of *Entomophthora floridana* with and in *Eutetranychus banksi* in Florida citrus groves. *J. Fla. agr. exp. Stn*, **3316**, 107–12.

(1970). Natural control potential of *Galendromus floridanus* (Acarina: Phytoseiidae) on Tetranychidae on Florida citrus trees. *Fla Entomol.*, **53**, 79–88.

*Murakami, Y. (1975). Biological and ecological studies on the parasites of *Unaspis yanonensis* (Kuwana) in Japan. In *Approaches to Biological to Control*, JIBP Synthesis Vol. 7 (eds. K. Yasumatsu & H. Mori), University of Tokyo Press, pp. 125–131.

*Murakami, Y., Uematsu, H., Ohga, S. & Kajita, H. (1972). Parasites of *Unaspis yanonensis* in Japan (Hymenoptera, Chalcidoidea). *Mushi*, **46**, 45–52.

Myburgh, A. C. (1962). Mating habits of the fruit flies *Ceratitis capitata* and *Pterandrus rosa*. *S. Afr. J. Sci.*, **5**, 457–63.

Nakagawa, S., Farias, G. J., Suda, D., Cunningham, R. T. & Chambers, D. L. (1971). Reproduction of the Mediterranean fruit fly: frequency of mating in the laboratory. *Ann. ent. Soc. Am.*, **64**, 949–50.

Nation, J. L. (1972). Courtship behaviour and evidence for sex attractant in the male Caribbean fruit fly, *Anastrepha suspensa*. *Ann. ent. Soc. Am.*, **65**, 1364–8.

Nault, L. R., Edwards, L. J. & Styer, W. E. (1973). Aphid alarm pheromones: secretion and reception. *Environ. Ent.*, **2**, 101–5.

Needham, N. V. (1930). A bacterial disease of *Aphis rumicis* Linn., apparently caused by *Bacillus lathyri* Manns and Taubenhaus. *Ann. appl. Biol.*, **24**, 144–7.

Neilson, W. T. A. (1971). Dispersal studies of a natural population of apple maggot adults. *J. econ. Ent.*, **64**, 648–53.

Neilson, W. T. A. & McAllan, J. W. (1964). Artificial diets for the apple maggot, *Rhagoletis pomonella*. II. Reproductive potential. *J. econ. Ent.*, **57**, 904–5.

Newcomer, E. J. & Yothers, M. A. (1929). Biology of the European red mite in the Pacific Northwest. *Tech. Bull. U.S. Dep. Agric.*, **89**, 3–70.

Newell, I. M. & Haramoto, F. H. (1968). Biotic factors influencing populations of *Dacus dorsalis* in Hawaii. *Proc. Hawaii ent. Soc.*, **20**, 81–139.

Newton, J. H., Palmer, M. A. & List, G. M. (1953). Fall migration of aphids with special reference to the green peach aphid. *J. econ. Ent.*, **46**, 667–70.

Nielson, M. W., Don, H., Schonhorst, M. H., Lehman, W. F. & Marble, V. L. (1970). Biotypes of the spotted alfalfa aphid in western United States. *J. econ. Ent.*, **63**, 1822–5.

Niemczyk, E. (1966). The occurrence of predacious insects in an apple orchard. *Pr. Inst. Sad.*, **10**, 331–57.

(1968). The occurrence of the bugs of the order Heteroptera in sprayed, and unsprayed apple orchards. *Pr. Inst. Sadow.*, **12**, 355–63.

(1973). Preliminary information on effectiveness of bark burg (*Anthocoris nemorum* L.: Heter.-Anthocoridae) in relation to red spider mite (*Panonychus ulmi* Koch.). *Zesz. Probl. Post. Nauk Roln.*, **129**, 243–9.

Nikolskaya, M. N. & Yasnosh, V. A. (1968). On the aphelinid fauna of the Caucasus. *Trud. Vses. Ent. O-va*, **52**, 3–42 (in Russian).

Niku, B. (1972). Der Einfluss räuberischer Feinde auf die Ausbreitung von Erbsenläusen (*Acyrthosiphon pisum* (Harr.)) im Bestand. *Z. angew. Ent.*, **70**, 359–64.

*Nishida, T. & Torii, T. (1970). A handbook of field methods for research on rice stem-borers and their natural enemies. *IBP Handbook* no. **14**. Blackwell, London, 132 pp.

*Nishida, T. & Wongsiri, T. (1972). Rice stem-borer population and biological control in Thailand. *Mushi*, **45**, *Suppl.*, 25–37.

Nohara, K. (1970). Study of the biological and chemical control of citrus pests, with special reference to the integrated control of *Unaspis yanonensis* (Kuwana) on *Panonychus citri* (McGregor). *Spec. Bull. Yamaguchi, agric. Exp. Stn*, 66–74.

Norris, D. O. (1943). Pea mosaic on *Lupinus varius* L. and other species in western Australia. *Bull. Coun. scient. ind. Res.*, Melbourne, **170**, 1–27.

Oatman, E. R. & McMurtry, J. A. (1966). Biological control of the two-spotted spider mite on strawberry in southern California. *J. econ. Ent.*, **59**, 433–9.

Oatman, E. R., McMurtry, J. A. & Voth, V. (1968). Suppression of the two-spotted spider mite on strawberry with mass releases of *Phytoseiulus persimilis*. *J. econ. Ent.*, **61**, 1517–21.

Oatman, E. R., McMurtry, J. A., Shorey, H. H. & Voth, V. (1967). Studies on integrating *Phytoseiulus persimilis* releases, chemical applications, cultural manipulations, and natural predation for control of the two-spotted spider mite on strawberry in southern California. *J. econ. Ent.*, **60**, 1344–51.

Oatman, E. R. & Platner, G. R. (1969). An ecological study of insect populations on cabbage in southern California. *Hilgardia*, **40**, 1–40.

Oatman, E. R. & Voth, V. (1972). An ecological study of the two-spotted spider mite on strawberry in southern California. *Environ. Ent.*, **1**, 34–9.

Ohinata, K. F., Fujimoto, M. S., Chambers, D. L., Jacobson, M. & Kanakaki, D. C. (1974). Bioassay techniques for investigating sex pheromones of the Mediterranean fruit fly. *J. econ. Ent.*, **66**, 812–14.

Olkowski, W., Pinnock, C., Toney, W., Mosher, G., Neasbitt, W., van den Bosch, R. & Olkowski, H. (1974). An integrated insect control program for street trees. *Calif. Agric.*, **28**, 3–4.

Onillon, J. C. & Onillon, J. (1975). Contribution à l'étude de la dynamique des populations d'homoptères inféodés aux agrumes. III-2-Modalités de la dispersion de *Cales noacki* How. (Hymenopt., Aphelinidae), parasite d'*Aleurothrixus floccosus* Mask. (Homopt., Aleurodidae). *Annls Inst. Phytopath. Benaki*, in press.

Orlob, G. B., Seto, D. & Sun, A. (1973). Effects of granulosis virus on aphids. *J. Invertebr. Path.*, **22**, 220–7.

Orphanidis, P. S. & Soultanopoulos, D. D. (1962). Some observations concerning the influence of the colour and the number of traps per tree on the captures of adult flies of *Dacus oleae* Gmel. *Annls Inst. Phytopath. Benaki*, **4**, 112–17.

Ossiannilsson, F. (1952). Bladlöss i växthus än en gång. *Växtskyddsnotiser*, **4**, 53–7.

(1959). Contributions to the knowledge of Swedish aphids. II. *LantbrHögsk. Annls*, **25**, 375–527.

Overmeer, W. P. J. & Harrison, R. A. (1969). Notes on the control of the sex ratio in populations of the two-spotted spider mite *Tetranychus urticae* Koch. *N.Z. J. Sci.*, **12**, 920–8.

Paillot, A. (1930). Parasitisme bactérien et symbiose chez *Aphis mali*. *C. r. Acad. Sci. Paris*, **190**, 895–6.

(1931). Parasitisme bactérien et symbiose chez *Aphis atriplicis*. *C. r. Acad. Sci. Paris*, **193**, 676.

Painter, R. H. (1968). Crops that resist insects provide a way to increase world food supply. *Bull. Kans. State agric. Exp. Stn*, **520**, 1–22.

Panis, A. (1975). Modalités de dispersion de *Metaphycus helvolus* Compere (Hymenoptera, Chalcidoidea, Encyrtidae) lâché en un point d'un verger d'argrumes. *Annls Inst. Phytopath. Benaki*, in press.

Parent, B. (1967). Population studies of phytophagous mites and predators in southwestern Quebec. *Can. Ent.*, **99**, 771–8.

Parent, B. & Lord, F. T. (1971). *Panonychus ulmi* (Koch), European red mite, (Acarina: Tetranychidae) and other phytophagous mites on fruit trees in Canada. In *Biological Control Programs against Insect and Weeds in Canada, 1959–1968*. Tech. Commun. CIBC, **4**, 28–30.

Parr, W. J. (1971). Integrated control of red spider mite. *Rep. Glasshouse Crops Res. Inst., 1970*, 119–20.

Parr, W. J. & Hussey, N. W. (1966). Diapause in the glasshouse red spider mite *Tetranychus urticae* Koch: a synthesis of present knowledge. *Hort. Res.*, 6, 1–21.

Parrish, W. B. & Briggs, J. D. (1966). Morphological identification of virus-like particles in the corn leaf aphid, *Rhopalosiphum maidis* (Fitch). *J. Invertebr. Path.*, 8, 122–3.

Parry, W. H. & Ford, J. B. (1969). The artificial feeding of phosphamidon to *Myzus persicae*: II. The effects of phosphamidon on liquid uptake through a parafilm membrane. *Entomologia exp. appl.*, 12, 1–18.

Pathak, M. D. (1970). Genetics of plants in pest management. In *Concepts of Pest Management* (eds. R. L. Rabb and F. E. Guthrie), pp. 138–57. North Carolina State University, Raleigh.

Persson, I. (1963). Studies on the biology and larval morphology of some Trypetidae (Dipt). *Opusc. ent.*, 28, 3–69.

Peska, W. (1931). Observations on biology of bark bug *Anthocoris nemorum* L. *Prace Wydz. Rosl. P.I.N.G.W.*, 10, 53–72.

Peterson, A. (1923). The pepper maggot, a new pest of peppers and egg-plants. *N.J. agric. Exp. Stn Bull.*, 373, 1–23.

Piasecka, I. (1969). Biology and behaviour of *Anthocoris gallarum-ulmi* Deg. (Heteroptera, Anthocoridae). *Annls Univ. Mariae Curie-Skłodowska*, Lublin–Polonia, 24, 269–77.

Pickett, A. D., Patterson, N. A., Stultz, H. T. & Lord, F. T. (1946). The influence of spray programs on the fauna of apple orchards in Nova Scotia. 1. An appraisal of the problem and a method of approach. *Scien. Agric.*, 26, 590–600.

Pickett, W. P., Fish, A. S. & Shan, K. S. (1951). The influence of certain organic sprays on the photosynthetic activity on peach and apple foliage. *Proc. Am. Soc. hort. Sci.*, 57, 111–14.

Pijnacker, L. P. & Ferwerda, M. A. (1972). Diffuse kinetochores in the chromosomes of the arrhenotokous spider mite *Tetranychus urticae* Koch. *Experientia*, 28, 354.

Pimentel, D. (1961). Natural control of aphid populations on cole crops. *J. econ. Ent.*, 54, 885–8.

Poe, S. L. & Enns, W. R. (1969). Predacious mites (Acarina: Phytoseiidae) associated with Missouri orchards. *Trans. Mo. Acad. Sci.*, 3, 69–82.

Poisson, R. (1940). Sur la réproduction agame de *Myzus persicae* (Sulz.) en Bretagne. *C. r. Soc. Biol.*, 133, 634–6.

Pollard, E. (1969). The effect of removal of arthropod predators on an infestation of *Brevicoryne brassicae* (Hemiptera, Aphididae) on Brussels sprouts. *Entomologia exp. appl.*, 12, 118–24.

(1971). Hedges. VI. Habitat diversity and crop pests: A study of *Brevicoryne brassicae* and its syrphid predators. *J. appl. Ecol.*, 8, 751–80.

Porath, A. (1969). On the establishment and distribution of *Casca smithi* Comp., a parasite of *Chrysomphalus aonidum* L. in Israel. *Israel J. Ent.*, 4, 41–4.

Post, A. (1962). Effect of cultural measures on the population density of the

References

fruit tree red spider mite, *Metatetranychus ulmi* Koch. (Acari, Tetranychidae). Diss. Rijks University, Leiden, 110 pp.

Prasad, V. (1967). Biology of the predatory mite *Phytoseiulus macropilis* in Hawaii. (Acarina: Phytoseiidae). *Ann. ent. Soc. Am.*, **60**, 905–8.

Pritchard, A. E. & Baker, E. W. (1955). A revision of the spider mite family Tetranychidae. *Mem. Pac. Coast ent. Soc.*, **2**, 472 pp.

Pritchard, G. (1969). The ecology of a natural population of Queensland fruit fly, *Dacus tryoni*. II. The distribution of eggs and its relation to behaviour. *Aust. J. Zool.*, **17**, 293–311.

Prokopy, R. J. (1968). Visual response of apple maggot flies, *Rhagoletis pomonella* (Diptera: Tephritidae): Orchard studies. *Entomologia exp. appl.*, **11**, 403–22.

(1972*a*). Evidence for a marking pheromone deterring repeated oviposition in apple maggot flies. *Environ. Ent.*, **1**, 326–32.

(1972*b*). Response of apple maggot flies to rectangles of different colours and shades. *Environ. Ent.*, **1**, 720–6.

Prokopy, R. J., Bennett, E. W. & Bush, G. L. (1971). Mating behaviour in *Rhagoletis pomonella* (Diptera: Tephritidae). I. Site of assembly. *Can. Ent.*, **103**, 1405–9.

(1972). Mating behaviour in *Rhagoletis pomonella*. II. Temporal organization. *Can. Ent.*, **104**, 97–104.

Prokopy, R. J. & Bush, G. L. (1972). Mating behaviour in *Rhagoletis pomonella* (Diptera: Tephritidae). III. Male aggregation in response to an arrestant. *Can. Ent.*, **104**, 275–83.

(1973). Mating behaviour in *Rhagoletis pomonella* (Diptera: Tephritidae). IV. Courtship. *Can. Ent.*, **105**, 873–91.

Putman, W. L. (1959). Hibernation sites of phytoseiids (Acarina: Phytoseiidae) in Ontario peach orchards. *Can. Ent.*, **91**, 735–41.

(1962). Life-history and behaviour of the predacious mite *Typhlodromus* (T.) *caudiglans* Schuster (Acarina: Phytoseiidae) in Ontario, with notes on the prey of related species. *Can. Ent.*, **94**, 163–77.

(1963). Lack of affect of DDT on fecundity and dispersion of the European red mite *Panonychus ulmi* Koch (Acarina: Tetranychidae) in peach orchards. *Can. J. Zool.*, **44**, 603–10.

(1967). Prevalence of spiders and their importance as predators in Ontario peach orchards. *Can. Ent.*, **99**, 160–70.

(1970). Occurrence and transmission of a virus disease of the European red mite, *Panonychus ulmi*. *Can. Ent.*, **102**, 305–21.

Putman, W. L. & Herne, D. C. (1966). The role of predators and other biotic factors in regulating the population density of phytophagous mites in Ontario peach orchards. *Can. Ent.*, **98**, 808–20.

Quayle, H. J. (1912). Red spiders and mites of citrus trees. *Bull. Calif. agric. Exp. Stn*, **234**, 483–530.

(1938). The development of resistance to hydrocyanic gas certain scale insects. *Hilgardia*, **11**, 183–210.

*Quezada, J. R. (1972). Biological control of citrus pests in El Salvador. *IOBC Newsletter*, **2**, 11.

*Quezada, J. R. & DeBach, P. (1971). Population interactions among the cottony cushion scale and its enemies. *Proceedings of the 12th Pacific Science Congress*, Canberra, **1**, 193.

* (1973). Bioecological and population studies of the cottony-cushion scale, *Icerya purchasi* Mask., and its natural enemies, *Rodolia cardinalis* Mul. and *Cryptochaetum iceryae* Will., in southern California. *Hilgardia*, **41**, 631–88.

*Quezada, J. R., DeBach, P. & Rosen, D. (1973). Biological and taxonomic studies of *Signiphora borinquensis*, new species (Hymenoptera: Signiphoridae), a primary parasite of diaspine scales. *Hilgardia*, **41**, 543–603.

Quilis Pérez, M. (1930). Los parásitos de los pulgones. Dos nuevas especies de '*Aphidius*'. *Bol. Patol. veg. Ent. agric.*, **4**, 49–64.

Quisumbing, A. R., Lauer, F. I. & Radcliffe, E. B. (1970). Resistance to green peach aphid, *Myzus persicae* (Sulz.), in hybrids from wild tuber bearing *Solanum* species. *Proc. N. cent. Brch ent. Soc. Am.*, **25**, 103–5.

Radcliffe, E. B. & Lauer, F. I. (1966). A survey of aphid resistance in the tuber-bearing *Solanum* (Tourn.) L. species. *Tech. Bull. Minn. agric. Exp. Stn*, **253**, 1–23.

(1967). Insect resistance in the wild *Solanum* species. *Proc. N. cent. Brch ent. Soc. Am.*, **22**, 165–7.

(1970). Further studies on resistance to green peach aphid and potato aphid in the wild tuber-bearing *Solanum* species. *J. econ. Ent.*, **63**, 110–14.

(1971 a). Resistance to green peach aphid in introductions of wild tuber-bearing *Solanum* species. *J. econ. Ent.*, **64**, 1260–6.

(1971 b). An appraisal of aphid resistant tuber-bearing *Solanum* germ plasm. *Tech. Bull. Minn. agric. Exp. Stn*, **281**, 1–24.

Rainey, R. C. (1972). Wind and the distribution of the desert locust, *Schistocerca gregaria* (Forsk.). *Proceedings of an international study conference on current and future problems of acridology*. London, 1970, pp. 229–37.

Rambier, A. (1972). Le *Phytoseiulus persimilis* Athias-Henriot dans le midi de la France. *Zesz. Probl. Post. Nauk Rolniczych*, Warsaw, **129**, 89–91.

*Rao, S. V. & DeBach, P. (1969 a). Experimental studies on hybridization and sexual isolation between some *Aphytis* species (Hymenoptera: Aphelinidae). I. Experimental hybridization and an interpretation of evolutionary relationships among the species. *Hilgardia*, **39**, 515–53.

(1969 b). Experimental studies on hybridization and sexual isolation between some *Aphytis* species (Hymenoptera: Aphelinidae). II. Experiments on sexual isolation. *Hilgardia*, **39**, 555–67.

* (1969 c). Experimental studies on hybridization and sexual isolation between some *Aphytis* species (Hymenoptera: Aphelinidae). III. The significance of reproductive isolation between interspecific hybrids and parental species. *Evolution*, **23**, 525–33.

Rao, V. P. (1965). Natural enemies of rice stem-borers and allied species in various parts of the world and possibilities of their use in biological control of rice stem-borers in Asia. *Tech. Bull. Commw. Inst. biol. Control*, **6**, 1–68.

References

Rasmy, A. H. (1970). A laboratory technique for mass rearing of a phytoseiid mite. *Z. angew. Ent.*, **65**, 159–61.

Razumova, Z. P. (1967). Variability of the photoperiodic reaction in a number of successive generations of spider mites (Acarina). *Ent. Obozr.*, **46**, 268–72.

Read, D. P., Feeny, P. P. & Root, R. B. (1970). Habitat selection by the aphid parasite *Diaeretiella rapae* (Hymenoptera: Braconidae) and hyperparasite *Charips brassicae* (Hymenoptera: Cynipidae). *Can. Ent.*, **102**, 1567–78.

Readshaw, J. L. (1971). An ecological approach to the control of mites in Australian orchards. *J. Aust. Inst. agric. Sci.*, **37**, 226–30.

(1972). Failure of lead arsentate in an ecological approach to the control of mites in Australian orchards. *J. Aust. Inst. agric. Sci.*, **38**, 308–9.

Remaudière, G. (1971). Vers l'utilisation pratique des Entomophthorales parasites de pucerons. *Parasitica*, **27**, 115–26.

Remaudière, G., Iperti, G., Leclant, F., Lyon, J. P. & Michel, M. F. (1973). Biologie et écologie des Aphides et de leurs ennemis naturels. Application à la lutte intégrée en vergers. *Entomophaga*, mém. hors. sér., **6**, 1–34.

Remaudière, G. & Michel, M. F. (1971). Première expérimentation écologique sur les Entomophthorales (Phycomycètes) parasites de pucerons en vergers de pêchers (*Myzus persicae*). *Entomophaga*, **16**, 75–94.

Remund, U. (1971). Anwendungsmöglichkeiten einer wirksamen visuellen Wegwerffalle für die Kirschenfliege (*Rhagoletis cerasi* L.). *Schweiz. Z. Obst- u. Weinbau*, **107**, 196–205.

Rice, R. E. & Moreno, D. S. (1969 *a*). Marking and recapture of California red scale for field studies. *Ann. ent. Soc. Am.*, **62**, 558–60.

(1969 *b*). Comparative production of pheromone by the California red scale reared on lemons and potatoes. *Ann. ent. Soc. Am.*, **62**, 958–9.

(1970). Flight of male California red scale. *Ann. ent. Soc. Am.*, **63**, 91–6.

Richardson, N. L. (1972). The ecology of oriental fruit moth (*Cydia molesta*) in relation to other pests of canning peaches in southern Australia. M.Sc. Thesis, University of Adelaide.

Richardson, H. P. & Westdal, P. H. (1965). Use of *Aphelinus semiflavus* Howard for control of aphids in a greenhouse. *Can. Ent.*, **97**, 110–11.

Ristich, S. S. (1956). Toxicity of pesticides to *Typhlodromus fallacis* (Gar.). *J. econ. Ent.*, **49**, 511–15.

Robert, Y., Rabasse, J. M. & Rouze-Jouan, J. (1972). Régulation naturelle des populations de *Capitophorus horni* Börner (Hom. Aphididae) par Hyménoptères Aphidiidae et Entomophthorales. *Entomophaga*, **17**, 59–69.

Robert, Y., Rabasse, J. M. & Scheltes, P. (1973). Facteurs de limitation des populations d'*Aphis fabae* Scop. dans l'Ouest de la France. 1. Epizootologie des maladies à Entomophthorales sur féverole de printemps. *Entomophaga*, **18**, 61–75.

Rock, G. C. & Yeargan, D. R. (1971). Relative toxicity of pesticides to organophosphorus-resistant orchard populations of *Neoseiulus fallacis* and its prey. *J. econ. Ent.*, **64**, 350–2.

(1972). Laboratory studies on toxicity of dinocap to *Neoseiulus fallacis* and its prey. *J. econ. Ent.*, **65**, 932–3.

Rock, G. C., Yeargan, D. R. & Rabb, R. L. (1971). Diapause in the phytoseiid mite *Neoseiulus* (T.) *fallacis*. *J. Insect Physiol.*, **17**, 1651–9.

Rodriguez, J. G. (1964). Nutritional studies in the acarina. *Acarologia*, **6**, 324–37.

Rodriguez, J. G., Chen, H. H. & Smith, W. T. (1960). Effect of soil insecticides on apple trees and resulting effect on mite nutrition. *J. econ. Ent.*, **53**, 487–90.

Rodriguez, J. G. & Hampton, R. E. (1966). Essential amino acids determined in the two-spotted spider mite, *Tetranychus urticae* Koch (Acarina, Tetranychidae) with glucose-U.C.[14] *J. Insect Physiol.*, **12**, 1209–16.

Rodriguez, J. G., Singh Pritam Seay, T. N. & Walling, M. V. (1967). Ingestion in the two-spotted spider mite, *Tetranychus urticae* Koch, as influenced by wavelength of light. *J. Insect Physiol.*, **13**, 925–32.

Rogers, A. W. (1967). *Techniques of Autoradiography*. Elsevier, Amsterdam, London and New York, 335 pp.

Rogers, C. E., Jackson, H. B. & Eikenbary, R. D. (1972*a*). Voracity and survival of *Propylaea 14-punctata* preying upon greenbugs. *J. econ. Ent.*, **65**, 1313–16.

Rogers, C. E., Jackson, H. B., Eikenbary, R. D. & Starks, K. J. (1972*b*). Host–parasitoid interaction of *Aphis helianthi* on sunflowers with introduced *Aphelinus asychis*, *Ephedrus plagiator*, and *Praon gallicum*, and native *Aphelinus nigritus* and *Lysiphlebus testaceipes*. *Ann. ent. Soc. Am.*, **65**, 38–41.

Rosen, D. (1967*a*). Effect of commercial pesticides on the fecundity and survival of *Aphytis holoxanthus* (Hymenoptera: Aphelinidae). *Israel J. Agric. Res.*, **17**, 42–52.

* (1967*b*). Biological and integrated control of citrus pests in Israel. *J. econ. Ent.*, **60**, 1422–7.

* (1973). Methodology for biological control of armored scale insects. *Phytoparasitica*, **1**, 47–54.

* (1975). Current status of integrated control of citrus pests in Israel. *Bulletin Eur. Med. Pl. Prot. Organ.*, in press.

*Rosen, D. & DeBach, P. (1970). Notes on the genus *Marlattiella* Howard (Hymenoptera: Aphelinidae). *Mushi*, **43**, 39–44.

* (1975*a*). Biosystematic studies on the species of *Aphytis*. *Annls Inst. Phytopath. Benaki*, in press.

(1975*b*). Diaspididae. In Biological Control of Insect Pests and Weeds – a Review (ed. C. P. Clausen). U.S. Department of Agriculture, in press.

(1973*c*). Systematics, morphology and biological control. *Entomophaga*, **18**, 215–22.

*Rössler, Y. & DeBach, P. (1972*a*). The biosystematic relations between a thelytokous and an arrhenotokous form of *Aphytis mytilaspidis* (Le Baron) (Hymenoptera: Aphelinidae). 1. The reproductive relations. *Entomophaga*, **17**, 391–423.

* (1972*b*). The biosystematic relations between a thelytokous and an

References

arrhenotokous form of *Aphytis mytilaspidis* (Le Baron) (Hymenoptera: Aphelinidae). 2. (Comparative biological and morphological studies. *Entomophaga*, **17**, 425–35.

Rothschild, G. H. L. (1970). Parasites of rice stem-borers in Sarawak (Malaysian Borneo). *Entomophaga*, **15**, 21–51.

(1971). The biology and ecology of rice stem-borers in Sarawak (Malaysian Borneo). *J. appl. Ecol.*, **8**, 287–322.

Russ, K., Boller, E., Vallo, V., Haisch, A. & Sezer, S. (1973). Development and application of visual traps for monitoring and control of populations of *Rhagoletis cerasi* L. *Entomophaga*, **18**, 103–16.

Russell, G. E. (1966). Preliminary studies in breeding for aphid-resistance in beet. *J. Inst. int. Rech. better.*, **1**, 117–25.

(1972). Inherited resistance to virus yellows in sugar beet. *Proc. R. Soc. Lond. (B)*, **181**, 267–79.

Sanford, K. H. (1967). The influence of spray programs on the fauna of apple orchards in Nova Scotia. XVII. Effects on some predacious mites. *Can. Ent.*, **99**, 197–201.

Sanford, K. H. & Herbert, H. J. (1967). The influence of spray programs on the fauna of apple orchards in Nova Scotia. XVIII. Predator and prey populations in relation to miticides. *Can. Ent.*, **99**, 689–96.

(1970). Influence of spray programs on the fauna of apple orchards in Nova Scotia. XX. Trends after altering levels of phytophagous mites or predators. *Can. Ent.*, **102**, 592–601.

Saini, R. S. & Cutkomp, L. K. (1966). The effect of DDT and sublethal doses of dicofol on reproduction of the spotted spider mite. *J. econ. Ent.*, **59**, 249–53.

Salt, G. (1937). The sense used by *Trichogramma* to distinguish between parasitized and unparasitized hosts. *Proc. R. ent. Soc. Lond. (B)*, **122**, 57–75.

Sapozhnikova, F. D. (1964). Photoperiodical response in the mite *Typhlodromus* (Amblyseius) *similis* (Koch) (Acarina: Phytoseiidae). *Zool. Zh.*, **43**, 1140–4.

Schneider, F. (1969). Bionomics and physiology of aphidophagous Syrphidae. *A. Rev. Ent.*, **14**, 103–24.

Schöll, S. E. & Daiber, C. C. (1958). Notes on the occurrence of holocyclic overwintering of the green peach aphid in South Africa. *J. ent. Soc. S. Afr.*, **21**, 315–22.

Schroeder, W. J., Chambers, D. L. & Miyabara, R. Y. (1973). Mediterranean fruit fly: propensity to flight of sterilized flies. *J. econ. Ent.*, **66**, 1261–2.

Schroeder, W. J., Mitchell, W. C. & Miyabara, R. Y. (1974). Dye-induced changes in melon fly behaviour. *Environ. Ent.*, **3**, 571.

Schultz, G. A. & Boush, G. M. (1971). Suspected sex pheromone glands in three economically important species of *Dacus*. *J. econ. Ent.*, **64**, 347–9.

Scopes, N. E. A. (1970). Control of *Myzus persicae* on year-round chrysanthemums by introducing aphids parasitized by *Aphidius matricariae* into boxes of rooted cuttings. *Ann. appl. Biol.*, **66**, 323–7.

Scopes, N. E. A. & Biggerstaff, S. M. (1973). Progress towards integrated pest

274

control on year-round chrysanthemums. *Proceedings of the 7th British Conference on Insects and fungi*, **1**, 227–33.

Severin, H. H. P. (1917). The currant fruit fly. *Maine agric. Exp. Stn Bull.*, **264**, 177–247.

Shands, W. A. & Simpson, G. W. (1969). Bioenvironmental control of the green peach aphid, *Myzus persicae. Am. Potato J.*, **46**, 56–8.

(1972). Insect predators for controlling aphids on potatoes. 7. A pilot test of spraying eggs of predators on potatoes in plots separated by bare fallow land. *J. econ. Ent.*, **65**, 1383–7.

Shands, W. A., Simpson, G. W. & Gordon, C. C. (1972*a*). Insect predators for controlling aphids on potatoes. 5. Numbers of eggs and schedules for introducing them in large field cages. *J. econ. Ent.*, **65**, 810–17.

Shands, W. A., Simpson, G. W., Hall, I. M. & Gordon, C. C. (1972*b*). Further evaluation of entomogenous fungi as a biological agent of aphid control in northeastern Maine. *Tech. Bull. Maine Life Sci. agric. Exp. Stn*, **58**, 1–33.

Shands, W. A., Simpson, G. W., Muesebeck, C. F. W. & Wave, H. E. (1965). Parasites of potato-infesting aphids in northeastern Maine. *Bull. Maine agric. Exp. Stn, Tech. Ser.*, **19**, 1–77.

Shands, W. A., Simpson, G. W. & Storch, R. H. (1972*c*). Insect predators for controlling aphids on potatoes. 3. In small plots separated by aluminium flashing strip-coated with a chemical barrier and in small fields. *J. econ. Ent.*, **65**, 799–805.

Shands, W. A., Simpson, G. W., Wave, H. E. & Gordon, C. C. (1972*d*). Importance of arthropod predators in controlling aphids on potatoes in northeastern Maine. *Tech. Bull. Univ. Maine*, **54**, 1–49.

Shaposhnikov, G. C. (1971). The principal trends and modes of evolution in aphids. *Proceedings of the 13th international congress on entomology*, Moscow, 1968, **1**, 196–7.

Sharp, J. L. (1972). Effects of increasing dosages of gamma irradiation on wingbeat frequencies of *Dacus dorsalis* Hendel males and females at different age levels. *Proc. Hawaii Ent. Soc.*, **21**, 257–62.

Sharp, J. L., Chambers, D. L. & Haramoto, F. H. (1975). Flight mill and stroboscope studies of oriental fruit flies and melon flies including observations on Mediterranean fruit flies. *J. econ. Ent.*, in press.

Shaw, J. G., Sanchez-Riviello, M., Spishakoff, M. L., Trujillo, G. & Lopez, D. F. (1967). Dispersal and migration of tepa-sterilized Mexican fruit flies. *J. econ. Ent.*, **60**, 992–4.

Shaw, J. G., Tashiro, H. & Dietrick, E. J. (1968). Infection of the citrus red mite with virus in central and southern California. *J. econ. Ent.*, **61**, 1492–5.

Shaw, M. J. P. (1970). Effect of population density on alienicolae of *Aphis fabae* Scop. II. The effect of crowding on the expression of migratory urge among alatae in the laboratory. *Ann. appl. Biol.*, **65**, 197–203.

Shaw, M. W. (1957). Aphids and seed-potato growing in Scotland. *Agric. Rev. Lond.*, 1957, pp. 28–36.

Shehata, K. K. & Weiseman, L. (1972). Rearing the predaceous mite *Phyto-*

10-2

seiulus persimilis Athias-Henriot (Acarina: Phytoseiidae) on artificial diet. *Biol. Czech. B*, **27**, 609–15.

Shiga, M. (1967). Ecological studies on the green peach aphid, *Myzus persicae* (Sulzer) and the cabbage aphid, *Brevicoryne brassicae* (Linnaeus) in Japan, with special reference to their biological control. *Mushi*, **41**, 75–89.

— (1968). The effect of aggregation of the green peach aphid, *Myzus persicae* (Sulzer), on the spatial distribution and parasitization of a hymenopterous parasite, *Aphidius gifuensis* Ashmead. *Sci. Bull. Fac. Agric. Kyushu Univ.*, **23**, 169–83.

Sidlyarevich, V. I. (1965). The importance of predacious mites and bugs in reducing the numbers of *Metatetranychus ulmi* Koch in the Byelorussia SSR. *Trudý vses. Inst. Zashch Rast.*, **24**, 240–7. (*Abst. Rev. appl. Ent.*, **56**, 1675).

Sigwalt, B. (1971). Les études de démographie chez les cochenilles diaspines. Applications à trois espèces nuisibles à l'oranger en Tunisie. Cas particulier d'une espèce à generations chevauchantes: *Parlatoria ziziphi* Lucas. *Annls zool. écol. anim.*, **3**, 5–15.

Sigwalt, B., Soria, F., Yana, A. & Baldy, C. (1968). Déplacement diurne apparent d'une population de Cératites sur une parcelle d'agrumes. *Annls épiphyt.*, **19**, 169–71.

Silvestri, F. (1920). La mosca della Brionia *Gonyglossum weidemanni* Meig. (Diptera: Trypaneidae). *Boll. Lab. Zool. gen. Agrar. Portici*, **14**, 205–15.

— (1922). Etat actuel de la lutte contre la mouche des olives. In *La lutte contre la mouche des olives dans les divers pays*. Annexe au Rapp. de M. Francisco Bilbao y Sevilla, 6e Ass. Générale, Inst. intern. agric., pp. 49–72.

Simmonds, P. (1972). Observations on the control of *Tetranychus urticae* on roses by *Phytoseiulus persimilis*. *Pl. Path.*, **21**, 163–5.

Simon, J. P. (1969). Esterase isozymes in the *Rhagoletis pomonella* species complex. *Isoenzyme Bull.*, **2**, 2–27.

Smirnoff, W. A. (1971). Effect of chitinase on the action of *Bacillus thuringienses*. *Can. Ent.*, **103**, 1829–31.

Smith, C. F., Martorelle, L. F. & Pérez-Escolar, M. E. (1958). *Myzus persicae* (Sulzer) in Puerto Rico. *J. Agric. Univ. P. Rico*, **42**, 63–6.

Smith, F. F., Boswell, A. L. & Webb, R. E. (1969). Segregation between strains of carmine and green two-spotted spider mites. *Proceedings of the 2nd international congress on acarology*, Sutton Bonnington (1967), 155–9.

Smith, F. F. & Henneberry, T. J. & Boswell, A. L. (1963). The pesticide tolerance of *Typhlodromus fallacis* (Garman) and *Phytoseiulus persimilis* A. H. with some observations on the predator efficiency of *P. persimilis*. *J. econ. Ent.*, **56**, 274–8.

Smith, I. (1968). *Chromatographic and Electrophoretic Techniques*. Vol. II. *Zone Electrophoresis*. John Wiley and Sons, Inc. N.Y., 2nd edn, 524 pp.

Smith, K. M., Hills, G. J., Munger, F. & Gilmore, J. E. (1959). A suspected virus disease of the citrus red mite, *Panonychus citri* (McG.). *Nature, Lond.*, **184**, 70.

*Snowball, G. J. (1972). Status of natural enemies of white wax scale, *Gascardia destructor* (Newst.) (Homoptera: Coccidae), in eastern Australia.

Proceedings of the 14th international congress on entomology, Canberra, p. 212 (abstract).

Solheim, W. G. (1971). A new light in a forgotten past. *Nat. Geog. Mag.*, Washington, D.C., **139**, 330–9.

Sologic, H. D. & Rodriguez, J. G. (1971). Microorganisms associated with the two-spotted spider mite. *Tetranychus urticae. J. Invertebr. Path.*, **17**, 48–52.

Solomon, M. G. (1972). Establishment of predators in orchards. *A. Rep. E. Malling Res. Stn*, Kent, 1971, 134.

Sonleitner, F. & Bateman, M. A. (1963). Mark-recapture analysis of a population of Queensland fruit fly, *Dacus tryoni* in an orchard. *J. Anim. Ecol.*, **32**, 259–69.

Specht, H. B. (1968). *Phytoseiidae Acarina: Mesostigmata* in the New Jersey apple orchard environment with description of spermathecae and three new species. *Can. Ent.*, **100**, 673–92.

Srivastava, P. N. & Rouatt, J. W. (1963). Bacteria from the alimentary canal of the pea aphid, *Acyrthosiphon pisum* (Harr.) (Homoptera, Aphididae). *J. Insect Physiol.*, **9**, 435–8.

Starks, K. J., Muniappan, R. & Eikenbary, R. D. (1972). Interaction between plant resistance and parasitism against the greenbug on barley and sorghum. *Ann. ent. Soc. Am.*, **65**, 650–5.

Starý, P. (1964*a*). Food specificity in the Aphidiidae (Hymenoptera). *Entomophaga*, **9**, 91–9.

   (1964*b*). Biological control of *Megoura viciae* Bckt. in Czechoslovakia. *Čas. Cesk. Spol. ent.*, **61**, 301–22.

   (1970). *Biology of Aphid Parasites*. Dr W. Junk N. V., The Hague. 643 pp. (Series Entomologica, no. 6).

Stathopoulos, D. G. (1967). Studies on the identification and bioecology of cotton pests. II. *Rep. Pl. Prot. agric. Res. Stn Thessaloniki*, 3, 41–9 (original in Greek).

Steffan, A. W. (1972). Möglichkeiten genetischer Bekämpfung von Blattläusen (Homoptera: Aphidina). *Z. angew. Ent.*, **70**, 267–77.

Steiner, H. & Baggiolini, M. (1968). *Anleitung zum integrierten Pflanzenschutz im Apfelanbau*. Landesanst. f. Pfl. schutz, Stuttgart, 64 pp.

Steiner, L. F. (1969). Mediterranean fruit fly research in Hawaii for the sterile fly release program. In *Insect ecology and the sterile male technique*. I.A.E.A., Vienna, pp. 73–82.

Steiner, L. F., Harris, E. J., Mitchell, W. C., Fujimoto, M. S. & Christenson, L. D. (1965*a*). Melon fly eradication by overflooding with sterile flies. *J. econ. Ent.*, **58**, 519–22.

Steiner, L. F., Hart, W. G., Harris, E. J., Cunningham, R. T., Ohinata, K. & Kamakaki, D. C. (1970). Eradication of the Oriental fruit fly from the Mariana Islands by the methods of male annihilation and sterile insect release. *J. econ. Ent.*, **63**, 131–5.

Steiner, L. F. & Lee, R. K. S. (1955). Large area tests of a male annihilation method for Oriental fruit fly control. *J. econ. Ent.*, **48**, 311–17.

Steiner, L. F., Mitchell, W. C., Harris, E. L., Kozuma, T. T. & Fujimoto,

References

M. S. (1965*b*). Oriental fruit fly eradication by male annihilation. *J. econ. Ent.*, **58**, 961–4.

Steinhaus, E. A. & Marsh, G. A. (1962). Report of diagnoses of diseased insects 1951–1961. *Hilgardia*, **33**, 349–490.

Steudel, M. (1952). Untersuchungen zur anholozyklischen Überwinterung der grünen Pfirsichblattlaus (*Myzodes persicae* Sulz.) an Brassicaceen. *Mitt. biol. Bundesanst. f. Land- u. Forstw.*, **73**, 1–32.

Stokes, G. W. & Valleau, W. D. (1968). Registration of Ky 10 and Ky 12. *Crop Sci.*, **8**, 131.

Stoltz, L. P., Chaplin, C. E., Lasheen, A. M. & Rodriguez, J. G. (1970). Mineral nutrition of strawberry plants in relation to mite injury. *J. Am. Soc. hort. Sci.*, **95**, 601–3.

Stoltzfus, W. B. & Foote, B. A. (1965). The use of froth masses in courtship of *Eutreta*. *Proc. Ent. Soc. Wash.*, **67**, 263–4.

Storms, J. J. H. (1965). Rearing methods for studying the effect of the physiological condition of the host plant on the population development of *Panonychus ulmi* (Koch). *Boll. Zool. agrar. Bachic.*, **7**, 80–5.

(1969). Observations on the relationship between mineral nutrition of apple rootstocks in gravel culture and the reproduction rate of *Tetranychus urticae* (Acarina: Tetranychidae). *Entomologia exp. appl.*, **12**, 297–311.

Storms, J. J. H., Harrewijn, P. & Noordink, J. Ph. W. (1967). A new approach to the physiological host plant–parasite relationship – a technique in the field of applied entomology. *Neth. J. Pl. Path.*, **73**, 165–9.

Storms, J. J. H. & Noordink, J. Ph. W. (1970). Nutritional requirements of the two-spotted spider mite, *Tetranychus urticae* (Acarina: Tetranychidae). VII. *Eur. Mite Symp. Pol. Acad.*, 29–37.

Sudderuddin, K. I. (1973). Studies of insecticide resistance in *Myzus persicae* (Sulz.) (Hem., Aphidoidea). *Bull. ent. Res.*, **62**, 549–56.

Summers, F. M. (1960). Several stigmaeid mites formerly included in *Mediolata* redescribed in *Zetzellia* Ouds., and *Agistemus*, new genus. *Proc. ent. Soc. Wash.*, **62**, 233–47.

Swift, F. C. (1968). Population densities of the European red mite and the predacious mite, *Typhlodromus* (*A.*) *fallacis* on apple foliage following treatment with various insecticides. *J. econ. Ent.*, **61**, 1489–91.

(1970). Predation of *Typhlodromus* (*A.*) *fallacis* on the European red mite as measured by the insecticidal check method. *J. econ. Ent.*, **63**, 1617–18.

Swirski, E., Amitai, S. & Dorzia, N. (1967). Laboratory studies on the feeding, development and reproduction of the predacious mites (*Amblyseius rubini* Swirski and Amitai and *Amblyseius swirskii* Athias (Acarina: Phytoseiidae) on various kinds of food substances. *Israel J. agric. Res.*, **17**, 101–19.

Swirski, E. & Dorzia, N. (1968). Studies on the feeding, development and oviposition of the predacious mite *Amblyseius limonicus* Garman and McGregor (Acarina: Phytoseiidae) on various kinds of food substances. *Israel J. agric. Res.*, **18**, 71–5.

Syed, R. A. (1972). Studies on overwintering in *Dacus* species (mimeo.).

Syed, R. A., Ghani, M. A. & Murtaza, M. (1970). Studies on trypetids and their natural enemies in West Pakistan. III. *Dacus* (*Strumeta*) *zonatus* (Saunders). *Tech. Bull., Commonw. Inst. biol. Control*, **13**, 1–16.

Tachikawa, T. (1970). A revised list of the hosts of encyrtid genera (Hymenoptera: Chalcidoidea). *Trans. Shikoku ent. Soc.*, **10**, 84–99.

Tachikawa, T. & Valentine, E. W. (1969a). A new species of *Aphycomorpha* (Hymenoptera: Encyrtidae) parasitic on a diaspidine scale from New Zealand. *N.Z. J. Sci.*, **12**, 535–40.

(1969b). A new genus of Encyrtidae from New Zealand (Hymenoptera: Chalcidoidea). *N.Z. J. Sci.*, **12**, 546–52.

Takaoka, I. (1960). Studies on the differentiation of morphological and ecological characters found in the life-cycle of the green peach aphid *Myzus persicae* (Sulzer). *Bull. Hatano Tob. Exp. Stn*, **48**, 1–95.

Tamaki, G. (1973). Spring populations of the green peach aphid on peach trees and the role of natural enemies in their control. *Environ. Ent.*, **2**, 186–91.

Tamaki, G., Halfhill, J. E. & Hathaway, D. O. (1970). Dispersal and reduction of colonies of pea aphids by *Aphidius smithi* (Hymenoptera: Aphidiidae). *Ann. ent. Soc. Am.*, **63**, 973–80.

Tamaki, G., Landis, B. J. & Weeks, R. E. (1967). Autumn populations of green peach aphid on peach trees and the role of syrphid flies in their control. *J. econ. Ent.*, **60**, 433–6.

Tamaki, G. & Powell, D. M. (1972). An insecticide and a defoliant evaluated for use in a program of integrated control designed to suppress the green peach aphid on peach trees. *J. econ. Ent.*, **65**, 271–5.

Tamaki, G. & Weeks, R. E. (1972a). Efficiency of three predators, *Geocoris bullatus*, *Nabis americoferus*, and *Coccinella transversoguttata*, used alone or in combination against three insect prey species, *Myzus persicae*, *Geramica picta*, and *Mamestra configurata*, in a greenhouse study. *Environ. Ent.*, **1**, 258–63.

(1972b). Biology and ecology of two predators, *Geocoris pallens* Stål and *G. bullatus* (Say). *Tech. Bull. U.S. Dep. Agric.*, **1446**, 1–46.

(1973). The impact of predators on populations of green peach aphids on field grown sugar beets. *Environ. Ent.*, **2**, 345–50.

Tambs-Lyche, H. (1950). Aphids on potato foliage in Norway. 1. With a supplement on aphids in glasshouses. *Norsk ent. Tidsskr.*, **8**, 17–46.

Tanaka, M. (1966). Fundamental studies on the utilization of natural enemies in the citrus grove in Japan. I. The bionomics of natural enemies of the most serious pests. II. *Stethorus japonicus* H. Kamiya (Coccinellidae) a predator of the citrus red mite, *Panonychus citri* (McGregor). *Bull. hort. Res. Stn, Japan, Ser. D.*, **4**, 22–42 (in Japanese, English summary).

Tanaka, M., Inoue, K. & Kita, K. (1969). Studies on the acaricide resistance of the citrus red mite, *Panonychus citri* (McG.). IV. Genetics of resistance to phenkapton. *Proc. Assoc. Pl. Prot.*, Kyushu, **15**, 151–5 (in Japanese, English summary).

Tanigoshi, L. K. (1973). Studies on the dynamics of predation of *Stethorus picipes* (Coleoptera: Coccinellidae) and *Typhlodromus floridanus* on the prey *Oligonychus punicae* (Acarina: Phytoseiidae, Tetranychidae). Ph.D. Thesis, University of California, Riverside.

References

Tao, C. & Chiu, S. (1971). Biological control of citrus, vegetables and tobacco aphids. *Spec. Publ. Taiwan agric. Res. Inst.*, Taipei, **10**, 1–110.

Tashiro, H. & Chambers, D. L. (1967). Reproduction in the California red scale, *Aonidiella aurantii* (Homoptera: Diaspididae). I. Discovery and extraction of a female sex pheromone. *Ann. ent. Soc. Am.*, **60**, 1166–70.

Tashiro, H., Chambers, D. L., Moreno, D. S. & Beavers, J. (1969). Reproduction in the California red scale, *Aonidiella aurantii* (Homoptera: Diaspididae). III. Development of an olfactometer for bioassay of the female sex pheromone. *Ann. ent. Soc. Am.*, **62**, 935–40.

Tauber, M. J. & Tauber, C. A. (1967). Reproductive behaviour and biology of the gall-former, *Aciurina ferruginea* (Doane) (Diptera: Tephritidae). *Can. J. Zool.* **45**, 907–13.

Tawfik, M. F. & Nagui, A. (1965). The biology of *Montandoniella moraguesi* Puton, a predator of *Cynaikothrips ficorum* Marchal in Egypt. *Bull. Soc. ent. Egypte*, **49**, 181–200.

Taylor, C. E. (1962). The population dynamics of aphids infesting the potato plant with particular reference to the susceptibility of certain varieties to infestation. *Eur. Potato J.*, **5**, 204–19.

Taylor, E. P. (1908). Life history notes and control of the green peach aphis, *Myzus persicae* L. *J. econ. Ent.*, **1**, 83–91.

Taylor, L. R. (1974). Monitoring changes in the distribution and abundance of insects. *A. Reps. Rothamsted Exp. Stn*, **2**, in press.

Thill, H. (1957). Untersuchungen über die Wirkung von Pflanzenschutzmassnahmen auf die Rote Spinne und ihre Feinde im Buhler Obstbaugebiet. *Verh. 4th int. Pflanzenschutzkongr.*, Hamburg, **1**, 941–5.

Thoizon, G. (1967). Contamination expérimentale d'Homoptères Aphididae par des souches d'*Entomophthora* (Phycomycètes) isolées de Lépidoptères et de Diptères. *C. r. Acad. Sci. Paris*, **265**, 2001–3.

(1970). Spécificité du parasitisme des Aphides par les Entomophthorales. *Annls Soc. ent. Fr.* (N.S.), **6**, 517–62.

Thompson, W. R. (1934). The development of a colony of *Aphelinus mali* Hald. *Parasitology*, **26**, 449–53.

Thurston, R. (1961). Resistance in *Nicotiana* to the green peach aphid and some other tobacco insect pests. *J. econ. Ent.*, **54**, 946–9.

Thurston, R., Smith, W. T. & Cooper, B. P. (1966). Alkaloid secretion by trichomes of *Nicotiana* species and resistance to aphids. *Entomologia exp. appl.*, **9**, 428–32.

Thurston, R. & Webster, J. A. (1962). Toxicity of *Nicotiana gossei* Domin to *Myzus persicae* (Sulzer). *Entomologia exp. appl.*, **5**, 233–8.

*Torii, T. (1970). Quantitative prediction of the rice stem borer based on the sequential sampling test. *Proc. Assoc. Pl. Prot. Kyushu*, **16**, 25–7 (in Japanese).

(1971*a*). Quantitative occurrence prediction based on the sequential sampling test of the degree of infestation by the rice stem-borer. *Sci. Bull. Fac. Agric., Kyushu Univ.*, Fukuoka, **25**, 103–12.

(1971*b*). The development of quantitative occurrence prediction of infesta tion by the rice stem-borer, *Chilo suppressalis* Walker in Japan. *Entomophaga*, **16**, 193–207.

(1971 *c*). The ecological studies of rice stem-borers in Japan: a review. *Mushi*, **45**, 1–49.

Traboulsi, R. (1968). Prédateurs et parasites d'*Aphytis* (Hym., Aphelinidae). *Entomophaga*, **13**, 345–55.

Trjapitzin, V. A. (1971). Review of genera of Palaearctic encyrtids (Hymenoptera, Encyrtidae). *Trudy Vses. Ent. Ova*, **54**, 68–155 (in Russian).

Tsakas, S. & Krimbas, C. B. (1970). The genetics of *Dacus oleae*. IV. Relation between adult esterase genotypes and survival to organophosphate insecticides. *Evolution*, **24**, 807–15.

Tsakas, S. & Zouros, E. (1969). Isozymes in olive fruit fly *Dacus oleae Isoenzyme Bull.*, **2**, 46 pp.

*Tunçyürek, M. & Oncuer, C. (1975). Studies on aphelinid parasites and their hosts, citrus diaspine scale insects, in citrus orchards of the Aegean region. *Annls Inst. Phytopath. Benaki*, in press.

Tychsen, P. H. (1972). Mating behaviour and the control of sexual responsiveness in the Queensland fruit fly, *Dacus tryoni* (Frogg). Ph.D. Thesis, University of Sydney, Australia.

Tychsen, P. H. & Fletcher, B. S. (1971). Studies on the rhythm of mating in the Queensland fruit fly, *Dacus tryoni*. *J. Insect Physiol.*, **17**, 2139–56.

Tzanakakis, M. E., Tsitsipis, J. A. & Economopoulos, A. P. (1968). Frequency of mating in females of the olive fruit fly under laboratory conditions. *J. econ. Ent.*, **61**, 1309–12.

Uhler, L. D. (1951). Biology and ecology of the goldenrod gall fly, *Eurosta solidaginis* (Fitch). *Cornell Univ. agric. Exp. Stn Mem.*, **300**, 1–51.

Uygun, V. N. (1971). Der Einfluss der Nahrungsmenge auf Fruchtbarkeit und Lebensdauer von *Aphidoletes aphidimyza* (Rond.) (Diptera: Itonididae). *Z. angew. Ent.*, **69**, 234–58.

Van de Vrie, M. (1962). The influence of spray chemicals on predators and phytophagous mites on apple trees in laboratory and field trials in the Netherlands. *Entomophaga*, **7**, 243–50.

(1965). *Acarogical Research*. Jaarverslag Proefstation voor de Fruitteelt in de Volle Grand, Wilhelminadorp, 162 pp.

(1974). Studies on prey–predator interactions between *Panonychus ulmi* and *Typhlodromus* (*A.*) *potentillae* (Acarina: Tetranychidae, Phytoseiidae) on apple in the Netherlands. *Proceedings of the FAO conference on ecology in relation to plant pest control* (Rome, 1972), pp. 145–50.

Van de Vrie, M. & Boersma, A. (1970). The influence of the predacious mite *Typhlodromus* (*A.*) *potentillae* (Garman) on the development of *Panonychus ulmi* Koch on apple grown under various nitrogen conditions. *Entomophaga*, **15**, 291–304.

Van de Vrie, M. & Fluiter, H. J. (1958). Some observations on the effect of insecticides and acaricides on the European red spider mite (*Metatetranychus ulmi* Koch) and its principal predators in commercial orchards in the Netherlands. *Proceedings 10th international congress on entomology*, Montreal, **4**, 603–6.

References

Van de Vrie, M. & Kropczynska, D. (1967). The influence of predatory mites on the population development of *Panonychus ulmi* Koch on apples. *Entomophaga*, **3**, 77–84.

Van de Vrie, M., McMurtry, J. A. & Huffaker, C. B. (1972). Ecology of tetranychid mites and their natural enemies: a review. III. Biology, ecology, and pest status, and host-plant relations of tetranychids. *Hilgardia*, **41**, 343–432.

van den Bosch, R. & Haramoto, F. H. (1953). Competition among parasites of the oriental fruit fly. *Proc. Hawaiian ent. Soc.*, **15**, 201–6.

van den Bosch, R., Lagace, C. F. & Stern, V. M. (1967). The interrelationship of the aphid, *Acyrthosiphon pisum*, and its parasite, *Aphidius smithi*, in a stable environment. *Ecology*, **48**, 993–1000.

van den Bosch, R., Schlinger, E. I., Dietrick, E. J., Hagen, K. S. & Holloway, J. K. (1959 a). The colonization and establishment of imported parasites of the spotted alfalfa aphid in California. *J. econ. Ent.*, **52**, 136–41.

van den Bosch, R., Schlinger, E. I., Dietrick, E. J. & Hall, J. C. (1959 b). The role of imported parasites in the biological control of the spotted alfalfa aphid in southern California. *J. econ. Ent.*, **52**, 142–54.

van den Bosch, R., Schlinger, E. I., Lagace, C. F. & Hall, J. C. (1966). Parasitization of *Acyrthosiphon pisum* by *Aphidius smithi*, a density-dependent process in nature (Homoptera: Aphididae) (Hymenoptera: Aphidiidae). *Ecology*, **47**, 1049–55.

van Emden, H. F. (1966). Studies on the relations of insect and host plant. III. A comparison of the reproduction of *Brevicoryne brassicae* and *Myzus persicae* (Hemiptera: Aphididae) on Brussels sprout plants supplied with different rates of nitrogen and potassium. *Entomologia exp. appl.*, **9**, 444–60.

(1969 a). The differing reactions of *Brevicoryne* and *Myzus* to leafage, turgidity and soluble nitrogen in brassicas. *Ann. appl. Biol.*, **63**, 324–6.

(1969 b). Plant resistance to aphids induced by chemicals. *J. Sci. Fd Agric.*, **20**, 385–7.

* (ed.) (1972 a). *Aphid Technology*. Academic Press, London and New York, 344 pp.

(1972 b). Aphids as phytochemists. In *Phytochemical Ecology* (ed. J. B. Harborne), pp. 25–43. Academic Press, London and New York.

(1973). Aphid–host plant relationships. Some recent studies. In *Perspectives in Aphid Biology* (ed. A. D. Lowe), pp. 54–64. *Bull. ent. Soc. N.Z.*, **2**.

van Emden, H. F. & Bashford, M. A. (1969). A comparison of the reproduction of *Brevicoryne brassicae* and *Myzus persicae* in relation to soluble nitrogen concentration and leaf age (leaf position) in the Brussels sprout plant. *Entomologia exp. appl.*, **12**, 351–64.

(1971). The performance of *Brevicoryne brassicae* and *Myzus persicae* in relation to plant age and leaf amino acids. *Entomologia exp. appl.*, **14**, 349–60.

*van Emden, H. F., Eastop, V. F., Hughes, R. D. & Way, M. J. (1969). The ecology of *Myzus persicae*. *A. Rev. Ent.*, **14**, 197–270.

References

van Emden, H. F. & Wearing, C. H. (1965). The role of the host plant in delaying economic damage levels in crops. *Ann. appl. Biol.*, **56**, 323–4.

van Emden, H. F. & Williams, G. F. (1974). Insect stability and diversity in *agro-ecosystems*. *A. Rev. Ent.*, **19**, 455–75.

Van Zon, A. Q. & Helle, W. (1967). Linkage studies in the pacific spider mite *Tetranychus pacificus*. I. Genes for pigmentless, white eye, stork, and organophosphate resistance. *Ent. exp. appl.*, **10**, 69–74.

Van Zon, A. Q. & Overmeer, W. P. J. (1972). Induction of chromosome mutations by X-irradiation in *Tetranychus urticae* Koch (Acarina: Tetranychidae) with respect to a possible method of control. *Entomologia exp. appl.*, **15**, 195–202.

Varley, G. C. & Gradwell, G. R. (1970). Recent advances in insect population dynamics. *A. Rev. Ent.*, **15**, 1–24.

*Vater, G. (1971). Über Ausbreitung und Orientierung von *Diaeretiella rapae* (Hymenoptera, Aphidiidae) unter Berücksichtigung der Hyperparasiten von *Brevicoryne brassicae* (Homoptera, Aphididae). *Z. angew. Ent.*, **68**, 187–225.

Vevai, E. J. (1942). On the bionomics of *Aphidius matricariae* Hal., a braconid parasite of *Myzus persicae* Sulz. *Parasitology*, **34**, 141–51.

Vinson, S. B. (1972). Competition and host discrimination between two species of tobacco budworm parasitoids. *Ann. ent. Soc. Am.*, **65**, 229–36.

Voronin, K. E. (1968). Acclimatation of the aphidophagous coccinellid *Leis axyridis* Pall. from Far East in the Ukraine south to Carpathians. Tr. Vses. n-i. *Inst. Zashch. Rast.*, **31**, 234–43 (original in Russian).

Voronina, E. G. (1964). Culture des champignons entomophthorales dans des milieux nutritifs. *Issled. biol. metod. bor'by s vredit, sel'sk lesn. khoz.*, Novosibirsk, 1964, pp. 92–4 (original in Russian).

(1971). Entomophthorosis epizootics of the pea aphid *Acyrthosiphon pisum* Harris (Homoptera, Aphidoidea). *Ent. Rev.*, Wash., **50**, 444–53.

Wald, A. (1947). *Sequential Analysis*, New York, London, Sydney, John Wiley & Sons, 212 pp.

Waldhauer, W. (1957). Untersuchungen an Klonen der grünen Pfirsichblattlaus *Myzodes persicae* (Sulz.) zur Frage ihrer virginogenen Überwinterung. Diss. Rhein. Friedrich-Wilhelms-Univ., Bonn, 115 pp.

Walling, M. V., White, D. C. & Rodriguez, J. G. (1968). Characterization, distribution, catabolism, and synthesis of the fatty acids of the two-spotted spider mite, *Tetranychus urticae*. *J. Insect Physiol.*, **14**, 1445–58.

Walton, R. R. (1954). Seasonal fluctuations of the green peach and turnip aphids on commercial greens crops in Oklahoma. *J. econ. Ent.*, **47**, 775–80.

Ward, K. M. (1934). The green peach aphid (*Myzus persicae* Sulzer) in relation to the peach in Victoria and the measures investigated for its control. *J. Dep. Agric. Vict.*, **32**, 97–104, 134–45, 256–68.

*Waterhouse, D. F. (1967). Interacting organisms and the balance of nature. In *Biology in the Modern World, Rep. Aust. Acad. Sci.*, **8**, 33–9.

Watson, T. F. (1964). Influence of plant condition on population increase of *Tetranychus telarius* L. (Acarina: Tetranychidae). *Hilgardia*, **35**, 273–32.

References

Way, M. J. (1973 *a*). Population structure in aphid colonies. In *Perspectives in Aphid Biology* (ed. A. D. Lowe), pp. 76–84. *Bull. ent. Soc. N.Z.*, **2**.

(1973 *b*). Objectives, methods and scope of integrated control. In *Insects: Studies in Population Management* (eds. P. W. Geier, L. R. Clark, D. J. Anderson and H. A. Nix), pp. 137–52. *Mem. ecol. Soc. Aust.* **1**.

Way, M. J. & Murdie, G. (1965). An example of varietal variations in resistance of Brussels sprouts. *Ann. appl. Biol.*, **56**, 326–8.

Wehrhahn, C. F. & Klassen, W. (1971). Genetic insect control methods involving the release of relatively few laboratory-reared insects. *Can. Ent.*, **103**, 1387–96.

Weiser, J. (1961). *Die Mikrosporidien als Parasiten der Insekten.* P. Parey, Hamburg und Berlin. 149 pp. *Monogr. angew. Ent.*, no. **17**.

Westigard, P. H. (1971). Integrated control of spider mites on pear. *J. econ. Ent.*, **64**, 496–501.

Wharten, J. D., Jr, Rudrum, M., Moreno, D. S. & Jacobsen, M. (1970). *Aonidiella aurantii* sex pheromone isolation. *J. Insect Physiol.*, **16**, 2207–9.

White, E. B., DeBach, P. & Garber, M. J. (1970). Artificial selection for genetic adaptation to temperature extremes in *Aphytis lingnanensis* Compere (Hymenoptera: Aphelinidae). *Hilgardia*, **40**, 161–92.

Whitten, M. J. (1970). Genetics of pests in their management. In *Concepts of Pest Management* (eds. R. L. Rabb and F. E. Guthrie), pp. 119–37. North Carolina State University, Raleigh.

Wiackowski, S. K. (1962). Studies on the biology and ecology of *Aphidius smithi* Sharma & Subba Rao (Hymenoptera, Braconidae), a parasite of the pea aphid, *Acyrthosiphon pisum* (Harr.) (Homoptera, Aphididae). *Pol. Pismo entomol.*, **32**, 253–310.

Wilbert, H. (1964). Das Ausleseverhalten von *Aphelinus semiflavus* Howard und die Abwehrreaktionen seiner Wirte (Hymenoptera: Aphelinidae). *Beitr. Ent.*, **14**, 159–221.

Wilding, N. (1968–72). Various reports. *Reps. Rothamsted Exp. Stn*, 1968–72.

(1969). Effect of humidity on the sporulation of *Entomophthora aphidis* and *E. thaxteriana*. *Trans. Br. mycol. Soc.*, **53**, 126–30.

(1970). *Entomophthora* conidia in the air-spora. *J. gen. Microbiol.*, **62**, 149–57.

(1971 *a*). The effect of temperature on the infectivity and incubation periods of *Entomophthora aphidis* and *E. thaxteriana* for the pea aphid *Acyrthosiphon pisum*. *Proceedings 4th international colloquium on insect pathology*, College Park, Maryland, 1970, pp. 84–8.

(1971 *b*). Discharge of conidia of *Entomophthora thaxteriana* Petch from the pea aphid *Acyrthosiphon pisum* Harris. *J. gen. Microbiol.*, **69**, 417–22.

(1972). The effect of systemic fungicides on the aphid pathogen, *Cephalosporium aphidicola*. *Pl. Path.*, **21**, 137–9.

(1973). The survival of *Entomophthora* spp. in mummified aphids at different temperatures and humidities. *J. Invertebr. Path.*, **21**, 309–11.

Willard, J. R. (1972 *a*). Studies on rates of development and reproduction of California red scale, *Aonidiella aurantii* (Mask.) (Homoptera: Diaspididae) on citrus. *Aust. J. Zool.*, **20**, 37–47.

(1972*b*). The rhythm of emergence of crawlers of California red scale, *Aonidiella aurantii* (Mask.), (Homoptera: Diaspididae). *Aust. J. Zool.*, **20**, 49–65.

Willcocks, F. C. & Bahgat, S. (1937). Insects and mites injurious to the cotton plant. In *Insects and Related Pests of Egypt*, vol. I, pt. 2. R. Agric. Soc., Cairo. 792 pp.

Williams, J. R. (1970). Studies on the biology, ecology and economic importance of the sugar-cane scale insect, *Aulacaspis tegalensis* (Zhnt.) (Diaspididae), in Mauritius. *Bull. ent. Res.*, **60**, 61–95.

Wilson, J. W., Kelsheimer, E. G., Griffiths, J. T. & Tissot, A. N. (1948). A preliminary report on the control of the green peach aphid on shade grown tobacco in Florida. *Can. Ent.*, **30**, 45, 47–56.

Wood, B. J. (1963). Imported and indigenous natural enemies of citrus coccids and aphids in Cyprus, and an assessment of their potential value in integrated control programmes. *Entomophaga*, **8**, 67–82.

Wood, T. G. (1967). New Zealand mites on the Family Stigmaeidae (Acari, Prostigmata). *Trans. Soc. N.Z.*, **9**, 93–139.

Wright, N. S., MacCarthy, H. R. & Forbes, A. R. (1970). Epidemiology of potato leafroll virus in the Fraser River delta of British Columbia. *Am. Potato J.*, **47**, 1–8.

Wyatt, I. J. (1965). The distribution of *Myzus persicae* (Sulz.) on year-round chrysanthemums. I. Summer season. *Ann. appl. Biol.*, **56**, 439–59.

(1970). The distribution of *Myzus persicae* (Sulz.) on year-round chrysanthemums. II. Winter season: the effect of parasitism by *Aphidius matricariae* Hal. *Ann. appl. Biol.*, **65**, 31–41.

Wysoki, M. & Swirski, E. (1971). Studies on overwintering of predacious mites of the genera *Amblyseius* Berlese, *Typhlodromus* Scheuten and *Iphiseius* Berlese (Acarina: Phytoseiidae) in Israel. In *Entomological Essays to Commemorate the Retirement of Professor K. Yasumatsu.* Shukosha Printing Co., Ltd, Fukuoka, pp. 265–92.

Yasumatsu, K. (1967*a*). Distribution and bionomics of natural enemies of rice stem-borers (Research on the natural enemies of rice stem-borers). *Mushi*, **39**, *Suppl.*, 33–44.

(1967*b*). The possible control of rice stem-borers by the use of natural enemies. In *The Major Insect Pests of the Rice Plant*, pp. 431–42. Baltimore: The Johns Hopkins Press.

(1971). On some biological control problems in Japan. *Israel J. Ent.*, **6**, 301–5.

(1972). Activity, scope and problems in rice stem-borer research. *Mushi*, **45**, *Suppl.*, 3–6.

*Yasumatsu, K. & Torii, T. (1968). Impact of parasites, predators, and diseases on rice pests. *A. Rev. Ent.*, **13**, 295–324.

Yates, R. W. (1969). Flight ability and flight metabolism with relation to age in the fruit fly *Dacus tryoni*. BSc. Thesis, University of Sydney, Australia, 45 pp.

Yendol, W. G. (1964). The pathogenesis of *Entomophthora coronata* (Costantin) Kevorkian and *Entomophthora virulenta* Hall & Dunn (Zygomycetes:

References

Entomophthorales) in termites and aphids. Ph.D. Thesis, Purdue University, Lafayette. 222 pp.

(1968). Factors affecting germination of *Entomophthora* conidia. *J. Invertebr. Path.*, **10**, 116–21.

Yinon, U. (1969). Food consumption of the armored scale lady beetle *Chilocorus bipustulatus* (Coccinellidae). *Entomologia exp. appl.*, **12**, 139–46.

Zaher, M. A., Wafa, A. K. & Shehata, K. K. (1969). Life history of the predatory mite *Phytoseius plumifer* and the effect of nutrition on its biology (Acarina: Phytoseiidae). *Entomologia exp. appl.*, **12**, 383–8.

Zimmermann-Gries, S. & Swirski, R. (1956). La diffusione degli afidi sulle patate in Israele. *Boll. Lab. Zool. gen. agric. Portici*, **33**, 303–11.

Zirnitis, J. (1944). Pflanzenpathologie im Ostland, 6. Mitteilung. *Z. angew. Ent.*, **30**, 381–90.

Zouros, E. (1969). On the role of female monogamy in the sterile-male technique of insect control. *Acta Inst. Phytopath. Benaki*, **9**, 20–9.

Zouros, E. & Krimbas, C. B. (1969). The genetics of *Dacus oleae*. III. Amount of variation of two esterase loci in a Greek population. *Genet. Res.*, **14**, 249–58.

(1970a). Frequency of female digamy in a natural population on the olive fly *Dacus oleae* as found by using enzyme polymorphism. *Entomologia exp. appl.*, **13**, 1–9.

(1970b). A case of duplication of an esterase locus in the olive fruit fly, *Dacus oleae*. *Isoenzyme Bull.*, **3**, 44 pp.

Zouros, E., Tsakas, S. & Krimbas, C. B. (1968). The genetics of *Dacus oleae*. II. The genetics of two adult esterases. *Genet. Res.*, **12**, 1–9.

Zuñiga, E. (1966). Los pulgones del duraznero en Chile Central. *Agric. téc.*, **27**, 32–9.

Zwick, R. W. (1972). Studies on the integrated control of spider mites on apple in Oregon's Hood River Valley. *Environ. Ent.*, **1**, 169–76.

Zwölfer, H. (1968). Untersuchungen zur biologischen Bekämpfung von *Centaurea solstitialis* L. Strukturmerkmale der Wirtspflanze als Auslöser des Eiablageverhaltens bei *Urophora siruna-seva* (H.G.). *Z. angew. Ent.*, **61**, 119–30.

(1969). *Urophora siruna-seva* (H.G.) (Dipt.: Trypetidae), a potential insect for the biological control of *Centaurea solstitialis* L. in California. *Tech. Bull., Commonw. Inst. Biol. Control*, **11**, 105–55.

Zwölfer, H., Englert, W. & Pattullo, W. (1970). Investigation on the biology, population ecology and the distribution of *Urophora cardui* L., *Commonw. Inst. Biol. Control, Weed Proj. Can. Prog. Rep.*, **28**, 17 pp.

Zwölfer, H. & Harris, P. (1971). Host specificity determinations of insects for biological control of weeds. *A. Rev. Ent.*, **16**, 159–78.

Index of animal names

Index of animal names

Aphidius rubifolii Mackauer, 98
Aphidius smithi Sharma & Subba Rao, 93–8
Aphidoletes, 101
Aphis, 105
Aphis craccivora Koch, 97
Aphis fabae Scopoli, 54, 88, 94, 101, 104, 107, 114
Aphis gossypii Glover, 101, 109
Aphis helianthi Monell, 93
Aphis nasturtii Kalt., 88, 105
Aphis rumicis L., 110
Aphytis, 141, 142, 143, 146, 148–51, 153, 154, 157, 160, 163, 165, 168, 172, 173, 174, 177, 178, 232, 234
Aphytis africanus Qued., 151, 152, 153, 169, 170, 176
Aphytis aonidiae (Merc.), 173, 176
Aphytis chilensis group, 150
Aphytis chilensis (How.), 151, 171, 176
Aphytis chrysomphali group, 150
Aphytis chrysomphali (Mercet), 144, 146, 151, 156, 157, 164, 165, 172, 173, 174, 176
Aphytis coheni DeBach, 151, 152, 153, 157, 168, 169, 170, 173, 176
Aphytis costalimai (Gomes), 176
Aphytis cyclindratus Comp., 177
Aphytis diaspidis (How.), 145, 173, 177, 178
Aphytis fisheri DeBach, 151, 152, 153, 173
Aphytis funicularis Comp., 162, 173
Aphytis hispanicus (Mercet), 177, 178
Aphytis holoxanthus DeBach, 144, 147, 151, 152, 153, 157, 161–4, 166, 170, 171, 172, 176
Aphytis 'khunti', 151, 152, 153
Aphytis lepidosaphes Comp., 144, 147, 151, 152, 153, 161–4, 168, 170–3, 177
Aphytis lingnanensis Comp., 144, 146, 151, 152, 153, 154, 156, 157, 160, 161, 162, 164, 167–73, 176, 177, 178
Aphytis lingnanensis group, 150, 152, 153
Aphytis maculicornis (Masi), 145, 147, 153, 161, 169, 172, 177
Aphytis melinus DeBach, 144, 146, 151, 152, 153, 156, 157, 158, 160, 161, 164–74, 176
Aphytis melinus group, 152
Aphytis mytilaspidis group, 150, 173
Aphytis mytilaspidis (LeBaron), 145, 147, 151, 153, 154, 172, 173, 176, 177, 178
Aphytis proclia group, 150
Aphytis proclia (Walk.), 151, 176, 178
Aphytis 'R-65-23', 151, 152, 153
Aphytis roseni DeBach & Gordh, 161, 163
Aphytis '2002', 151, 152, 153
Aphytis vittatus (Comp.), 151
Aphytis vittatus group, 150
apicalis Holm., *Goryphus*, 131, 132
apterus (Fab.), *Himacerus*, 220
arboreus Chant, *Typhlodromus*, 216
articulatus (Morg.), *Selenaspidus*, 148, 161, 163, 175, 178
asiatica Arch., *Tecaspis*, 175
aspidioti Comp. & Ann., *Habrolepis*, 176
aspidiotinorum Comp., *Metaphycus*, 163, 178
Aspidiotiphagus, 142, 177

Aspidiotiphagus citrinus (Craw.), 144, 162, 167, 173, 176
Aspidiotus destructor Sign., 5, 142, 144, 146, 166, 175, 176
Aspidiotus hederae (Vallot), 175, 176
Aspidiotus nerii Bouché, 144, 161, 167, 169, 175, 176
aspidistrae (Sign.), *Pinnaspis*, 175, 176
asychis Walk, *Aphelinus*, 92, 94, 95
Atomus, 218
Atractotomus mali (Meyer), 218, 220
Aulacaspis rosae (Bouché), 175, 176
Aulacorthum solani (Kalt.), 88
aurantii (Mask.), *Aonidiella*, 142, 143, 144, 146, 148, 149, 150, 155–60, 164–71, 175, 176,
aurescens Athias-Henriot, *Amblyseius*, 216
auricilia (Dudgeon), *Chilo*, 122
australicum Gir., *Trichogramma*, 131, 132
australis Aldrich, *Procecidochares*, 45, 47
Austrodacus cucumis (French), 40, 45
axyridis Pallas, *Harmonia*, 100, 102
ayyari Rohw., *Tetrastichus*, 131, 132
Azotus separaspidis Ann. & Ins., 176
Azya trinitatis Marsh., 166

balteata DeG., *Epistrophe*, 92, 101
banski (McGregor), *Eotetranychus*, 227, 228
basicincta Gah., *Encarsia*, 174
basilaris Holm., *Goryphus*, 132
basiola (Osten Sacken), *Rhagoletis*, 23, 45
Bathytricha truncata (Walk.), 122
beckii (Newm.), *Lepidosaphes*, 143, 144, 147, 148, 150, 157, 159, 161–4, 166, 168, 170, 175, 177
bella Gah., *Prospaltella*, 177
beneficus (Hiura), *Lyptocoris*, 133
berberidis Kand., *Rhagoletis*, 45
berlesei (How.), *Prospaltella*, 145, 147, 177
bifasciata How., *Comperiella*, 142, 144, 146, 155–8, 160, 164–8, 170, 172, 173, 174, 176, 178
biguttula (Munak.), *Temelucha*, 131, 132
bipunctata L., *Adalia*, 103
bipustulatus (L.), *Chilocorus*, 156, 177
blanchardi (T.T.), *Parlatoria*, 5, 145, 147, 150, 156, 175, 177
Blepharidopterus angulatus, (Fall.) 218, 220
boisduvalii Sign., *Diaspis*, 175, 176
boringuensis Quez., DeBach & Rosen, *Signiphora*, 155, 172
boycei Cresson, *Rhagoletis*, 45
Bracon, 132
Bracon chinensis Szépl., 131, 132
Bracon onukii Watanabe, 131, 132
brassicae L., *Brevicoryne*, 2, 66–72, 79, 88, 97, 101, 111, 115
braziliensis (Hemp.), *Prospaltella*, 171
Brevicoryne brassicae L., 2, 66–72, 79, 88, 97, 101, 111, 115
Brevipalpus phoenicis Geijskes, 192
brevis Uhl., *Deraeocoris*, 220
brevispinus (Kennett), *Amblyseius*, 216
Bryobia, 187, 197
Bryobia rubrioculus (Scheut.), 197, 221, 222

Index of animal names

Index of animal names

Lepidosaphes conchiformis (Gmel.), 175
Lepidosaphes ficus (Sign.), 145, 175, 177
Lepidosaphes gloverii (Pack.), 175, 177
Lepidosaphes malicola Borkh., 175, 177
Lepidosaphes newsteadi (Sulc), 144, 147
Lepidosaphes pistaciae Arch., 175
Lepidosaphes pistacicola Borkh., 175
Lepidosaphes ulmi (L.), 145, 147, 175, 177
Leptomastix dactylopii How., 172, 174
Leptothrips mali (Fitch), 218, 219
limonicus Garm., *Amblyseius*, 191, 192
lindi Blackb., *Rhizobius*, 165
Lindorus, 142
Lindorus lophantae (Blaisd.), 163, 164, 165, 176
lindquisti (Shust. & Pritch.), *Amblyseius*, 216
lingnanesis Comp., *Aphytis*, 144, 146, 151, 152, 153, 156, 157, 160, 161, 162, 164, 167–73, 176, 177, 178
lingnanensis group, *Aphytis*, 150, 153, 154
Liriomyza bryoniae Kalt., 215
livida Reut., *Campylomma*, 197
lobosus Boudr., *Tetranychus*, 217
lombardini Baker & Prichard, *Tetranychus*, 181
longipennis (Haan), *Conocephalus*, 133
longipilis (Nesbitt), *Galendromus*, 219
longipilus Nesbitt, *Typhlodromus*, 212
longirostris (Sign.), *Ischnaspis*, 144, 146
longisetus Gonzales, *Agistemus*, 196
longispinus (T.T.), *Pseudococcus*, 167
lophantae (Blaisd.), *Lindorus*, 163, 164, 165, 176
loxtoni Britton & Lee, *Stethorus*, 198
ludens (Loew), *Anastrepha*, 33, 34, 36, 37, 38, 42, 45
luteator (Fab.), *Ischnojoppa*, 131, 132
Lyptocoris beneficus (Hiura), 133
Lysiphlebus fabarum Quilis, 93

macropilis (Banks), *Phytoseiulus*, 194, 209, 218, 220
Macrosiphum euphorbiae Thomas, 2, 77, 81, 88, 105, 107, 113, 114
Macrosiphum (*Sitobion*) *fragrariae* Walk., 95
maculicornis (Masi), *Aphytis*, 145, 147, 153, 161, 169, 172, 177
maiusculus Reut., *Orius*, 220
Malacarcoris chlorezins (Panz.), 220
maleti (Vayss.), *Quadraspidiotus*, 149
mali Haldemann, *Aphelinus*, 92, 94
mali (Meyer), *Atractotomus*, 218, 220
mali (Fitch), *Leptothrips*, 218, 219
mali Ewing, *Zetzellia*, 196, 218, 219, 220
Maliarpha separatella Ragonot, 122
malicola Borkh., *Lepidosaphes*, 175, 177
malus (Shimer), *Hemisarcoptes*, 145, 147
mandschurica (Uchida), *Lampronota*, 131, 132
mangiferus R. & P., *Oligonychus*, 191
maraquesi Puton, *Montandoniola*, 202
marcescens Bizio, *Serratia*, 109, 110
marginalis Reut., *Orthotylus*, 220
Marietta, 149, 150
Marietta exitiosa Comp., 157
Marlattiella, 149, 150
Masonaphis maxima Mason, 97, 98
matricariae Hal., *Aphidius*, 89, 116, 117, 215

maxima Mason, *Masonaphis*, 97, 98
mcdanieli McGregor, *Tetranychus*, 193, 212 217, 219
Mediolata, 195
meigeni Loew, *Rhagoletis*, 45
Melanaphis donacis Passerini, 103
melinus DeBach, *Aphytis*, 144, 146, 151, 152, 153, 156, 157, 158, 160, 161, 164–74, 176
melinus group, *Aphytis*, 152
mendax Curran, *Rhagoletis*, 39, 45
Mesidia nigra Lagace, 94, 96
mesoxanthus maculipennis (Cam.), *Goryphus*, 131, 132
Metaphycus, 167
Metaphycus aspidiotinorum Comp., 163, 174
Metaphycus helvolus (Comp.), 161, 167, 168, 169, 172
Metarrhizium anisopliae Metsch., 109
Metaseiulus floridanus (Muma), 227
Micrapis discolor (Fab.), 133
Micrapis vincta (Gorham), 133
Microgaster russata Hal., 131, 132
Microterys okitsuensis Comp., 159
minima (T.T.), *Carulaspis*, 144, 147, 175, 176
minki Dohrn, *Anthocoris*, 202
minor (Mask.), *Pinnaspis*, 145, 175
minutus (L.), *Orius*, 200, 202, 220
Miscanthaspis tegalensis (Zehnt.), 149, 175, 177
miyarai Tach., *Adelencyrtus*, 177
modesta (Smith), *Xanthopimpla*, 131, 132
Monoctonus crepidis Hal., 95
Monoctonus paulensis Ashm., 95
Monotarsobius crassipes L., 21
Montandoniola maraquesi Puton, 202
montrouzieri Muls., *Cryptolaemus*, 172
munakatae (Munak.), *Chelonus*, 127, 131, 132
muscarium Petch., *Cephalosporium*, 108
musculus Say, *Anthocoris*, 200, 218
Mycodiplosis acarivora (Felt), 216
Myrmica laevinodis Nyl., 21
mytilaspidis group, *Aphytis*, 150, 173
mytilaspidis (LeBaron), *Aphytis*, 145, 147, 151, 153, 154, 172, 173, 176, 177, 178
Myzus, 105
Myzus persicae (Sulz.), 2, 51–63, 65, 67–74, 77, 78, 80, 81, 83–91, 95, 96, 99, 101, 102, 105, 109, 111, 113–19, 215, 223, 230, 231, 235

Nabis americoferus Car., 101
narangae (Ashm.), *Itoplectis*, 131, 132
Nasonovia, 95
Nasonovia ribisnigri Mosley, 67, 95
nasturtii Kalt., *Aphis*, 88, 105
nebulosus (Uhl.), *Deraeocoris*, 218, 219
nematophilus Poinar & Thomas, *Achromobacter*, 129
nemoralis Fab., *Anthocoris*, 202, 220
nemorum L., *Anthocoris*, 103, 200, 201, 202, 220
Neoaplectana carpocapsae Weiser, 129
neocaledonicus André, *Tetranychus*, 181
neohumeralis (Perkins), *Dacus*, 40, 45
Neomyzus circumflexus Buckton, 96

Index of animal names

Pilophorus perplexus (D. & S.), 218, 220
pilosus (Jak), *Orius*, 202, 220
pinifoliae (Fitch), *Phenacaspis*, 175, 177
pini-radiatae Davidson, *Schizolachnus*, 107
Pinnaspis aspidistrae (Sign.), 175, 177
Pinnaspis buxi (Bouché), 145, 146
Pinnaspis minor (Mask.), 145, 175
Pinnaspis strachani (Cool.), 175
pistaciae Arch., *Lepidosaphes*, 175
pistacicola Borkh., *Lepidosaphes*, 175
pisum Harris, *Acyrthosiphon*, 66, 97, 104, 105, 106, 109, 110
plagiator Nees, *Ephedrus*, 89
Plagiognathus obscurus (Uhl.), 218
Plagiognathus politus Uhl., 219
planchoniana Cornu, *Entomophthora*, 104–8
plantaginea Passerini, *Dysaphis*, 103
platanoidis Schr., *Drepanosiphum*, 103
Platyparea poiciloptera Schr., 28–32
poiciloptera Schr., *Platyparea*, 28–32
politus Muls., *Chilocorus*, 144
politus Uhl., *Plagiognathus*, 219
polychrysa (Meyrick), *Chilo*, 121, 122, 127, 128
pomi (Parrot), *Typhlodromus*, 219
pomonella (Walsh), *Rhagoletis*, 15, 17, 18, 19, 21–33, 38, 39, 41, 45
Popillia japonica Newman, 204
pornia (Walk.), *Dirioxa*, 38
potentillae Garm., *Typhlodromus*, 204, 220
Praon exsoletum Nees, 92
Praon pequodorum Vier., 94
Procecidochares, 47
Procecidochares australis Aldrich, 45, 47
Procecidochares sp. A, 45
proclia group, *Aphytis*, 150
proclia (Walk.), *Aphytis*, 151, 176, 178
Pronematus, 224
Pronematus anconai Baker, 225
Pronematus ubiquitus (McGregor), 225
Propylaea quatuordecimpunctata Ganglb., 102
Prospaltella, 142, 162, 165, 171–4, 177
Prospaltella bella Gah., 177
Prospaltella berlesei (How.), 145, 147, 177
Prospaltella braziliensis (Hemp.), 171
Prospaltella elongata Dozier, 177
Prospaltella fasciata Malen., 163
Prospaltella inquirenda Silv., 177
Prospaltella lahorensis How., 161, 162, 174
Prospaltella opulenta Silv., 160, 163
Prospaltella perniciosi Tower, 144, 145, 146, 148, 156, 164, 167, 168, 171, 172, 176, 178
prunorum Borkh., *Tecaspis*, 175
Psallus ambiguus (Fall.), 220
Pseudaonidia duplex (Ckll.), 175, 177
Pseudaulacaspis pentagona (T.T.), 142, 145, 147, 175, 177
pseudococci (Gir.), *Anagyrus*, 172
Pseudococcus longispinus (T.T.), 167
pseudomagnoliarum Kuw., *Coccus*, 159
Pseudoscymnus, 166
Pteroptrix, 142
Pteroptrix chinensis (How.), 142, 173
Pteroptrix smithi (Comp.), 147, 166
Pteroptrix wanhsiensis (Comp.), 173

pulchellus Montr., *Rhizobius*, 146, 166
pulcher (C. & F.), *Genopalpus*, 192
punctata HRL., *Iziphya*, 96
punctata (Fab.), *Xanthopimpla*, 131, 132
punctillum Weise, *Stethorus*, 200, 222
punctipes (Stål), *Geocoris*, 216
punctum LeConte, *Stethorus*, 200, 219
punicae (Hirst), *Oligonychus*, 192, 200, 228
purchasi (Mask.), *Icerya*, 4, 157, 227
pyri (Licht.), *Quadraspidiotus*, 175
pyri Scheut., *Typhlodromus*, 194, 211, 218, 220, 223

Quadraspidiotus forbesi (Johns.), 175, 177
Quadraspidiotus juglansregiae (Comst.), 175, 178
Quadraspidiotus maleti (Vayss.), 149
Quadraspidiotus ostreaeformis (Curt.), 175, 178
Quadraspidiotus perniciosus (Comst.), 142, 145, 148, 150, 165, 166, 168, 175, 178
Quadraspidiotus pyri (Licht.), 175
quatuordecimpunctata Ganglb., *Propylaea*, 102
quinquesignata punctata Kirby, *Hippodamia*, 101

'R-65-23', *Aphytis*, 151, 152, 153
rapae L. *Pieris*, 109
rapae M'Intosh, *Diaeretiella*, 89, 95, 96, 98
rapax (Comst.) *Hemiberlesia*, 175, 177
reticulatus Oud., *Typhlodromus*, 216
Rhaconotus oryzae Wilk., 131, 132
Rhaconotus schoenobivorus (Rohw.), 131, 132
Rhagoletis, 41, 47
Rhagoletis alternata Fall., 23
Rhagoletis basiola (Osten Sacken), 23, 45
Rhagoletis berberidis Kand., 45
Rhagoletis boycei Cresson, 45
Rhagoletis cerasi (L.), 15–24, 27–31, 33, 35, 38, 41, 45, 46, 47
Rhagoletis cingulata (Loew), 23, 45
Rhagoletis completa Cresson, 23–7, 45
Rhagoletis cornivora Bush, 45
Rhagoletis fausta (Osten Sacken), 23, 24, 45
Rhagoletis juglandis Cresson, 45
Rhagoletis meigeni Loew, 45
Rhagoletis mendax Curran, 39, 45
Rhagoletis pomonella (Walsh), 15, 17, 18, 19, 21–33, 38, 39, 41, 45
Rhagoletis ribicola Doane, 23
Rhagoletis suavis (Loew), 23, 24, 45
Rhagoletis tabellaria Fitch, 45
Rhagoletis zephyria Bush, 45
rhenanus (Oud.), *Typhlodromus*, 194, 218, 220, 222
Rhizobius lindi Blackb., 165
Rhizobius pulchellus Montr., 146, 166
Rhopalosiphum insertum Walk., 101
Rhopalosiphum padi L., 95, 96, 109
ribicola Doane, *Rhagoletis*, 23
 ibisnigri Mosley, *Nasonovia*, 67, 95
Rodolia cardinalis (Muls.), 157
rosa Karsh, *Ceratitis*, 33
rosae (Bouché), *Aulacaspis*, 175, 176
roseni DeBach & Gordh, *Aphytis*, 161, 163

294

Index of animal names

rouxi Comp., *Habrolepis*, 163, 176
rowani (Gah.), *Telenomus*, 131, 132
rubifolii Mackauer, *Aphidius*, 98
rubrioculus (Scheut.), *Bryobia*, 197, 221, 222
ruficoxatus (Son.), *Gambrus*, 131, 132
rufilabris Burm., *Chrysopa*, 219
Rugaspidiotus tamaricicola Mal., 175
rumicis L., *Aphis*, 110
russata Hal., *Microgaster*, 131, 132

sabalis (Comst.), *Comstockiella*, 144
sakaguchii (Mats. et Uch.), *Enicospilus*, 131, 132
Saissetia oleae (Bern), 161 167, 168, 169
salicellus (H. & S.), *Coniortodes*, 218
salicis (L.), *Chionaspis*, 175, 176
sarothamni Douglas & Scott, *Anthocoris*, 201, 202
Saula japonica Garm., 227
Scambus annulitarsis (Ashm.), 131, 132
Schizaphis graminum Rondani, 56, 63, 93, 100
Schizolachnus pini-radiatae Davidson, 107
schlechtendali (Nal.), *Aculus*, 217, 218, 219
schoenobii Wilk., *Apanteles*, 131, 132
schoenobii Ferr., *Tetrastichus*, 127, 131, 132
schoenobii (Vier.), *Tropobracon*, 131, 132
schoenobivorus (Rohw.), *Rhaconotus*, 131, 132
Scirpophaga chrysorrhoa Zeller, 122
Scolothrips sexmaculatus (Pergande), 216, 224
scripta L., *Sphaerophoria*, 101
scutellaris Bezzi, *Dacus*, 38
Scutellista, 167
Scymnus, 171
Selenaspidus, 163
Selenaspidus articulatus (Morg.), 148, 161, 163 175, 178
seminotus Silv., *Physcus*, 177
separaspidis Ann. & Ins., *Azotus*, 176
separatella Ragonot, *Maliarpha*, 122
septempunctata L., *Coccinella*, 101, 102
Serratia marcescens Bizio, 109, 110
Sesamia inferens (Walk.), 121, 122, 127, 130
sesamiae Yosh., *Tetrastichus*, 132
sexmaculatus (Riley), *Eotetranychus*, 227
sexmaculatus (Pergande), *Scolothrips*, 216, 224
Signiphora, 171, 177
Signiphora boringuensis Quez., DeBach & Rosen, 155, 172
Signiphora coquilletti Ashm., 176
similoides Buch. & Pritch., *Amblyseius*, 216
sinicus (Holm.), *Eriborus*, 131, 132
siruna-seva (Hering), *Urophora*, 24
Sitobion, 105
smithi Sharma & Subba Rao, *Aphidius*, 93–8
smithi (Comp.), *Pteroptrix*, 166
solani (Kalt.), *Aulacorthum*, 88
soleiger (Rib.), *Typhlodromus*, 194, 220
solidaginis (Fitch), *Eurosta*, 23, 24
Spathius helle Nix., 131, 132
sphaegiformes (Rossi), *Globiceps*, 220
Sphaerophoria scripta L., 101
sphaerosperma Fres., *Entomophthora*, 104, 105, 107, 108
spiniferus Brèth., *Amitus*, 159, 160, 164, 168, 172, 174

stangli (Ashm.), *Temelucha*, 131, 132
stemmator (Thunb.), *Xanthopimpla*, 131, 132
Steneotarsonemus pallidus (Banks), 215
Stenobracon nicevillei (Bingh.), 131, 132
Stethorus, 196–200, 218, 221, 227
Stethorus nilvifrons Muls., 222
Stethorus japonicus Kamiya, 200, 227
Stethorus loxtoni Britton & Lee, 198
Stethorus nigripes Kapur, 198
Stethorus picipes Casey, 199, 200, 216, 220, 228
Stethorus punctillum Weise, 200, 222
Stethorus punctum LeConte, 200, 219
Stethorus tetranychi Kapur, 222
Stethorus vagans Blackb., 198
strachani (Cool.), *Pinnaspis*, 175
suavis (Loew), *Rhagoletis*, 23, 24, 45
subflavus Ann. & Ins., *Physcus*, 177
subsolidus Begl., *Typhlodromus*, 220
subtilisetosus Begl., *Typhlodromus*, 220
suppressalis (Walk.), *Chilo*, 121, 122, 127–30
suspensa (Loew), *Anastrepha*, 36, 42, 45
sydneyensis Timb., *Anarhopus*, 167
Sympiesomorpha chilonis Ishii, 131, 132
Syntomosphyrum israeli Kur., 131, 132

tabellaria Fitch, *Rhagoletis*, 45
tamaricicola Mal., *Rugaspidiotus*, 175
Tecaspis asiatica Arch., 175
Tecaspis prunorum Borkh., 175
tegalensis (Zehnt.), *Miscanthaspis*, 149, 175, 177
telarius (L.), *Tetranychus*, 206
Telenomus dignoides Nix., 131, 132
Telenomus dignus (Gah.), 131, 132
Telenomus rowani (Gah.), 131, 132
Telsimia, 163
Telsimia nitida Chapin, 145
Temelucha biguttula (Munak.), 131, 132
Temelucha philippinensis (Ashm.), 131, 132
Temelucha stangli (Ashm.), 131, 132
terebrans (Grav.), *Eriborus*, 131, 132
terresiana Tillyard, *Ischnura*, 133
testaceus Masi, *Physcus*, 177
tetranychi Kapur, *Stethorus*, 222
Tetranychus, 180, 187
Tetranychus cinnabarinus (Bois.), 191, 192, 217
Tetranychus lobosus Boudr., 217
Tetranychus lombardini Baker & Prichard, 181
Tetranychus mcdanieli McGregor, 193, 212, 217, 219
Tetranychus neocaledonicus André, 181
Tetranychus pacificus McGregor, 180, 199, 224, 225, 226
Tetranychus telarius (L.), 206
Tetranychus tumidus Banks, 180
Tetranychus urticae (Koch), 179–86, 189, 197–200, 202, 203, 205, 206, 211, 212, 213, 215, 216, 217, 219, 221, 222, 224, 225
Tetranychus viennensis Zacher, 217
Tetrastichus, 126
Tetrastichus ayyari Rohw., 131, 132
Tetrastichus israeli (Mani & Kur.), 131, 132
Tetrastichus schoenobii Ferr., 127, 131, 132
Tetrastichus sesamiae Yosh., 132

295

Subject index

Abate, 170
Acarina (acarids), 124, 156, 179, 187, 196
acid phosphatase, 45
adenylate kinase, 45
African countries, 1, 66, 144, 163, 167, 169
age-structure of populations, 72–5, 129
Agrionidae, 133
agro-ecosystems, 143
alanine, 182, 185
alcaline phosphatase, 45
alcohol dehydrogenase, 45
aldehyde oxidase, 45
aldicarb, 208, 215
aldolase, 45
aldrin, 208
aleyrodids, 141
alfalfa, 105
Algeria, 145
alternative host (food), 103, 130, 219, 221–6
aluminium foil trap, 28, 29
amide, 70
amino acids, 68–71, 182–5
γ-aminobutyric acid, 71
p-aminobenzoic acid, 185
androcycly, 60–3
anholocycly, 56, 58–63, 65, 66, 90, 106, 111, 113
Anthio, 209
anthocorids, 81, 83, 84, 124, 133, 196, 200–2, 218–22
ants, 21, 22, 124
Anystidae (anystid mites), 80, 217
aphelinids, 5, 92, 99, 142, 143, 149, 155
aphidicides, use of, 114, 115
aphidiids, 92, 99
aphids, 2, 3, 8, 51–119, 141, 196, 215, 230, 235
apple, 101, 187, 193, 200–3, 210–12, 217–21, 22
apple rust mite, 217
Arachnoidea, 100
Aramite, 208
Araneida, 187
Argentina, 140, 145, 160, 162, 164, 172
arginine, 182, 184, 185
arid zones, 5
armoured scale insects, 4, 5, 139–78, 232, 233
arrhenotoky, 153, 154
Arrowhead scale, 171, 175
ascorbic acid, 185
Asia, Asian countries, 4, 66, 121, 130, 133, 135, 136, 148, 231
aspartic acid, 182, 184, 185
Aspidiotini, 175
Aspidistra scale, 175

Aster tripolium, 109
attractants, 14, 24, 230
Australia, 7, 9, 15–17, 20, 28, 30, 38, 66, 72, 89–91, 97, 98, 122, 123, 133, 134, 140, 144, 148, 149, 158, 160–2, 164, 165, 167, 170–2, 174, 176, 197–9, 217, 221, 232
Austria, 145
avocado, 187, 200, 228
avocado brown mite, 228
azinphosmethyl, 196, 199, 208, 210–12

bacteria, 109, 110
bait sprays, 12
Bali, 144
Bangladesh, 124
Barbados, 163
Bdellidae (bdellids), 217, 219
bean, 109, 182, 183, 184, 203
benomyl, 205, 214, 215
Bermuda, 144, 145, 147, 176
Bermuda cedar, 144
Bermuda cedar scale, 144, 147
Beta vulgaris, 114
Beta vulgaris ssp. *maritima*, 115
Bidrin, 208
binapacryl, 204, 207, 209
biotin, 185
biotypes, 44, 46, 52, 53, 56, 58, 63, 66, 67, 116, 118
black parlatoria scale, 175
black scale, 161, 167
Boiduval scale, 175, 176
bottlebrush, 225
brassicas, 61, 67, 68, 71, 86, 115
Brazil, 76, 81, 145, 161, 162, 164, 171, 176
Britain, 57, 59, 62, 63, 67, 75, 76, 78, 79, 80–2, 101, 109, 114, 115, 176, 193, 201, 209, 213, 220
brown mite, 217, 223
Brussels sprouts, 61, 66, 68, 69, 71, 86, 87, 101
Burma, 129

cabbage, 86–8
cactus scale, 175
calcium, 203
California, 5, 92, 95, 139, 140, 144–8, 155–64, 167, 169, 171–4, 177, 178, 199, 212, 215–17, 220, 222, 224–8
California red scale, 143, 144, 148, 160, 175, 176
Cambodia, 129
campesterol, 183

Subject index

298

Subject index

Holland, 89, 101, 105, 213
holocycly, 56, 58–63, 65, 66, 90, 111
homeostasis, 156
homogamic cross, 153, 154
homologues, ecological, 156
Homoptera, 141
honeydew, artificial, 103, 192
honeysuckle, 46
host
  age, influence of, 94
  discrimination, 150
  feeding, 94, 151, 170
  infection, 105
  plant interaction, 67–71, 95, 97, 202, 217, 230
  preference, 105, 118, 151, 155
  range, 95, 127
  selection, 56, 94, 95, 150
  size, influence of, 94, 107
  specifity, 48, 127, 142, 232
humidity, influence of, 105, 106, 157, 194, 226, 228
hybridisation, 151, 152, 153, 154, 232
Hymenoptera, 124, 149

IABCR, 122
Ichneumonoidea, 123, 131, 132
Imidan, 208, 210, 211
immigration, effect of, 117
immune reaction, 95
India, 92, 122, 123, 128, 129, 144, 145, 161, 162, 173, 176
Indonesia, 124
inoculative release of parasites, 119
inoculum, 106, 107
inositol, 185
insecticidal check method, 158
insecticides, use and effect of, 44, 48, 119, 129, 130
integrated control (pest management), 3, 23, 51, 52, 56, 67, 100, 102, 116, 117, 119, 121, 133, 143, 147, 167, 169, 170, 200, 204, 206, 210–12, 214, 218, 219, 223, 225, 229, 231, 233–5
international coordination, 1–3, 5, 7, 8, 11, 51, 84, 139, 229, 234, 236
Intration (thiometon), 209
intrinsic rate of increase, 96, 188, 189, 200
IOBC, 3, 140
Iran, 76, 81, 83, 92, 128, 145, 172, 177
iron, 203
irradiation, influence of, 34, 35, 154, 180
island situations, 12
isocitrate dehydrogenase, 45
Isolan, 208, 210
isoleucine, 182, 185
Israel, 140, 143, 144, 147, 149, 156–8, 160, 166, 167, 171, 172, 174, 176, 232
Italian pear scale, 175
Italy, 2, 11, 15, 17, 21, 28, 30, 36, 75, 81, 83, 85, 145, 160, 161, 169, 172, 174, 177

Jamaica, 162, 174
Japan, 9, 36, 37, 101, 122, 123, 128, 130, 133, 140, 144, 146, 155, 159, 162, 171, 174, 176, 217, 220, 221, 227

Java, 144, 176
Johnson grass, 225

k-strategists, 13, 14
kale, 61
karathane, 196, 204
karyotype, 43, 60, 63
key-factor analysis, 130

lacewings, 84, 85, 88, 89
latania scale, 175, 177
Lathridiidae, 197
lead arsenate, 196, 208
leafhoppers of rice, 126
leaf-miner, 215
Lebanon, 28, 30, 31
lemon, 144
Lepidoptera, 112
lesser snow scale, 175
lettuce, 115
leucine, 182, 185
leucine amino peptidase, 45
levulose, 185
lichi, 17
life cycle
  of *Aphytis*, 150
  of *Myzus persicae*, 57
life tables, 11, 15, 19, 22, 130, 189, 200, 230
light, influence of, 184
lime-sulphur, 207, 209
lindane, 208, 209
linoleic acid, 183, 185
linolenic acid, 183, 185
linyphiid spiders, 80
lipids, 183, 185
losses of crop
  with armoured scale insects, 139, 148, 162
  with spider mites, 213
lysine, 182, 185

magnesium, 203
Malachiidae, 100
malate dehydrogenase, 45
malathion, 165, 198, 205, 208, 209
Malaysia, 100, 122, 133
maneb, 204, 207, 209
mango, 17
marsh flies, 123
mating behaviour, 23, 26, 39, 40, 42, 44, 49, 180, 181
Mauritius, 144
McPhail trap, 32, 33
mealybugs, 5, 148, 166
medfly, 2
Mediterranean area, 140, 166, 176, 177, 194
meetings, 9, 11, 51–3, 121, 139, 140, 179
menazon, 208
metasystox, 206, 208, 209
methionine, 69, 70, 182, 185
methodologies, 236
  artificial feeding system for mites, 182
  assessment of damage (rice stem-borers), 136–7

Subject index

paratin, 208
Parlatoria date scale, 145, 147, 175, 177
Parlatoria tea scale, 175
Parlatorini, 175
parthenogenesis, 58, 59, 61, 62, 67, 111
pathogens, 3, 52, 72, 104, 129, 163, 221, 230, 231
peach, 71, 89, 106, 145, 198, 210, 221, 223, 231
peach silver mite, 222, 223
peroxydase, 45
'Persian strain' (of *A. maculicornis*), 147, 169
Peru, 144, 145, 160–4, 172, 178, 227, 232
pesticide use and effect, 4, 5, 101, 134, 141, 143,
    148, 158, 159, 162, 179, 187, 196–9, 202–4,
    206–12, 217–21, 223, 224, 226, 227, 229,
    231, 233, 234
Petunia, 67
Phaseolus, 203
phencapton, 208, 209
phenology of coccinellids, 156
phenology of mites, 196
phenology of rice, 126
phenotypes, 62
phenotypic polymorphism, 112
phenylalanine, 71, 182, 185
pheromones, 11, 12, 14, 24–7, 40–2, 97, 149,
    154
Philippines, 121, 123, 135, 174, 193
phleothripids, 219
phosalone, 208, 209, 211
Phosdrin, 208
phosphamidon, 208, 209
phosphoglucomutase, 45
6-phosphogluconate dehydrogenase, 45
phosphoglucose isomerase, 45
phosphorus, 182, 203
photoperiod, effect of, 59, 186, 189, 195, 199,
    224
photosynthesis, effect of, 203
phytoseiids, 179–228
phytosterols, 183
pine needle scale, 175, 177
pinimor, 215
planning of project, 7, 11
plant age, effect of, 126
Plictran, 208, 210, 211
Poland, 76, 83, 194, 196, 209, 210, 220
pollen, 192, 196, 197, 225–8
population genetics, 12, 14, 43, 46, 47, 61, 230
potatoes, 53, 60, 61, 67, 71, 72, 75, 77, 79, 85,
    87, 88, 101, 102, 105, 112, 113, 230
potassium, 69, 182, 203
predators
    of aphids, 100–4
    of armoured scale insects, 156–9
    of rice stem-borers, 133
    of spider mites, 187–96
    of tephritids, 16–18, 20–2
prediction of damage, 136, 137
Principe, 144, 146
proline, 182, 185
propensity test for flight, 34, 35
protein hydrolystates, 12
Protozoa, 109
Prunus nigra, 65

Psocoptera, 197
Puerto Rico, 145, 177
pupal mortality, of tephritids, 14–16, 18–22,
    230
purple scale, 144, 147, 161, 175, 177

races, 12, 44, 46–8, 56, 58, 59, 84, 85, 146, 181
radish, 86
radioactive labelling studies, 182, 184
rearings, 52
red scale, 146
'red scale strain'
    of *P. perniciosi*, 146, 155, 171
    of *C. bifasciata*, 170
red spider mite, 186, 204, 215, 223, 234
reproductive isolation, 46–8, 67, 151, 152, 153
reservoir of wild hosts, 2, 3
resistance of plant, 4, 52, 53, 56, 66, 113–16,
    118, 119, 127, 135, 181, 229, 233–5
resistance to pesticides, 2–4, 6, 52, 63, 113, 181,
    196, 197, 199, 210–12, 214, 215, 218–20,
    223, 231, 233
response to colour, 27, 30, 33
resting spores, 104, 105, 107, 108
resurgence, insecticide-induced, 52, 231
Rhodesia, 178
Rhotane, 208
riboflavin, 185
rice, 3, 4, 8, 121–37, 231
rice ecosystem, 124, 126
rice stem-borers, 3, 121–37, 231
    distribution, 122, 128
    free areas, 127
    origin, 125
RNA, 185
rose, 215
rose scale, 175, 176
r-strategists, 13, 14
rufous scale, 148, 161, 175, 178
Ryania, 207, 208, 210

sampling, 57, 72, 89, 96, 121, 134
San Jose scale, 145, 147, 175, 178, 223
Sarawak, 123, 128, 130, 133
scale insects, 5, 192
Scandinavia, 215
scelionids, 131, 132
schradan, 208, 210
SCIBP, 8, 9
Sciomyzidae, 123
sea beet, 115
selbar, 204
selective insecticides (including spray and use),
    117, 134, 143, 167, 170, 219, 231, 235
selective pressure, 58, 111, 154, 213, 214
'semispecies', 153
serine, 182, 183, 185
sex determination, 180
sexual behaviour, 12, 39, 41, 230
sexual isolation, 151, 232
Seychelles, 144, 145
sibling species, 44, 46, 47, 147, 151, 153, 154
Signiphoridae, 142, 155
silver mite, 223

Subject index